Carbon Capture and Storage

Carbon Capture and Storage

Efficient Legal Policies for Risk Governance and Compensation

Michael G. Faure and Roy A. Partain

The MIT Press
Cambridge, Massachusetts
London, England

This book was set in Stone Sans and Stone Serif by Toppan Best-set Premedia Limited. Printed on recycled paper and bound in the United States of America.

Library of Congress Cataloging-in-Publication Data

Names: Faure, Michael (Michael G.) | Partain, Roy.
Title: Carbon capture and storage : efficient legal policies for risk governance and
    compensation / Michael G. Faure and Roy A. Partain.
Description: Cambridge, MA : The MIT Press, [2017] | Includes bibliographical
    references and index.
Identifiers: LCCN 2016026784 | ISBN 9780262035590 (hardcover : alk. paper)
Subjects: LCSH: Carbon sequestration. | Carbon sequestration--Economic aspects. |
    Climate change mitigation.
Classification: LCC SD387.C37 F37 2017 | DDC 628.5/3--dc23 LC record available
    at https://lccn.loc.gov/2016026784

10 9 8 7 6 5 4 3 2 1

Contents

List of Abbreviations and Unique Terms

AGI	Acid gas injection
ANI	American Nuclear Insurers
Ann.	Annotated
API	American Petroleum Institute
Ar	Argon gas
ART	Alternative risk transfer
BBC	British Broadcasting Corporation
B USD	Billions of US dollars
CAT Bonds	Catastrophe bonds
CCR	Caisse Centrale de Réassurance
CCS	Carbon capture and storage, carbon capture and sequestration
CERCLA	Comprehensive Environmental Response, Compensation, and Liability Act of 1980 ("SuperFund")
CEREL	Centre for Environmental Resources and Energy Law
CH_4	Methane, commonly known as natural gas
christmas tree	A collection of values at the top of a well that regulate the flow and pressure within the well apparatus
CO_2	Carbon dioxide
DOE	US Department of Energy
EC	European Community
ECBM	Enhanced coal bed methane recovery
EEA	European Economic Area
EEC	European Economic Community
ELD	Environmental liability directive
EMANI	European Mutual Association for Nuclear Insurance
EMS	Environmental management system
EOR	Enhanced oil recovery
EPA	US Environmental Protection Agency

EU	European Union
ExxonMobil	Exxon Mobil Corporation
FEMA	Federal Emergency Management Agency
GAREAT	Gestion de l'Assurance et de la Réassurance des Risques Attentats et Actes de Terrorisme
GGGSA	Victoria Greenhouse Gas Geological Sequestration Act
GHG	Greenhouse gas (sing. and pl.)
GMOs	Genetically modified organisms
GS	Geological sequestration
GSU	Geologic sequestration utility
$GtCO_2$	Gigatonnes of carbon dioxide
HB	House bill: proposed, but not yet enacted, legislation in the lower house
He	Helium gas
H.R.	House of Representatives of the US Congress; usually used to denote proposed legislation
H_2	Hydrogen gas
H_2S	Hydrogen sulfide
IEA	International Energy Agency
IEEE	Institute of Electrical and Electronics Engineers
IEEE-SA	Institute of Electrical and Electronics Engineers Standards Association
IOGCC	US Interstate Oil and Gas Compact Commission
IPCC	Intergovernmental Panel on Climate Change
IPCC CCS	IPCCs report on the potential to implement CCS technologies
ISO	International Organisation for Standardization
kg/sec	Kilogram per second
kg/day	Kilogram per day
km	Kilometers
Kr	Krypton gas
MGA	Midwestern Governors Association
Mont.	Montana
M USD	Million US dollars
MW	Megawatt
Ne	Neon gas
NEIL	Nuclear Electric Insurance Limited
NFIP	National Flood Insurance Plan
NGO	Nongovernmental organization

NHT	Nederlandse Herverzekeringsmaatschappij voor Terrorismeschade
NO_x	Any of the compounds from the nitrogen oxide group (e.g., nitrogen dioxide)
NRC	Nuclear Regulatory Commission
OBO	Operated-by-other
OECD	Organisation for Economic Co-operation and Development
OPA	Oil Pollution Act
OPOL	Offshore Pollution Liability Association
OPGGSA	Offshore Petroleum and Greenhouse Gas Storage Act
OSLTF	Oil Spill Liability Trust Fund
OSPAR	Oslo Paris Convention
p.a.	Per annum
PAC	Protective action criteria
ppm	Parts per million
P&I Clubs	Protection and Indemnity Clubs
RCRA	Resource Conservation and Recovery Act of 1976
SACROC	Scurry Area Canyon Reef Operators Committee; legacy name of an oil field located in the Permian Basin in Scurry County, Texas
SAFETY Act	Support Anti-Terrorism by Fostering Effective Technologies Act of 2002
SB	Senate Bill, proposed legislation, but not yet enacted, in the upper house
SCAPA	US Department of Energy Subcommittee on Consequence Assessment and Protective Actions
SMCRA	US Surface Mining Control and Reclamation Act
SO_x	Any of the compounds from the sulfurous oxide group (e.g., sulfur dioxide)
SWDA	US Safe Water Drinking Act
SwissRe	Swiss Reinsurance Company
TEEL	Temporary emergency exposure limits
TEPCO	Tokyo Electrical Power Company
TFEU	Treaty on the Functioning of the European Union
TRIA	Terrorism Risk Insurance Act
UIC	Underground injection control
UK	United Kingdom

UMTRCA	Uranium Mill Tailings Radiation Control Act of 1978
UNFCCC	United Nations Framework Convention on Climate Change
US	United States of America
USD	US dollar
USGS	US Geological Service

1 Introduction

Carbon capture and storage (CCS) is a potential solution to a devious problem,[1] but it has not yet become widely used to fight this problem. Why? Several problems are believed to be holding the development of CCS back, but one of the main reasons is thought to be that the risks of faulty implementation could lead to great liabilities.[2] Accidents from failed CCS containment systems are known to potentially cause a wide variety of damages.[3] But it is also a problem that exactly what liabilities, and thus legal damages, might result are also unclear because the underlying liability rules remain cloudy or confused. Thus, it is difficult for potential operators to accurately scope and forecast their business plans, and so this poorly defined liability exposure substantially hampers CCS rollout.

This book endeavors to resolve that problem by addressing the complexity and confusion on CCS-related liability rules and attempting to provide a rigorous framework to identify efficient or robust policy recommendations.[4] It also expands its inquiry into how regulations might also enable efficient governance of the risks from CCS activities.[5] Finally, it addresses how compensation for CCS-related accidents might be handled. The challenge is to design compensation systems that neither interfere with the primary goal of accident prevention nor create moral hazards for either the operator or the potential victims.[6] This book does find its target. The summary of policy recommendations is set out in the final chapter.[7]

1.1 Carbon Capture and Storage: Risks and Rewards

The world is confronted by a substantial problem: anthropogenic climate change. If too much carbon dioxide (CO_2) is emitted into the atmosphere, a chain of chemical and physical consequences could ensue that would lead to increased capture of solar radiation and thus increase the overall energy levels of the global climate.[8] This problem is commonly called either global warming or climate change. The results of global climate change could include adverse impacts to a wide variety of ecologies and communities, potentially threatening the survival of a great many species, including humans. Lesser risks include agricultural problems, major shifts in weather and climatic conditions, and the potential loss of land masses due to rising ocean waters. Thus, a consensus is emerging that CO_2 emissions, and other greenhouse gases, should be prevented or otherwise sequestered in carbon sinks.

One of the main sources of emitted CO_2 is the combustion of fossil fuels for energy production or heat production.[9] Fuels such as coal, crude oil, and natural gas release gaseous volumes of CO_2 as they combust; usually those volumes are vented and emitted directly to the atmosphere. To reduce the levels of CO_2 emissions, there are choices to consider: (1) reduce emissions by reducing the amount of fuel combusted, or (2) replace those carbon-rich fuels with carbon-free sources of energy.[10] However, despite technological innovations and rapid production levels, it would appear that carbon-free energy sources may take decades to supply a sufficient level of energy supplies to obviate the need for fossil fuels. And it would appear that very few leaders or consumers are eager to reduce energy budgets, so with no other recourse, they are equally hesitant to substantially reduce the levels of fossil fuel consumption. The world clearly needs a third choice.

CCS might be that third option: it might enable continued fossil fuel consumption while reducing CO_2 emissions.[11] The world has taken notice of this option. Some even argue that CCS is unavoidable, especially given the deep commitment of the world's largest economies to fossil fuels.[12] While many remain adverse to the long-term use of fossil fuels, there does appear to be a consensus that CCS should be used at least as long as it takes to displace fossil fuels.[13] This perspective is recognized in the approval of CCS as a clean development mechanism under the UN Framework Convention on Climate Change (UNFCCC).

Yet the development of CCS remains a lackluster option. It is not due to a lack of commercially feasible technologies, as those are already in place and operational.[14] Nor is it due to a lack of urgency: the risks of climate change are well known, and the scientific consensus is strong. It does, however, appear to suffer from a few major factors; among leading concerns are an inefficient market price for carbon and/or its injection services and a variety of concerns on existing liability rules and uncertainty in expected damages.[15]

This book focuses on the latter two concerns: that clear and efficient standards need to be set for liability and compensation policy if policymakers were to want to encourage the broader development of CCS projects.[16] In addition, these types of standards need to be set even if policymakers do not want to encourage CCS projects. The actual underlying goal is to set clear legal standards that facilitate the creation of incentives that guide operators and other members of the community to efficiently reach the community's welfare goals.[17]

There are good reasons to pay attention to the potential liability issues and the related issues of compensation for CCS-related damages. The basic process of CCS is to inject highly compressed CO_2 gas down injection wells into deep geological formations. The gas, which remains under great pressure and is buoyant, will seek to migrate within the rock and escape to the surface and back into the atmosphere unless impeded by a secure geological barrier or seal.[18] A variety of risks and hazards arise during the development and operational phases of such facilities; suffice it to say, the health and safety of humans, animals, and broader ecological communities could be substantially impaired.[19]

But the current liability rules are unclear in their application to the facts and circumstances of CCS activities.[20] There is an overlap of various liability rules, along with confusing property law rules and unclear assignment of actor liability. Due to the exotic location and nature of the torts and injuries involved, causation is complicated.[21] Also, the time line of potential injuries engages a truly unique sense of the phrase *long run*; this sense of long run extends potentially multiple millennia, beyond the likely lifetimes of either humans or even corporate entities.[22] Thus, there is a strong need for theoretic frameworks that can assist policymakers in finding their way through this liability and compensation wicket.

The development of CCS activities faces a substantial problem in that the activity itself is recognized as risky, but the rules and regulations remain unclear in many jurisdictions. Equally challenging is the question of who might pay compensation for damages once injuries occur; especially challenging is who might pay if the injuries arise after the operator has gone insolvent or otherwise been extinguished due to the passage of centuries.[23] Could the solution be found in the marketplace, or would public authorities of some kind need to become responsible for CCS storage activities at some point? These are the types of questions and concerns impeding the development of CCS activities.

But the delay in expanding CCS storage options and capacity almost surely means that carbon emissions continue with insufficient sinks, and thus the threat of anthropogenic climate change looms ever nearer. Thus, this book focuses on providing some solutions to these liability and compensation questions.

1.2   Goals and Methodology of This Book

The goals of this book are to research the liability questions raised by the development of CCS storage operations with particular regard to the risk and harms that might arise from accidental releases or subterranean movements of the stored CO_2. What means of risk governance would best ensure the optimal level of CCS activities to ensure the minimal costs from accident prevention and of damages incurred?

It investigates whether rules of civil liability should be employed to govern these risks. Furthermore, we ask which particular rules of civil liability should be chosen: a fault-based rule of negligence, with its attendant notions of sufficient due care, or the nonfault methodologies of strict liability, trespass, and nuisance? We provide research into the subsidiary liability policies that need to be developed to support rules of civil liability, such as to whom to assign liability and the question of whether to channel liability, among other affiliated questions.

This book recognizes that one of the unique liability questions from CCS-related accidents is how to address liability when the time is so extremely open ended, as it would be for certain potential CCS-related accidents. The very nature of CCS storage implies that the CO_2 storage activity would necessarily continue for hundreds, if not thousands, of years, and technically accidents could emerge centuries after the final act of injection. Rarely have liability systems needed to address such extreme latency considerations. This book investigates how a liability system might best address such latency problems. It also investigates appeals from the literature and from recently enacted legislation to examine the policy option of transferring responsibility for later-emerging accidents from operators to some form of public authority as a means to provide a respondent for those far future events.

Given that liability systems are concerned with the management of risk posed by risky activities and recognizing that regulatory measures can also perform similar services, this book investigates if regulatory measures might also be part of the risk governance mechanism applied to the risks from CCS storage activities. This book not only explores the potential role of public regulatory measures but also those of private regulatory measures themselves. Indeed, the book takes up a particular research question of whether public and private regulatory efforts could be coordinated to jointly provide superior standards more efficiently than what might be obtained by either means of regulation separately.

This book also sets as its goal to research the compensation questions that necessarily follow questions of liability. If injuries were indeed to occur from CCS-related accidents, what would be the best way to provide compensation? And how might the need for compensation be best coordinated with policies that prioritize the prevention of accidents? This book investigates whether compensation should be provided primarily by the commercial operators of CCS storage sites, in application of various financial tools available on the private market, or be derived from public resources as a function of local government.

In exploring the potential of privately providing compensation, this book provides research into the feasibility of providing insurance to CCS operators to address their liability exposures. It investigates more traditional forms of insurance as well as innovative alternatives such as self-insurance, captured agencies, and questions on the need for governmentally supported reinsurance policies. Similarly, the book investigates the utility of other financial market tools that could provide contingent amounts of capital, such as guarantees and deposits, letters of credit, performance bonds, and catastrophe bonds. Private arrangements to share risk among operators, known as risk-sharing agreements, are investigated as an alternative to insurance for high-risk activities.

This book also researches the efficiency and efficacy of compensation measures that public resources support. Should governments provide direct payments of compensation, or should they be engaged in stewarding capital from the operators? And if the stewardship option were to be found to be a better choice, how best might that be done: with a managed compensation fund, compulsory financial guarantees, or publicly underwriting reinsurance? Or should governments instead consider the development of public CCS utilities, as might be done for power, water, or public wastes? This book endeavors to compare these public compensation measures with the private means of compensation to find the optimally efficient solution.

Ultimately, the goal of this book is to provide practical recommendations for policymakers, lawmakers, and community leaders to consider in their development of CCS policies and legislation. To that end, this book provides a summary of our research-based recommendations in accessible language.

Methodologically, our book provides its analysis on the basis of the law and economics literature; in particular, it is founded on the approaches developed to address research questions on tort law and accident law.[24] There are sound reasons to employ a law and economics approach to address liability and compensation for damage resulting from CCS storage sites. A first reason is that although other researchers have investigated the design of liability and compensation schemes for CCS-related damage,[25] they did not approach the issues from an economic analysis of law perspective. The law and economics methodology is distinct from other legal schools in that it investigates the comparable economic efficiency of legal policies to attain their goals. Thus, investigating these liability and compensation matters from a law and economics method provides fresh insights for policymakers.

A second, and perhaps more important, reason is that by employing the methodology of the law and economics literature, we gain insight into how legal measures provide incentives that motivate behavioral changes and affect decisions on risky activities. The methodology reveals how legal measures directly affect operators of CCS storage sites with particular regard to liability rules and compensation measures.[26]

A deeper review of the merits, and perhaps necessity, of employing the law and economics methodology is provided internally for each research question; however, we provide an expanded introduction to the law and economics paradigm and its methodological approach in chapter 3.[27]

1.3   Limits of Its Goals

The goal of the study is to provide a theoretical analysis of liability and compensation for damage resulting from CSS storage sites. We hope to provide policymakers with some insights on potential policy options and on the efficacy and efficiency of those recommendations.

A body of legal literature has grown to address the many legal issues facing CCS development. Bankes has distinguished three clusters of legal literature on CCS policies.[28] The first strand focuses on the coordination of offshore CCS storage systems with preexisting international conventions.[29] The second focuses on the legal concerns of CCS storage and long-term liabilities.[30] The third has focused on the coordination of CCS efforts and UNFCCC-related measures, such as carbon trading systems.[31] This study stems from the second group, focused on liabilities associated with CCS projects, especially liabilities that face long-term latency problems.

The goal of this study, however, is not to review the actual liability of operators in particular legal systems.[32] That would go beyond its scope. It also does not address specific legal questions limited to specific jurisdictions, for example, how to address liability for CCS within the framework of the European environmental liability or emissions trading directives.[33]

A variety of international legal conventions might interface with the activities contemplated within CCS storage activities, and questions could be asked to what degree CCS could be a useful instrument for states to implement those obligations under those conventions. More squarely put, a question might be asked as to how the implementation of CCS storage plans might interact with international law. However, those research questions have been discussed in previous literature;[34] our study thus remains limited to the scope we have already identified.

Finally, we stress that the goal of this study is obviously not to repeat or summarize what earlier studies on this topic have already provided, although we do survey and discuss the outstanding literature. The added value of this study is that it addresses the issue from a law and economics perspective and of issues that have been less, or less explicitly, addressed thus far.

1.4 Structure of This Book

This book is presented in three basic stages, followed by a summary and conclusion.

First, a scientific and technological background is provided for the later discussions. Second, research into the governance of the risks from CCS storage sites is presented. Third, research into compensation for CCS-related damages is presented. Finally, the book concludes with an executive summary of the policy recommendations developed within the previous three stages.

The first stage is set out in chapter 2, which addresses a range of scientific and technological background materials related to CCS storage activities. The natural phenomena that occur underground when CCS storage is undertaken are explained with reference to their geochemical, geophysical, and geological conditions. The concepts of pore space, storage, and storage reservoir are explained and thus support a discussion on potential storage capacity. Next, the technologies and procedures of CCS injection and sequestration are reviewed, as are the historical antecedents of modern CCS technologies. The century of experiences gained from natural gas storage systems is a key focus, along with the recent decades of injection technologies affiliated with enhanced oil recovery (EOR) and acid gas injection (AGI) systems.

Historically, there are relevant examples of natural CO_2 venting accidents and of man-made CO_2 accidents that provide insight into what might occur from CCS-related accidents. From that beginning, a characterization of the potential accidents from CCS storage sites finds two basic groups: accidents at ground level or above and accidents that occur underground. The potential injuries and damages from both sets of risks and hazards are explored and detailed. The consequences of the very long storage periods necessary to protect the atmosphere also engage the potential of accidents emerging long after the cessation of on-site activities. After the review of potential risks and hazards is completed, a discussion is presented on the potential to observe and monitor the risky activities as they progress and on the potential to prevent, mitigate, and remedy and restore whatever risks or harms ensue.

Chapters 3 through 6 address the second stage. Chapter 3 provides a review of the scholastic calls for clarity in liability rules for CCS storage projects. It demonstrates that multiple rules of liability currently could apply in potentially confusing overlap. It then explores and explains the paradigm of law and economics, especially in its application to tort law and accident law. It first evaluates the fitness of the rule of strict liability for the circumstances of CCS storage operations and then similarly evaluates the rule of negligence. Each of these investigations combines theoretical analysis with specific application to CCS storage activities. Strict liability will be shown to be more robust than negligence for the circumstances of CCS storage activities.

Chapter 4 evaluates a variety of liability policy options for the context of CCS operator liability concerns. Clarity in the assignment of liability is investigated, with attention brought to bear on notions of stakeholders, liability channeling, and the potential application of joint and several liability. The potential interactions of regulatory measures on implementation of civil liability rules are reviewed. The difficulties of causation in complex environmental settings, such as found with CCS storage systems, are discussed. Also, research is provided on the question of whether liability should be capped in order to better encourage the development of the CCS industry.

Chapter 5 focuses exclusively on the research question of how best to address the very long time frame of certain latent risks from CCS storage activities and investigates in particular the efficiency and efficacy of transferring responsibility for the CCS storage activities from private hands to public hands at some point after the storage facilities are closed. It provides insight into several rationales offered in support of such a strategy and a survey of existing legislative acts on the issue. It then develops a recommendation for the transfer of liability after successful closure of the storage site.

The potential application of regulatory measures to address the risks and hazards of CCS storage activities is investigated in chapter 6. The law and economics literature provides a framework to analyze under what circumstances public regulations can offer improvements on the performance of civil liability systems and when they cannot. The literature also provides insights into when private regulatory and self-regulatory efforts might improve on public regulation and civil liability. This chapter demonstrates that both public and private forms of regulation could and should be brought to complement the function of a civil liability system in governing the risks and hazards of CCS storage activities.

The third stage is explored in chapters 7 and 8. Chapter 7 focuses on the potential to find compensation measures implementable from private sources of capital and in the public market. Primary among those means is private insurance. Research is provided that supports the argument that insurance will likely be available for CCS-related risks, and related findings provide insight into how best to provide that insurance. Alternative insurance forms are also investigated, such as self-insurance. In the alternative to insurance, measures that ensure the availability of capital funds at future points or under contingent events are reviewed. Risk-sharing agreements, guarantees and deposits, letters of credit, performance bonds, trust funds, and escrow accounts are all evaluated. Finally, the chapter investigates the novel financial tool of catastrophe bonds. Ultimately a conclusion is reached that private compensation measures are feasible and would be best undertaken by using a mixture of the financial instruments reviewed in the chapter.

Chapter 8 evaluates the policy option of financing compensation schemes with public resources or, alternatively, having public authorities provide stewardship of privately sourced capital resources. It also evaluates direct compensation plans and investigates the potential role of public authorities in overseeing compulsory financial guarantees, compensation funds, and acting as a reinsurer of last resort.

The book concludes with a summary of the recommendations accumulated during the research presented over the previous chapters. It also touches on the topics of liability for climate change injuries and how developing countries might adjust our recommendations to better fit their needs.

2 Survey of CCS Technologies and Risks

The idea of CCS is for the most part novel to the general public, but its technological history extends a century back. CCS engages the potential to place gaseous volumes into geological storage by injecting them into the deep earth. In many ways, it resembles the reverse of producing natural gas. CCS storage artificially replicates the natural means of storing gases deep in geological reservoirs; CO_2 reservoirs exist in nature and have been artificially produced for a number of decades.

Current CCS technologies build on decades of experience in related technological feats. Natural gas has been stored in geological systems for over a century, enabling cities to provide continuity of supplies. CO_2 injection technologies have enabled greater recoveries of oil and natural gas for decades. Even the situations of ruptured CO_2 wells and of venting, even of worst-case events, have been encountered by the energy industry in preceding decades, providing a historical perspective on potential risks and hazards and also on the potential lack thereof.

This chapter provides an overview of the geology, geochemistry, and geophysics of CCS injection and storage systems. It explains and provides detail on the technologies and methods in CCS injection and storage systems. It also examines natural occurrences that resemble the processes and risks of CCS storage systems.

This chapter then characterizes a variety of risks and hazards and classifies them in two basic categories: risks at or above ground level and risks and hazards underground. The centrality of the site selection decision and the decisions on daily operational safety are presented as determinative of most of the key risks and hazards. The chapter also reviews the potential to identify risks and prevent them from occurring and the potential to remedy and restore injuries that were not preventable.

These observations are precedent to the arguments made in the subsequent chapters on the effectiveness of civil liability rules to govern the risks and hazards effectively. But these observations place great support on implementing regulations, setting standards, and monitoring and governing behaviors before accidents and injuries occur. The central finding of this chapter is that although CCS operations pose foreseeable risks and hazards, they would likely be capable of safe and efficient operation given the proper incentives.

2.1   Nature and Scale of Subterranean Storage Structures

2.1.1   Geology, Geochemistry, and Geophysics

Even in nature, there is carbon storage: the largest natural carbon reservoirs are within subterranean structures.[1] Geological storage of hydrocarbons, coals, methane (CH_4), and carbon dioxide (CO_2) gases has occurred for hundreds of millions of years, primarily within the upper crust of the earth.[2] Other forms of carbon are trapped in rock formations as carbonate materials.[3] CO_2 can be produced from chemical and volcanic reactions from those materials and migrate to form CO_2 reservoirs, trapped under subterranean structures.[4]

Natural accumulations of CO_2 exist as both gas reserves and solutions in aqueous modes.[5] Such systems are found around the globe, usually in sedimentary basins or volcanic-rich areas; research has focused on how these natural CO_2 "storage systems" might provide insight into how to best establish man-made CO_2 storage systems.[6] Several major CO_2 deposits in the United States have provided examples of large CO_2 reservoirs that exist with some amount of natural leakage.[7] Smaller CO_2 deposits have also been examined in Australia, France, Germany, Greece, and Hungary.[8] Evidence from these and other petroleum fields with associated stored gases suggests that gases, including CO_2, can be trapped and stored for millions of years.[9]

Thus, the concept of injection anthropogenic carbon fluids into subterranean structures conceptually follows natural events.

2.1.1.1   Pore Space and Storage CO_2 storage can be achieved by injecting the gas into sedimentary basins containing oil fields, depleted natural gas reservoirs, deep coal seams, or underground saline formations.[10] The preferred basins for CCS storage generally feature thick sedimentary layers, large saline reservoirs within permeable rock layers, overlaying impermeable rock to act as barriers, and, in general, simple geological structures.[11]

Geological storage of CO_2 and other gases is possible because of the granular structure of certain rock types.[12] A pore space between grains of rock material enables gases and fluids to move in between the grains of rock.[13]

There might already be a gas or fluid in the pore space;[14] the injection of CO_2 can either displace those volumes or conjoin them in solution.[15]

The storage of CO_2 in those intragranular pore spaces is achieved by injecting fluidized CO_2 down an injection well into a suitable rock layer with sufficient granular pore space available.[16] In order to transmit the fluid CO_2 into the rock layer, the bottom of the well is perforated with small holes that lead directly to the well bed.[17] As the CO_2 fluid enters the geological zone of injection, the pressure inside that zone increases.[18] That pressure plus the difference of fluid density, as affected by gravity, can enable both the injected volumes and the original volumes to move and migrate within the pore space.[19]

2.1.1.2 Attaining Storage and Permanence The CO_2 becomes stored when its exit from the subterranean depths is blocked by a "thick low-permeability seal," a conversion into solid minerals, or adsorption onto the surfaces of the underground structures.[20] A seal could be formed by cap rocks, low-permeability shale rocks, or shallow beds of methane hydrates in certain locations.[21] Various faults or fractured or folded rock layers could also potentially serve as seal layers.[22]

It is not necessary to have a hard physical barrier for the seal, as a seal might be created by hydraulic forces that would strongly limit the ability of the injected CO_2 fluids to migrate.[23] The migratory behavior of injected CO_2 depends on its density, viscosity, and the pressures present in the reservoirs. Such a barrier can also involve dissolution into a nonmobile water body.[24] If the overlaying seal is wide enough, even as a hydraulic feature, when the mobility of the injected CO_2 is thus constrained, it might take millennia for the fluid CO_2 to reach beyond the barrier's edge to leak toward the ground surface.[25]

During those millennia of "temporary" storage, under either hard or hydraulic seal, the supercritical CO_2 would undergo several changes to attain permanent sequestration.[26] While the gas would be in solution, a certain amount of solubility trapping would occur, fixing that volume in solution on a permanent basis.[27] As CO_2 dissolves into the water, carbonic acid is formed.[28] The carbonic acid will dissolve rock, and ionic variants of CO_2 would become abundant: sodium silicate, potassium silicate, magnesium carbonate, iron carbonate, and other silicate minerals.[29] Those ions and the dissolved rock fluids would form into carbonate rock accretions.[30] Estimates place the total time to convert the injected CO_2 into rocky carbonates at approximately 5,000 years.[31] Other comparable models suggest that 60 percent of the injected CO_2 fluids could be mineralized by the end of the injection years with close to 100 percent mineralization after the first millennium.[32] Thus, if carefully sited, most injection locations would rapidly attain large percentages of permanent mineralization of the injected CO_2; even in cases where the CO_2 fluids buoyed toward the cap rock, they would likely become mineralized at that upper boundary prior to escape.

Similarly, the transition from temporary to permanent storage can be achieved by the adsorption of the CO_2 onto the organic carbon surfaces of coal or organic-rich shales.[33] And yet another form of semipermanent sequestration can be achieved by the absorption of the CO_2 volumes into hydrates structures as can be found offshore, in arctic areas, and in the Tibetan plateau, for example.[34]

2.1.1.3 Storage in Saline, Coal, Shale, and Salt CO_2 storage can be accomplished within structures both on land and offshore.[35] The offshore locations suitable for CCS storage are primarily on the continental shelf; most sedimentary areas farther offshore are either too thin or insufficiently impermeable for safe storage.[36] Somewhat fortunately, the needs of early industry to be located close to fuel sources has left many, if not most, industrial areas adjacent to those sedimentary basins with suitable CCS storage locations.[37]

Saline formations occur when water reservoirs, rich in dissolved salts, are present within sedimentary rock.[38] Because of that salt content, the waters are generally unusable for human or livestock drinking waters and are also limited in agricultural application due to potential damage to topsoil from the salts.[39] Because of the properties of water to hold CO_2 in solution and because these saline waters are of limited human use, they are considered one of the premier targets for CO_2 injection storage.[40] The potential to inject CO_2 safely into saline reservoirs has been studied in Norwegian waters at the Sleipner Project in the North Sea.[41] Sleipner offers very large storage capacity underneath a clay and shale barrier; the injection zone is approximately 1,000 meters below the seabed.[42] Despite observed migration, data showed that the seal has prevented the CO_2 fluids from leaving the storage zone.[43] Furthermore, studies have suggested that the future of the injected fluids is to become denser and fall deeper into the storage zone, effectively preventing their escape.[44]

Coals that lay deep underground and are otherwise expected to be of no economic value are another potential storage option.[45] Cleats are the naturally occurring fractures within coal seams and provide ample surfaces for the injected CO_2 to flow onto even smaller surfaces and become permanently adsorbed.[46] However, as CO_2 is injected into coal seams, the coal can become rubberized; attention needs to be provided during the injection period to ensure that the coal remains adsorbent.[47] Another potential side effect is that the injection of fluidized CO_2 could wash and displace whatever volumes of methane had previously been adsorbed onto the coal faces; thus coal bed methane recovery could be facilitated,[48] or, if the methane is undesired for production, then caution would merit ensuring that both the methane and the CO_2 could be securely sequestered within that coal seam.[49] Barring those two concerns, the placement of CO_2 into coal seams is seen as a means to quickly stabilize the CO_2 into permanent storage for "geological time" periods.[50]

Other minerals that might facilitate rapid geological permanence are basalts, which can react with the CO_2 to form carbonates, and hydrocarbon-rich shales, which could provide adsorption akin to that of coal seams.[51] Also, mines that once exploited sedimentary rocks in access to minerals such as potash, lead, or zinc might be serviceable for storage.[52]

Salt caverns exist that were created by either natural or artificial flooding and washing away of the salts.[53] The storage capacity of such caverns has been exploited for the storage of both liquid hydrocarbons and gaseous volumes of methane; thus, it is expected that the same technologies and caverns could be used to store fluidized CO_2.[54] Due to the reduced likelihood of a transition to permanence by means of mineralization, salt caverns retain certain risks that other storage systems might lack.[55] However, given the vast experience from short-term storage of natural gas in such caverns, salt caverns might see active use as intermediate storage sites for CO_2 en route to locations of permanent storage.[56]

2.1.2 Global Storage Capacity

Estimates have suggested that "the global capacity to store CO_2 deep underground is large."[57] Caution does bear noting that the terms *reserves* and *proven reserves* can vary; generally the term *reserves* or *potential storage* indicates a scientific estimate of carrying capacity without economic considerations, whereas *proven reserves* tends to reflect reservoir capacity that would be supported by means of contemporary economic considerations.[58]

2.1.2.1 Factors Determining Capacity

Capacity in underground storage is driven by several factors: (1) volumetric trapping, the available pore space in the rock layers; (2) solubility trapping , the potential to store CO_2 in the in situ liquids such as water or crude oil; (3) adsorption trapping, or available surface areas within coal and shale deposits; and (4) mineral trapping, the potential chemical matrix to enable carbonate precipitation.[59]

Site selection is critical. Good sites should provide adequate capacity for injection and storage, sufficient means of containing the injected volumes, and a stable geological environment to ensure that long-term planning can be successful.[60] The primary settings of such conditions are midcontinent sedimentary basins and sedimentary basins offshore on the continental shelf.[61] In order to ensure sufficient capacity, adequate porosity and injection layer thickness are necessary.[62] In all likelihood, the most advantageous locations for storage are likely to be fields that have already been explored for oil and gas or coal development because they have already been exposed to intense scientific characterization studies; hydrocarbons or coal reservoirs were either economic and depleted[63] or found to be uneconomic and thus employable in both cases;[64] and much of the necessary infrastructure, such as a well, might already be in place and thus reduce costs of CCS installation and developments.[65] Another set of considerations is the absence of subterranean water flows, major gas reserves, and major natural fractures.[66]

Basins located in tectonically active areas are inappropriate, as are basins in the vicinity of subducting continental plates, which would bear ongoing structural changes likely to result in new faults and fractures.[67] Generally there are a half-dozen particular risks to avoid in site selection: (1) basins with less than a 1,000 meter depth; (2) basins with poor seal structures; (3) basins that contain many faults, folds, and fractures; (4) basins that feature many discontinuities in geological structures; (5) basins with significant diagenesis; and (6) basins that already bear excessively high internal pressures.[68]

2.1.2.2 Storage Capacity Estimates   One thing is clear: there are a lot of places that might safely store CO_2, although estimates do vary. A rough estimate of European and North American capacity alone suggests approximately 5,000 $GtCO_2$ in storage capacity. Globally, storage capacity in hydrocarbon reservoirs is estimated at 926 to 1,200 gigatonnes of CO_2 ($GtCO_2$).[69] Global saline reservoirs might offer from 200 to 56,000 $GtCO_2$, reflecting a variety of modeling techniques; a consensus view is that 1,000 $GtCO_2$ should be a sure minimum of potential storage.[70] Global potential storage within coal seams has been estimated at 60 to 200 $GtCO_2$.[71] This provides a scoping estimate that would more than likely suffice for several decades of technology bridging to greener forms of power generation, if also for longer time periods.

Most major industrial areas have access to local massive storage options. The EPA has estimated that 3,000 GtCO_2 could be safely stored in CCS storage sites within the United States;[72] this is effectively 1,000 years of 1,000 coal-fired energy plants. Other industrialized regions have similar storage capacities.[73]

2.2 CO_2 Injection and Storage

The discussion in this section covers the technology and engineering of CCS injection and storage systems. While the idea of injecting high-pressure CO_2 into subterranean storage zones might appear as an exciting technology of tomorrow, it is important to begin this discussion with the clarification that CCS storage technology exists as an off-the-shelf technology today; within certain conditions, CCS storage is already commercially viable.[74]

In fact, the injection of CO_2 into geological formations is a long-established industrial practice.[75] Even the idea of injecting CO_2 to ameliorate climate change concerns is decades old.[76] In addition, injection has been used to address unwanted chemicals, pollutants, and by-products.[77] There are also a number of CCS injection testing locations, either operational or planned, around the world.[78] It is useful to keep in mind that natural storage systems have contained CO_2 underground for hundreds of millions of years without significant leaks,[79] so the physical concept is indeed reliable if done in the correct circumstances.

2.2.1 Based on Preexisting Technologies

While CCS is often discussed as a novel technology or industry, there are already in excess of 13,000 permitted CO_2 injection wells in the United States, at least 6,000 of which have been recently used.[80] Since 1972, the United States has injected over 600 million tons of CO_2.[81] The current levels of injection are volume-equivalent to the volume of CO_2 emitted from twenty 500 megawatt coal-fired electrical plants.[82] And while the vast bulk of that injection activity is enhanced oil recovery (EOR) related, and generally thought not to be equivalent to permanent storage, the evidence suggests that approximately 50 percent of those injected volumes are never recovered; they become permanently sequestered.[83] Thus, while the CO_2 volumes are generally not captured from industrial sources and these operations are not aimed at reducing anthropogenic emissions of greenhouse gases, it would be misleading to suggest that CCS has no preexisting technology base or that there is a substantial lack of industrial experience to draw learnings and from which to draw estimates of future risks.[84] Indeed, the decades of EOR and other forms of gas injection and storage have relied on a variety of well-established technologies and a variety of legal frameworks. In essence future legal frameworks for CCS benefit from an *"acquis* of state and federal statutes, judicial precedent, regulatory rules, and commercial practices."[85]

Acid gas injection (AGI) is a currently practiced form of CO_2 injection.[86] The first AGI project began twenty-five years ago.[87] This process returns noncommercial and potentially polluting by-product gases from natural gas field production, such as CO_2 and hydrogen sulfide (H_2S), back into the reservoir.[88] It thus facilitates the goals of reservoir pressure maintenance, emissions management, and by-product disposal.[89]

AGI is widely practiced at substantial volumes comparable to CCS storage needs,[90] particularly with regard to the surface facilities and injection wells. AGI also injects into similar structures as CCS, in that the preferred targets for injection are deep saline formations, depleted oil and gas fields, and water zones underlying petroleum fields.[91] The capstones are also similar, with AGI projects preferring carbonates and quartz-rich sandstones.[92]

AGI does provide for the long-term replacement of CO_2 back into reservoirs, so while not abating anthropogenic greenhouse gas emissions from industrial sources, AGI does provide a strong comparable for future CCS technological and costs concerns. It therefore provides strong reassurance that field experience does exist with the risks associated with CCS, even those for projects with impure CO_2 streams. In fact, in the scores of AGI projects in Canada, no leaks have been detected at any of the facilities.[93]

Historically natural gas and helium have also been injected for storage purposes, albeit not for the time frames envisioned for CCS storage needs.[94] Approximately 470 natural gas storage facilities are operating in the United States.[95] Natural gas has primarily been stored to provide supply-smoothing potential for domestic suppliers, so its storage time frame is set in years, not decades or longer, although certainly some natural gas has remained in storage for decades.[96] The first injection to support a natural gas storage facility began in 1915.[97] That first gas storage effort was an injection back into a partially depleted gas field,[98] similar to a current option for CCS storage. Injection into saline reservoirs, another injection option for CCS storage, began in the 1950s.[99]

Thus, there exists over a century of industry experience in injecting methane into depleted gas reserves and over a lifetime of experience in injecting into saline and water reservoirs, demonstrating that a range of evidence can be drawn from in forecasting future CCS risk profiles. There have been reported leaks, primarily due to poor completion jobs on the injection well, improperly plugged and abandoned wells on site of the injection facilities, and leaky faults; however, the general observation is that such problems were remediated, and such operations can be undertaken safely and effectively.[100]

While Benson and Cook report the total expected retained CO_2 to remain in excess of 99 percent for a thousand years, they also report that the actual historical record suggests that the total expected accidental release volume is approximately 00.001 percent of the stored volumes.[101] That number must be taken as a high-end estimate for potential losses from CO_2 storage efforts, as methane is much harder than CO_2 to trap and retain;[102] comparable modeling estimates for CO_2 events limit the percentage lost to 00.00001 percent,[103] a tenth of that lost for methane. Even lower levels of risk are modeled for storage within saline systems, as dissolution occurs fairly quickly on a geological scale, and the migratory speeds then slow the escape to millions of years.[104] Thus, the expected or modeled volumes that might be released are very small percentages, even over very large time frames.

EOR also provides another example of current industrial practices that inject large volumes of CO_2 into subterranean reservoirs.[105] CO_2-assisted EOR is a widespread field practice, with close to a hundred such operations worldwide.[106] In just the United States, it was reported in 2005 that seventy-three EOR projects injected CO_2 as part of their operations.[107] However, the primary goal of CO_2 injection with EOR is to increase the production of crude oil, and a portion of the CO_2 often returns in stream with the produced crude oil.[108] While not designed for storage purposes, EOR injections can provide a variety of useful data for CCS model building.[109]

2.2.2 Methods and Techniques

CCS technology exists in an off-the-shelf state: practically all of the necessary mechanical and technological elements are already in service in the petroleum and related industries.[110] That includes technology and solutions for drilling, injecting, and otherwise handling the wells.[111]

CCS storage begins with the arrival of fluid CO_2 to the injection site.[112] The CO_2 is likely to have been captured from a variety of industrial sources— for example, power plants, petroleum and gas processing facilities, and brewing and fermentation facilities.[113] This capture process has some cost complexity; it is difficult to estimate the impact of CO_2 capture on the energy output of power plants connected to CCS systems.[114]

In most cases, the CO_2 would be transported from capture facilities to the injection fields through transport pipelines[115] to ensure sufficient volumes on a steady basis. In the United States, for example, there are already approximately 5,000 kilometers of CO_2 pipelines in operation,[116] predominantly for the support of EOR activities.[117] It is not a technological requirement that the CO_2 arrive through pipelines. Indeed it is possible that some scenarios might envisage shipments of CO_2 using boats to either offshore disposal locations or to affordable disposal options abroad.[118] Natural gas has long been transported in bulk by vessel, and new hydrate-based storage technologies could reduce the costs of shipping for both natural gas and CO_2 transports.[119]

Once the CO_2 arrives at the injection field, several steps need to be undertaken to prepare the gas fluids for injection. The gas would need to be scrubbed to remove impurities.[120] The CO_2 would need to be compressed to a state known as "supercritical" to facilitate injection down-hole.[121] Being in a supercritical state enables the CO_2 fluids to compress further as they reach deeper levels of injection.[122] This is because a supercritical fluid can behave as a gas and flow through porous rocks while simultaneously behaving as a liquid and enable dissolution and other liquid operations.[123]

Common injection wells are fitted with redundancy—for example, one valve for routine flow control and another for emergency cutoff.[124] It has been recommended that CO_2 injection wells be fitted with automatic shutoff valves to prevent accidental releases;[125] there could be multiple such devices per well depending on cost structures and regulatory requirements. The number of injection wells needed for any given storage facility would depend on five factors: (1) total injection rate, (2) permeability of the injection zone, (3) thickness of the storage zone, (4) the maximum pressure allowances of the injection zone, and (5) the availability of surface lands to support injection well facilities.[126]

The safety of the injection well systems would depend on routine maintenance and operational safeguards.[127] Such procedures could reduce the probabilities of blowouts or mitigate any injuries from a blowout.[128]

Injection into the formation would require the fluidized CO_2 to become more highly pressured than the fluids within the injection zone, yet artificially increasing the down-hole pressure could lead to fracturing or other disturbances to the injection zone that could lead to fluid migration.[129] Research, however, demonstrates that injection can be safely accomplished.[130]

Once the injection activities are completed, the injection well would need to be plugged and abandoned to permanently seal the well and prevent the injected fluids from returning up the wellbore.[131] Such technology already exists for EOR and AGI injection wells and should be extendable to CCS injection wells.[132] To facilitate the safety of the plugs from acids, hydraulic barriers can be put in place down-hole to aid in protecting the well structure.[133]

2.3 Historical CO_2 Venting from Subterranean Reservoirs

There are documented eruptions, ventings, and seepings of CO_2 from both natural and anthropogenic events.[134] Lewicki, Birkholzer, and Tsang listed thirteen examples of natural events and eight instances of man-made CO_2 leakages events.[135] They include eruption events from the geothermal fields in Italy; CO_2 wells in Oklahoma, Utah, and Colorado; and subterranean gas storage facilities in Kansas and Wyoming.[136] While the data from naturally occurring CO_2 vents and seeps can provide insight into potential storage problems, it is generally held that accidents from anthropogenic gas storage facilities, mostly methane storage, would be the closest analog for CO_2 storage events.[137] Implicitly, the natural events lack the potential in most cases to experience mitigation efforts that are available and required for anthropogenic gas leak events.[138]

What is now a tourist scenic spot in Utah began in 1936 when an oil well drilling effort failed by breaching a saline reservoir that then prevented casing and securing of the well.[139] The saline was pressurized with dissolved CO_2 gas arising from a carbon dioxide reservoir.[140] Since the failure of the drilling attempt, the well has continuously erupted CO_2 and brine into the air, and plumes of CO_2 have vented into the atmosphere.[141] The well erupts directly at ground level, and tourists can visit what is now known as the Crystal Geyser.[142]

Due to the design of the wellbore and the lack of control valves, the well can be considered equivalent to a maximum leakage rate for similar storage structures.[143] The well erupts approximately 11,000 tons per year of CO_2.[144] However, the eruptions are not continuous but intermittent in bimodal eruptions with resting periods, all in natural flow.[145] It appears that the eruptions need to recharge in between, perhaps correlated with depth to reservoir.[146] Near the well, the CO_2 saturation levels reach 6,000 to 7,000 parts per million (ppm) and have not exceeded 21,500 ppm.[147] In a concluding remark, the overall dangers of the situation were described:

For such scenarios, it appears that in many circumstances the instantaneous leakage rates *do not present a substantial risk to human health*. It also appears that leakage of this kind, even in such spectacular cases, may not be detected more than 100 m from the vent due to atmospheric mixing [italics added].[148]

Whereas the Crystal Geyser was drilled into a CO_2 reservoir that was in communication with a saline aquifer reservoir, such drilling accidents have occurred with CO_2 reservoirs lacking communication with aquifers; that is, the drilling accident led to dry seeping and venting accidents.[149] The Sheep Mountain Drill Well accident resulted from an operational failure,[150] a blowout, in 1982 after safely operating for seven years.[151] The well had produced CO_2 from the Sheep Hill Carbon Dome, located in Colorado, to provide CO_2 streams for enhanced oil recovery projects.[152] This carbon dome is an apt model of a CCS sequestration injection well failure.

The damage to the Sheep Hill well appeared to be subsurface loss of well integrity because the CO_2 seeped from the ground near the well and also vented directly from the well.[153] During the venting period, which went uncontrolled for seventeen days, the well vented approximately 7,000 to 11,000 tons of CO_2 per day.[154] Note that this eruption rate contrasts the same volumes as the Crystal Geyser, but Sheep Hill erupted as much CO_2 in a day as Crystal Geyser does in a year.[155] The failed well was successfully resealed and plugged after several attempts. Three decades later, it remains safely sealed and intact.[156] That no loss of life resulted is ascribed to both the topology of the wellsite and the local winds. These observations speak to the importance of careful selection of well sites for carbon storage projects.[157]

The Torre Alfina geothermal field is located within the Latera Caldera in the Monti Vulsini area of central Italy, near Lazio, approximately 80 to 100 kilometers northwest of Rome.[158] The underlying reservoir is geothermically heated saline waters bearing a substantial amount of CO_2,[159] which in many ways is a suitable representation of CO_2 stored in a saline reservoir. In 1973, exploration activities led to a well blowout and an eruption of CO_2.[160] The result was free-flowing gas from the mouth of the well;[161] the flow rate was measured at 76 kilograms per second, equivalent to 300 tons per hour or 7,200 tons per day, in comparison with Sheep Hill volumes.[162] There was also limited seepage around the wellhead.[163] After several days and approximately 25,000 tons of CO_2 emissions, the well was plugged and fitted with a tall venting pipe to enable excess CO_2 to vent at a height in the atmosphere designed to keep ground conditions safe.[164]

The Travale geothermal field, near Siena, Italy, suffered a similar blowout event to the Torre Alfina.[165] It produced 450 tons of fluid per hour, yielding

113 kilograms per second of CO_2.[166] The well was allowed to discharge and followed up with subsequent monitoring.[167]

A gas well that blew out in Kingfisher, Oklahoma, resulted in "geysering gas and water through surface water bodies."[168] The gas vented from December 2005 to January 2006, but the well was capped and plugged to eliminate the continued venting of gas.[169] By the time the well was plugged, it had produced at a rate of 77,760 kilograms per day,[170] or approximately 86 tons daily.

At the Leroy Storage Facility in Uinta County, Wyoming, gas that was stored underground began to migrate and then seep to the surface.[171] It seeped through the ground and vented to the atmosphere by bubbling though surface waters.[172] The migration was mitigated, if not perfectly resolved, by reducing the injection pressure within the storage reservoir.[173] Approximately 18 million cubic meters of gas were vented prior to resolution;[174] however, this was a very slow release against a large amount of venting surface.[175]

These examples had no reported deaths associated with the loss of gas volumes, but two human fatalities were associated with the leaks from the Yaggy Gas Storage Facility in Kansas.[176] Yaggy presented injection of gas into a salt reservoir.[177] After the onset of injection activities, the well casing of an injection well cracked, enabling free-flowing gas to create surface geysers.[178] The well was plugged and sealed to eliminate additional losses.[179] It is important to note that the accident at Yaggy Gas Storage Facility involved methane in addition to CO_2 and that methane is explosive while CO_2 is not; the cause of the fatalities was linked to the explosions caused by the leakages. Thus, concerns on mortal risk factors should retain that distinguishing aspect of the Yaggy events: it was not a pure CO_2 event but one present with substantial volumes of explosive methane gas.

Aines and colleagues found that while energetic releases of CO_2 could be comparable to similar events from conventional hydrocarbon operations, there are substantial differences.[180] CO_2 is neither flammable nor explosive;[181] it is not caustic or corrosive;[182] and although it could cause injury by asphyxiation, almost no such injuries have ever occurred with CO_2 events due to the likelihood of the CO_2 rising high above the field and dissipating to harmless levels.[183]

Lewicki, Birkholzer, and Tsang offered several broad conclusions on the lessons learned from historical events of gas ventings.[184] One concern raised is that many natural events of CO_2 release were associated with geological events. To the extent that injection activities could either stimulate seismic events or be exposed to natural seismic events could increase the riskiness of the injection operations.[185] Second, unsealed faults or fracture zones could provide high-speed pathways for the escape of CO_2.[186] Third, wells with engineering defects or otherwise structurally unsound could enable the rapid release of large amounts of CO_2.[187] Fourth, certain rupture and venting events could become "self-enhancing" and enable deteriorating conditions.[188] In a very important fifth observation, the historical risk to human health and mortality from CO_2 leaks or eruptions has been very low.[189] One reason is that ambient air conditions appear to have rarely threatened local observers of the events; either air quality remained safe enough or at least did not present a broad and challenging risk to safe escape. Another reason is that even when CO_2 made subterranean contact with water reserves, the waters remained safely potable for humans, despite scientific concerns that such events could lead to chemical hazards to the water.[190]

Thus, despite large-volume venting events and contamination of subterranean waters by CO_2 migration, both air and water volumes have remained safe for humans.

2.4   Risks from CCS Activities

While the evidence is strong from natural storage reservoirs that CO_2 can be securely stored for very long periods exceeding millennia, it is also clear that even those reservoirs can suffer from certain circumstances that enable migration and sometimes escape of those CO_2 volumes. Furthermore, while a century of industrial experience provides a wide set of data from which to take actuarial measurements, the risks of large-scale CCS storage projects would retain certain novel elements that need to be taken into consideration. These risks are reviewed and evaluated below, but in general it will be demonstrated that the risks are identifiable, measurable, and for the most part can be prevented or mitigated with proper planning if undertaken in advance.

The risks from CCS projects have been identified as arising from three primary concerns; the massive quantities of CO_2 that would need to be injected, the long time periods of sequestration required, and the buoyancy of subterranean CO_2.[191] The premier risk is identified as leaks and ventings close to or in close integration with the well assembly.[192]

Six risks are commonly referred to in the literature on CCS storage:[193] (1) groundwater contamination, (2) anthropogenically induced seismicity, (3) risks of asphyxiation to humans and animals, (4) risks of anthropogenic climate injuries, (5) property damage to subterranean assets, and (6) general environmental degradation. These categories can be clustered into three sets: surface-related events, subsurface contaminations, and geological impacts,[194] which is how the subsequent study of risks will be organized in this chapter.

The forecasting of hazards presents complexities. Swayne and Phillips listed four main practical problems with determining the probabilities of expected risk and hazards of CCS operations.[195] First, it remains difficult to accurately model the effects of the geomechanical effects from injection on faults and fractures, particularly on the potential to reopen such weak points.[196] Second, it remains challenging to predict where precipitation or dissolution into brines might occur.[197] Third, it is complex to model wells for their geomechanical stresses as injection continues.[198] And fourth, the mechanics of induced seismicity remain somewhat unclear.[199] Thus, while practical knowledge and experience have been gained in the field over decades, the scientific models of geological, geophysical, geochemical, and geoseismic forces will need ongoing research and development to better enable effective decision procedures for CSS injection and storage activities.

However, it is important to emphasize that CO_2 is not per se a pure waste product, but rather a unique resource that can eventually become inert and embedded in geological structures. Thus, a key factor that distinguishes CO_2 storage from other waste storage paradigms is that subterranean CO_2 fluids become safer over time as they chemically change into either less dangerous solutions or carbonates or become otherwise permanently sequestered.[200] They do not retain their chemical hazard in the way other hazardous wastes might.

Before engaging in more detailed discussions of the risks of CCS, we again stress the importance of focusing on the combination of the probability that certain risks could precipitate into accidents and the scale of the resultant harms if those events were to occur in order to determine the expected impact on welfare. The potential, or probabilistic, damage that could result from CO_2 storage obviously corresponds to a large extent to the risk categories already indicated. However, the expected risks are primarily focused on potential contamination of water reserves, and even in those cases, the damages are not expected to be huge.[201] In other words, most of the damaging effects of CCS can, according to the literature, be estimated and valued in monetary terms and addressed with policy tools such as liability rules, regulations, and certain compensation schemes, as opposed to outright avoidance measures.[202]

2.4.1 Fugitive CO_2 and Other Migrating Fluids

It is clear that despite the limits to engineering models, the greatest risks are during the active periods of fluid injection, with those risks substantially tapering off once injections stop.[203] Once the fluids are injected and stabilized, most events of risk or accident would need novel additions of energy to enable events temporally removed from original injection activities.[204]

After injection, CO_2 fluids can retain mobility and migrate within subterranean structures. There are nine cataloged means of transport:[205]

1. Pressure gradients created by the injection of the fluids.
2. Natural hydraulic pressure gradients already present before injecting the fluids.
3. Fluid CO_2 has a density different from fluids and gases already present within the reservoir structure.
4. The fluid CO_2 can be motivated by diffusion.
5. Because the subterranean surface of the reservoir is uneven and inconsistent, dispersion and "fingering" will result from the contrasting characteristics of the injected CO_2 fluids and the in situ fluids and gases. ("Fingering" is the movement of a fluid as driven by viscosity and other capillary physical forces present in micro-tubular conditions.)[206]
6. Some of the CO_2 would dissolve into the in situ fluids,[207] such as into saline volumes.
7. Mineralization would occur when the CO_2 reacts to form carbonates and other materials.
8. Certain pore spaces themselves resemble microtraps and would enable trapping of the CO_2 fluids.
9. The CO_2 can become adsorbed onto the surfaces of organic materials such as coal.

The behavior of injected CO_2 fluids will greatly depend on their miscibility with native fluids and their buoyancy, which could enable the fluids to return to the surface and leak. When CO_2 is injected into deep saline reservoirs, the gas usually remains immiscible for a certain period.[208] The CO_2 acts buoyant[209] and rises through the earthen layers toward the surface.[210] If the capstone or seal is impermeable, then the CO_2 plume would widen as it encounters that barrier on its vertical trajectory.[211]

CO_2 injected into a petroleum reservoir could be miscible or immiscible depending on the characteristics of the in-place minerals.[212] If the conditions are miscible, there would be no resultant buoyancy of the CO_2;[213]

otherwise, the CO_2 would be less dense than the crude oil and display buoyancy.[214]

Generally, supercritical CO_2, as found at the bottom of the injection well, is less viscous than crude oil or water and would be buoyant in response, that is, the CO_2 fluid would want to float above the other two fluids. Due to the mobility of the CO_2, its saturation into the crude oil or water would be expected to be limited to about 30 to 60 percent.[215] However, a certain volume of CO_2 would become permanently trapped by fingering within the pore structure.[216]

In natural gas reservoirs, CO_2 would be more viscous than the methane volumes,[217] and thus the CO_2 fluids would be expected to sink within the injection zone.[218] Because of the tendency to drop within natural gas, limited opportunity would exist for the CO_2 to become trapped by fingering,[219] but the negative buoyancy of the CO_2 fluids would greatly limit their capacity for mischief.

When CO_2 is injected into coal beds, miscibility is not considered given the solid nature of the coal face. However, the CO_2 would adsorb onto the organic faces of the coal seams;[220] in that process, methane could be desorbed from the coal bed.[221]

Once the CO_2 begins to migrate within the subterranean water or saline reserves, some of the CO_2 fluids would dissolve in the aqueous volumes.[222] Given a couple of centuries, it would be expected that all of the CO_2 could dissolve in this way.[223] Once dissolved, the CO_2 would travel amid the subterranean water flow.[224] However, this should not cause alarm because the speed of water flow, especially saline within deep sedimentary basins, is generally believed to be less than 1 meter for every century passed.[225] If the water volumes have sufficient vertical permeability, then a percolation cycle could enable the CO_2-rich waters to sink and be replaced with unsaturated waters, aiding in quicker dissolution of the CO_2 fluids.[226]

The functional plan of CO_2 storage is to place the CO_2 underground by injecting the fluids into saline aquifers, depleted hydrocarbon fields, or nonproducing coal seams.[227] Therefore, it is likely that the subterranean storage areas would overlap with previous man-made wellbores and testing shafts that could provide potential pathways for the injected volumes to travel.[228]

If the injected CO_2 is not pure or if it contains contaminants,[229] then additional considerations are required.[230] With regard to injection and long-term storage,[231] the inclusion of such contaminants could render the injection stream legally hazardous and become subject to additional regulations beyond the scope of CO_2 injection regulation.[232] In underground storage, impurities affect the compressibility of the fluid CO_2 and reduce the effective storage capacity of reservoirs.[233] Such impurities can also affect the rate of dissolution and precipitation as the CO_2 mineralizes; the impurities might also affect the leaching of heavy metals.[234] However, previous experience with AGI facilities has provided data suggesting that such effects might be minimal.[235]

Researchers have learned that even properly plugged and abandoned wells of recent vintage could be at risk to reopen and provide transit for the injected CO_2 fluids.[236] The injected fluids could encounter water volumes and combine to create carbonic acid, which could then enact corrosion on hydrated cements.[237] There are particular concerns for orphaned wells and abandoned wells that were sealed earlier with poorer-quality cements and for those plugged without cement.[238] There are also a great number of wells from much earlier time frames, but it is generally held that the drill depths of most of those wells would prevent much likelihood of their interference with injection activities.[239] However, most wells properly plugged and abandoned after the adoption of the American Petroleum Institute's (API) Code 32[240] on plugging and proper cements have used cements that match the depths and conditions of each well type.[241] Additional improvements in plugging requirements were developed to protect water supplies after the United States enacted the 1974 Safe Water Drinking Act.[242]

The importance of site selection is a frequent theme in the literature.[243] The size, porosity, and geological and demographic setting of the storage facility are highly determinative of potential risks and resultant harms.[244] Ingelson, Kleffner, and Neilson opined that there exists a general conclusion that while certain risks did exist, the risk of leaks was "very low" if CCS storage facilities were properly sited and properly operated.[245] "A key premise, however, is that storage sites are carefully selected."[246] Also, they concluded that the initial behaviors of the operators would determine the ultimate risk levels during the long-term postclosure period.[247] Thus, the incentives to govern operator behavior need to be both sufficient and in place before the start of any CCS storage activity, including site selection.[248] Furthermore, they found previous research that reservoir seals could securely confine CO_2 for "millions of years and longer," and, barring intruding or failing wells, statistical models demonstrated probabilities very close to zero.[249]

Thus, it might be reasonable to conclude that CCS storage operations have risks but that such risks are readily modeled and potentially preventable or remediable. However, it might be useful to consider the potential impact of a catastrophic failure event. Aines et al. provided a model of what might occur in the worst case: free flow at the wellhead to the atmosphere.[250] Such an event could occur under several scenarios, from a misdirected bulldozer removing the christmas tree of the injection well, or from a poorly secured crown valve after a reworking, or from a failure in the tubing of the well.[251] (The "christmas tree" is the collection of valves immediately following the mouth of the well. Legend has it that the assembly was historically refered to as a "Texas X-mas Tree.") Calculations for such events can be based on the (1) the pressure of the subterranean reservoir, (2) the diameter/radius of the well, (3) the amount of friction and heat transferred from the inner walls of the well, and (4) the maximum escape velocity as determined by atmospheric conditions at the top of the wellbore.[252] This modeling capacity supports the idea that actuarial models might be compliable prior to actual accidents.[253] Their models resulted in free flow rates at the mouth of the well that closely matched the historical data observed from the Sheep Mountain accident.[254] It is worth remembering that Sheep Mountain was remediated and successfully resealed and thus was a contained accident despite the high volumes of CO_2 present at the beginning of the accident.

2.4.2   Risks from Surface and Atmospheric Events

One of the most obvious concerns in the public mind is drawn from ordinary observations of balloons: if you inflate one, then you can expect that the gases inside the balloon will try to find a way to escape. In CCS, the concern is that the stored CO_2 volumes might escape from their storage systems and wreak havoc and harm to surrounding areas, ecologies, and communities. Within these CO_2 escape processes, two behaviors can be distinguished. *Leakage* has been used to describe movement of the injected CO_2 away from the primary formation of original injection, whereas *seepage* has been used to describe migration of the CO_2 out of the ground and into the atmosphere.[255]

The historical experience with engineered gas storage systems suggests that the vast bulk of gases could be safely stored and permanently sequestered, but some small amount is expected to escape to the atmosphere.[256] Similarly, the EPA concluded that there were only minimal risks to ecosystems of asphyxiation or respiratory distress and to agriculture from surface or atmospheric releases.[257]

Surface leakage could occur because injected CO_2 will be less dense and thus possess a buoyant character underground.[258] Because there have been observations of both natural and industrial sources of CO_2 leaks to the atmosphere, there is a strong scientific basis from which to model and forecast the impacts of such events on health, safety, and environmental impacts.[259] In particular, experiences with natural gas storage facilities, CO_2 production for EOR injection, and AGI activities have provided substantial opportunity to develop these tool sets.[260]

Ingelson, Kleffner, and Neilson detail that while CO_2 is generally not toxic at regular atmospheric levels, it can become toxic at certain densities.[261] Ordinary atmospheric conditions present primarily nitrogen (78 percent) and oxygen (21 percent), with carbon dioxide (0.1 percent to 0.4 percent) represented as one of the dominant trace gases.[262]

Sufficiently high levels of CO_2 are known to be dangerous to humans. Flatt reported on the potential toxicity of airborne CO_2 plumes; the primary risk to humans is asphyxiation from prolonged exposure to sufficient levels of the gas.[263] General discomfort at onset, such as headaches, can be low as 1 to 5 percent of atmospheric CO_2.[264] Benson and Cook provided that at atmospheric concentrations of 2 percent, respiratory problems could start and at or above 7 to 10 percent could cause loss of consciousness or death.[265] Humans can begin to pass out at levels above 10 percent, and levels above 30 percent are generally lethal.[266]

On the contrary, no adverse affects at all had been found for concentration levels at or below 1 percent.[267] Routine ambient levels of CO_2 are at 370 ppm, and it has been reported that humans can safely handle CO_2 levels up to 10,000 ppm (i.e., 1 percent), so the risk is presented only from very high-density CO_2 plumes.[268] Note that these levels at 10,000 ppm exceed historical experiences with CO_2 well rupture events, so even the Sheep Mountain and the Aines et al. free-flow catastrophe model fit within this range of safety.[269] Furthermore, temporary emergency exposure limits (TEEL) of TEEL-2 and TEEL-3 for CO_2 require much higher levels of ambient CO_2; thus it is unlikely that a leak would present much hazard to ground personnel.[270]

Furthermore, Ingelson, Kleffner, and Neilson report that the risks to human asphyxiation are well understood and that industrial practices and regulations exist for comparable oil and gas industrial practices to enable CCS operations to be "carried out safely."[271] Thus, the risks of asphyxiation might be present only within a very narrow range of CO_2 accidents and would not be expected at most CCS storage locations, given proper siting methods. While death can occur, it is generally modeled that deaths from ambient releases of CO_2 from CCS storage would be improbable.[272]

Aines et al. provided models that revealed the impact of wind speeds on the ambient safety levels around wells leaking CO_2 at maximum free-flow rates.[273] The safety of the event would be determined by the wind speed, wind shear, and buoyancy of the winds at the time of the leakage.[274] In the worst case, where those three factors were only minimally present and the land surface provides no downflow to drain away accumulated emissions, a danger zone can be calculated.[275] The danger zone for free-flowing CO_2 wells would be limited to a radius of 274 meters from the open mouth of the well.[276] This distance should enable the majority of ground personnel to retreat to safety without major health incident; a six-minute stroll should enable an adequate evacuation distance well within the safe time limits.[277]

There is also little to no chance that a major CO_2 eruption to the surface would be a stealth killer.[278] The CO_2 would exit the well at or at close to the speed of sound; a large roar would likely accompany that atmospheric displacement of the air volumes near the wellhead.[279] It would also be expected that snow would immediately precipitate around the well site due to the chilling effects of the vented CO_2.[280] Thus, the roar of the vented CO_2, the rush of chilled air, and the sight of falling snow and icing of local equipment should give clear notice of the events at hand for ground personnel. Indeed, no special safety measures were required when recent blowouts of CO_2 injection wells occurred, albeit not in a CCS context.[281] Aines et al. found that their models strongly supported the practical observations that such eruptions pose minimal to no risk at all at distances sufficiently away from the well site.[282] Additionally, they found that even close to the well, the risks were limited to a close adjacency to the wellhead; even in that case, the risks were minimal if an effort was made to walk away from the wellhead as soon as the eruption was noticed. Hazards that you can walk away from, which are also incredibly noisy and result in obvious icing and snow precipitates, are not likely to be sudden stealth killers.

Lakes near volcanoes are known to emit large volumes of CO_2. These events have been observed to create such thick blankets of CO_2 close to the ground that loss of life occurred.[283] In northwestern Cameroon, Lake Nyos presented such a venting event in 1986; reports are that approximately 1,700 people perished from asphyxiation.[284] However, the scientific consensus is that the waterborne CO_2 that erupted through a broad lake surface is not analogous to the storage conditions of subterranean CO_2, and thus such worries of public asphyxiation would need better evidence of a feasible pathway before substantial concerns should be raised.[285]

Finally, it should be mentioned that the emission of CO_2 can also cause local negative environmental effects.[286] Seepage of the CO_2 can disrupt the quality of surface soils and reduce the viability of plants in those soils.[287]

However, such damage is limited to the plants thus exposed, and remediation would be feasible in most circumstances.

2.4.3 Risks from Subterranean Events

Smaller volumes of the injected CO_2 fluids would likely become trapped by residual trapping, chemical dissolution, and mineralization; however, the remainder of the fluids is expected to remain in a mobile phase for early periods following injection activities.[288]

Regional impacts of CCS injection include several hazards: induced seismicity, basin-scale displacement of subsurface fluids, and increased risks of seepage to the ground level.[289]

Management of the risks from CCS storage activities can be accomplished; there are four primary concerns.[290] First, a careful site selection process must be implemented with concerns addressed for injection fitness, socioeconomic factors, and environmental factors.[291] Second, scientific and engineering monitors should already be in place to enable observation of the injection activities and the movement of the injected volumes.[292] Third, an effective legal regime should be in place to enable oversight of the CCS activities.[293] Fourth, remedial measures need to be identified prior to the development of the injection wells and storage areas; furthermore, not only should the means by identified but also enabled and put on standby to minimize "the causes and impacts of leakages."[294]

While legal scholars have found that there is little overall risk of CO_2 leakage of any type,[295] the potential for harm remains. The most likely form of harm is to be found in the threat to freshwater supplies for human and agricultural purposes.[296] However, not all scholars share the belief that this potential harm is limited to groundwater and drinking water.[297]

2.4.3.1 Site Selection

For CCS storage projects, few other decisions affect as much subsequent risk exposure as the choice of site selection. Site selection determines the location not only of the surface facilities, where at-well safety considerations are a factor, but also the subterranean location and migratory zones of the postinjection CO_2 fluids. Site selection also engages the geochemistry and geophysics of the storage area. Thus, site selection is one of the most important decisions to be made with regard to safety and risk prevention.

Site selection for the injection activities and the corresponding zones of geological storage is one of the most critical elements determinative of both storage capacity and storage permanence.[298] Various scholars have provided alternative listings of preferred characteristics; often the same scholars have provided alternative lists.[299] Thus, while the centrality of site selection is agreed on, clarity on precisely what will be preferred might remain locally regulated.[300]

Benson and Cook list thirteen factors that should be considered in site selection:[301]

"(1) volume, purity and rate of the CO_2 stream; (2) suitability of the storage sites, including the seal; (3) proximity of the source and storage sites; (4) infrastructure for the capture and delivery of CO_2; (5) existence of a large number of storage sites to allow diversification; (6) known or undiscovered energy, mineral or groundwater resources that might be compromised; (7) existing wells and infrastructure; (8) viability and safety of the storage site; (9) injection strategies and, in the case of EOR and ECBM, production strategies, which together affect the number of wells and their spacing; (10) terrain and right of way; (11) location of population centres; (12) local expertise; and (13) overall costs and economics.[302] [Enumeration added.]

Benson and Cook then provide the caveat that those thirteen characteristics should be employed to provide risk evaluations for five aspects of the injection plans: (1) storage capacity, (2) injectivity, (3) containment, (4) site resources, and (5) natural resources.[303]

In evaluating the cap rock and the seal that it might provide, the cap rock itself should be "regional in nature and uniform in lithology."[304] Given those characteristics, a secondary stage of review should confirm that there is sufficient strength in the rock, a lack of natural penetrations of the rock, a lack of "anthropomorphic penetrations"[305] of the rock, and a minimal amount of "potential CO_2-water-rock reactions that could weaken" the overall sealing capacity of the rock.[306]

Geomechanical factors can affect storage integrity.[307] As the CO_2 fluids are injected, they could increase the pressure in the storage zone and as a result increase the pressure on the cap rock or seal and lead to either new fractures or the reopening of old but closed fractures.[308] Also, depletion of hydrocarbon fields could induce new fractures or faults, leading to potential migration pathways.[309] However, those fractures might also be resealed by carbonate precipitation; thus, careful modeling of the particular site would be required.[310]

Geochemical factors could result in three events that would reduce the storage capacity: the CO_2 would become dissolved in the in situ waters, carbonic acid would be formed, and bicarbonate ions would be released.[311] Additionally, the acidic fluids and molecules could engage in the cement and well structures, leading to migration opportunities.[312]

Thus, the centrality of site selection for operational and environmental safety is paramount. If a site were to fail to meet the necessary characteristics, then operational and safety problems would be readily fore-seeable. To the extent that civil liability rules could ascertain the economic impact of eventual and expected damages, site selection decision processes could be subject to mechanisms designed to incentivize selection of a site with a full set of desirable characteristics. Similarly, regulations could engage in the decision-making process and set explicit and publicized stan-dards for site selection.

2.4.3.2 Induced Seismic Events   There are concerns that injection of compressed CO_2 into subterranean storage could cause incidents of ground tremors or earthquakes, that is, what is described in the literature as seis-micity.[313] The US Geological Service (USGS) found that such injections could cause anthropogenic seismic events and that certain controls on the injection process could mitigate risks and injuries.[314] Injected waters have been identified as the cause of induced seismic events in both the United States and the United Kingdom.[315]

The research literature provides evidence that the intensity of anthropo-genic seismic events was correlated with the volume of injected fluids,[316] which necessarily correlate with the amount of injected energy given the volumes and pressure levels of the reservoir. However, it would take a large volume to create events that ordinary observers would detect.[317]

Cessation of injections would lead to a gradual decrease in seismic events; however, because the subsurface plates of the earth are not smooth and uniform, the intensity of quakes over time after injection might not decline smoothly but unevenly toward none.[318] Models can be and are built to forecast the incidence of anthropogenic seismic events, but accuracy is generally limited in time frame.[319] It is clear from research that the most critical time to affect the ultimate risk of anthropogenic seismic events is during the early years of injection.[320] Siting decisions thus can play a central and determinative role in managing the overall risk profile of CCS storage projects. The risks of anthropogenic seismic events can be lessened by siting injection wells away from sensitive locations such as urban areas or factories.[321]

The USGS report called for the development of a "seismic network" that could quickly identify tremors smaller than 2 on a Richter scale.[322] With such a network in place, the precise locations of tremors could be con-nected to the data on injection wells to enable a better means of identifying risks such as the reactivation of old faults and forecasting future events and impact zones.[323] An early experimental version of a seismic network, deployed near Greeley, Colorado, in 2014, enabled injectors to govern injection rates and depths to mitigate seismic risks.[324] The USGS advocated two policy positions on such a seismic network: that there be a "traffic light" system that responds to risk factors to provide injectors with real-time feedback on operational conditions and that the public have similar real-time access to the seismic data collected in coordination with CO_2 injection activities.[325] Bommer et al.[326] described the three signals of their seismic traffic light system in this way:

The boundaries on the "traffic light" were then interpreted as follows in terms of guiding decisions regarding the pumping operations:

- Red: The lower magnitude bound of the Red zone is the level of ground shaking at which damage to buildings in the area is expected to set in.
- Amber: The Orange zone was defined by ground motion levels at which people would be aware of the seismic activity associated with the hydraulic stimulation but damage would be unlikely.
- Green: The Green zone was defined by levels of ground motion which are either below the threshold of general detectability or, at higher ground motion levels, at occurrence rates which are lower than the already established background activity level in the area.[327]

Bommer et al. explained that the impact of anthropogenic seismic events on structures is difficult to predict because of two factors: (1) "the relationship between the natural vibration frequency of the structure and the frequency content of the motion is a key factor in determining the response of buildings to earthquake shaking"[328] and (2) "the rate at which this energy is imparted by the ground motion is as important as the total energy carried by the seismic waves."[329] Thus, the forecast would need to be both building specific and aware of the energy transfer potential of the land near the structure; both of these data types are likely to be beyond the scope of CCS injection permitting requirements and thus might need ancillary data collection to better ensure awareness of risk for structures sufficiently local to injection fields.

The impact of field seismicity on humans and their health is also complicated.[330] Humans, similar to structures, have a unique frequency sensitivity to movement,[331] so "eggshell client" rules might be a concern for liability rules. However, a guideline to reasonable expectations of tolerable vibrations and seismic impacts might be gained from preexisting rules, both public and private, on acceptable vibrations and shocks.[332]

This risk of displacement of brines into potable aquifers and the potential of contamination of hydrocarbon resources are also mentioned as risks in other literature that points equally at the danger of pressure changes causing ground heave, even triggering potentially seismic events.[333] However, Wilson, Klass, and Bergan argue that "these risks likely will be small with properly managed sites."[334] Therefore, once again the centrality of careful site determination is invoked. By careful selection of injection site and storage zones, the risks of anthropogenically induced seismic events could be substantially reduced.

2.4.3.3 Underground Displacements of Gases and Liquids Subterranean storage of CO_2 will result in fugitive behavior of the CO_2.[335] This is not a statement of risk or of safety; it is the simple fact that the injected pressurized CO_2 fluids would be expected to display kinetic activity to some extent. In certain circumstances, that movement could be harmful, but in other circumstances, it may be of no substantial consequence or, given negative buoyancy conditions, even helpful in terms of stabilizing sequestration efforts. The review of the consequences of CO_2 fluid mobility results in a focus on two key factors; site selection and operational procedures at the time of injection.

The simple fact that water reserves are inundated with dissolved CO_2 does not necessarily imply a catastrophe for drinking water. Many readers of this book have likely enjoyed bottled beverages that were naturally carbonated, as Perrier is from its spring in Vergèze, France. However, the combination of CO_2 with water can result in carbonic acid, which can aid in the dissolution of potentially adverse or hazardous metals and minerals and thus make such contaminated waters unsuitable for drinking water or agricultural purposes.[336] This section reviews the scientific understanding of how such damages might arise and what probabilities of damage face policymakers.

In most onshore settings, subterranean CO_2 is gravitationally buoyant;[337] unless otherwise prohibited, it will migrate through the pore space.[338] Movement of the subterranean CO_2 could result in the gas moving beyond the property limits of the operator. It could also relocate to areas not foreseen as part of the injection reservoir system under earlier permitting review procedures.[339] Simulations of CCS storage for a 1 gigawatt coal-powered coal plant would spread out from the initial injection point to a radius of kilometers after thirty years.[340]

As a result of this fugitive nature and the inherent solvent nature of carbonic acid,[341] the postinjection CO_2 would possess an ability to mobilize and relocate both organic and inorganic compounds, providing a risk to underground biota and water resources.[342] The buoyant CO_2 volumes could also displace other underground fluids, such as brine, and force the movement of those fluids into new locations.[343] Dissolved CO_2 could have an impact on the geochemistry of the subterranean waters and lead to changes in flavor, odor, and taste, as well as provide potential contamination from metals, sulfates, or chlorides.[344]

Clearly the protection of subterranean water supplies is a central concern for policymakers.[345] The EPA identified four main risks to drinking water: (1) contamination of water supplies from co-injectants within the injected CO_2 fluids, (2) migration of stored CO_2 into drinking water supplies through damaged cap rocks, (3) contamination of freshwater supplies from hydraulic-motivated neighboring brines, and (4) the potential contamination from metals and other contaminates from the carbonic acid created by CO_2 dissolving into freshwater supplies.[346]

In a US Department of Energy (DOE) study, 1,800 tons of CO_2 were injected into a subterranean reservoir at the South Liberty Pilot Project in Texas.[347] The reservoir was a saline rich formation.[348] Dissolved metals were discovered to be flowing along with the displaced saline, with the plume dissolving metallic oxides from the mineral substrate.[349] These results were noted within the first week of the test injections, raising concerns that CO_2 injections could rapidly alter groundwater conditions.[350] The injection of CO_2 might enable the CO_2 injection plume to mix with underground aquifers, alter the composition of those aquifers, and displace those aquifers, resulting in a deleterious impact on freshwater supplies and to the local environment.[351]

The range of the damage to water could be enabled by two distinct processes, both from direct contact with the injected CO_2 fluids and elevated pressures resulting from the CO_2 injection.[352] The nature of those risks is, however, different: CO_2 plumes in sequestration reservoirs can extend several kilometers from a CO_2 injection well, but the area of elevated pressure in which brines could be forced into aquifers can extend across tens of kilometers.[353]

Apart from the movement of in-place subterranean minerals, CO_2 injection streams might not always be pure streams of CO_2; they might contain other molecules such as H_2S, which pose health and environmental risks of their own once injected underground.[354] If the CO_2 fluids contained SO_2 when injected, the SO_2 could increase the resulting acidity levels and thus enable greater dissolution of dangerous earthen metals.[355]

There is potential for the injection of CO_2 to stimulate certain geological events, such as breaching of the reservoir's cap rock, inducement of ground heaving, and the inducement of seismicity.[356] The choice of injection site could determine the level of risk of such events; injection into more porous reservoirs appears to present less evidence of such events.[357]

Recognition of these risks can be found in early CCS-related regulations, such as in the EU's CCS Directive.[358] There are repeated lists throughout the directive that draw attention to precisely these subterranean risks.[359] In particular, the directive provides explicit guidance on the selection of storage site locations.[360]

Thus, while the injection of CO_2 fluids into waters, both fresh and saline, is not hazardous in itself, the resultant acidification of those watery targets could result in demineralization and the leaching of metals from the underground storage structures. Even beyond the immediate range of the injected CO_2 fluids, the pressure waves from that injection could propagate underground and enable additional geochemical results at some distance from the injection site. Local conditions and circumstances, plus operational behaviors during the injection period, would be determinative of a CCS injection facility's risk profile.

Again, proper siting would potentially prevent or reduce these risks, as mineral content of the storage structures and nearby geologic areas could be reviewed to minimize the at-risk ore contents of those subterranean structures. Proper attention to injection pressures and monitoring of water migration and those migratory liquids' chemical composition could also enable damage prevention and mitigation. Thus, although risks are present, they can be addressed with rules of civil liability and regulatory measures.

2.4.3.4 Centrality of Site Choice and Risk Creation

As the discussion in this chapter has noted,[361] much of the literature holds that provided that a location site is correctly selected, the CCS risk profile is in fact rather limited, especially when the oft-feared long-tail risks are considered. In this respect, the previous quote, often repeated in the literature, from the IPCC study showing that the risk of leakage of CO_2 from a well-managed sequestration site is "likely to be small in magnitude and distant in time"[362] is important. For sites that are well selected, designed, operated, and monitored, it is likely that the vast bulk of stored CO_2 would remain sequestered for centuries.[363]

The literature holds that significant scientific research has been done concerning the risks associated with CCS. The key issue is that the literature not only holds that although the nature of those risks remains basically the same during the CO_2 injection and operation, as well as after the closure of the site, the probabilities of those risks of carbon leakage after the injection phase are expected to diminish over time.[364] One study holds that "the good news is that the likelihood that the risk of some events occurring that result in an unexpected release of carbon dioxide more than 10 years after termination of injection will become increasingly remote due to geochemistry conditions."[365] Other studies also mention that CO_2 storage indicates a potentially declining risk profile.[366]

Many studies have indicated that notwithstanding remaining uncertainties, it is possible to quantify the risks involved in CCS, since estimates can be made of probabilities of releases as well as of the potential damage that could occur as a result.[367] Such a monetization of the expected losses can be done in a site-specific manner, based on an assessment of the probability of the risk materializing and the potential damage that could result.

Some scholars also stress that there has already been practical experience with CO_2 sequestration, which would also confirm the low probability of the risk. CO_2 was used to facilitate oil extraction, which involves pumping large volumes of the gas underground. Experience would indicate that the safety record of CO_2 in those cases was excellent.[368] Other studies too refer to experience with the injection of fluids that suggest that long-term containment of CO_2 can be achieved in sites that are appropriately chosen, constructed, operated, and monitored.[369]

These results strongly suggest that if the behavior of the operators can be incentivized to select the injection site carefully and to govern their operational behaviors during the injection phase, most of the risks could be addressed. Thus, one might reasonably argue that the risks of CCS storage are mostly confined to acts well bounded in time and correspond well to the time periods in which operators might be most responsive to both rules of civil liability and forms of regulation.

The risks of CCS storage facilities in all likelihood could be squarely addressed with existing forms of mechanism design, including rules of civil liability. But even if these efforts to prevent CCS-related accidents were to fail to prevent all accidents, it would appear that many, if not most, of the remaining accidents could be cured by remediation efforts, which is detailed in the following section.[370]

2.5 Of Permanence, Risk, and Remedies

The important question from a liability and insurance perspective is to what extent it is possible to assess the long-term risks of a CCS project, especially those that occur after the cessation of injection activities and after the operator has no active operations at the site. This is particularly important since liability and insurance are often well suited for sudden and accidental events but less for so-called long-tail risks. Hence, understandably, a lot of the debate focuses on the nature of the potential long-tail character of risks related to CCS.[371] Therefore, in this study, we strongly focus on the final phase of CCS, the geological storage of CO_2, since that is the most interesting phase from a liability and compensation perspective.[372]

The concept of CCS-related CO_2 storage has been described as a sequence of life cycle events, each with distinguishable risky activities and resultant hazards; broadly speaking, they range from exploration, development, operation, closure, to postclosure.[373] A similar pattern has been found for the nonconventional fracturing extraction technologies.[374] It is said to be necessary for any effective CCS regulatory regime to cover the whole project life cycle.[375]

2.5.1 Identification of Long-Run Risks

Beck and colleagues documented a five-period model of the CCS life cycle: (1) exploration authorization phase, (2) storage authorization phase, (3) injection notification phase, (4) cessation of injection notification, and (5) closure authorization and transfer of responsibility.[376] These five phases underlay the development of the Model Framework of the International Energy Agency (IEA).[377] Similarly, Havercroft and Macrory described a life cycle system of five periods: (1) exploration and site selection, (2) active injection, (3) closure and removal of the injection facilities, (4) postclosure observation and monitoring, and (5) a long-term period after the transfer of responsibility to a public actor.[378]

Other sources in the literature find fewer periods but multiple additional dimensions or activity, increasing the complexity of life cycle analysis. Ingelson, Kleffner, and Neilson describe a three-period model of the CCS life cycle: before start-up, operations and closure, and afterclosure.[379] They focused on the liabilities associated with long-term storage, so they provided a 3-by-5 matrix of life cycle concerns for that stage of operations.[380] First, they broke out the types of postclosure events as geographically delimited to local to the well site or injection site, regional to the areas surrounding the storage facility, and global impacts to climate stability from large-scale greenhouse gas releases.[381] Those regional characteristics were set against five types of injuries or foreseeable hazards: (1) potential hazards to human health and safety, (2) hazards to subterranean water reservoirs, (3) hazards to terrestrial and marine ecologies, (4) increased risks from induced earthquakes, and (5) the potential for other chemical impurities to manifest during the storage process.[382]

Aldrich, Koerner, and Sloan presented a similar three-stage life cycle: (1) preinjection and operation, (2) closure, and (3) postclosure.[383] They state that the preinjection and operation periods could last about two to three decades at a given well site.[384] They provided a listing of short-term operational liabilities,[385] that is, prior to closure and a list of the long-term concerns at closure and postclosure.[386] Thus, the liabilities appeared to fall into two camps: near term and long-term latency.

These issues would need to be addressed as the storage site proceeds from closure to postclosure.[387] Current scientific consensus holds that such goals should be readily obtainable in certain circumstances.

2.5.2 Remedies for Migration Problems

Even if one were to accept the idea that CCS storage has manageable risks, one would still need to be prepared to resolve accidents that did occur despite the preventative efforts. A prudent operator would need to be able to demonstrate that mitigation and remediation techniques were available and feasible.

There are several known methods for addressing problems with migrating CO_2 or other fluids caused by the CCS-related injections.[388] The primary potential incidents are listed as (1) leaks within a storage reservoir, (2) leakage away from the storage area through faults and fractures, (3) shallow groundwater problems, (4) vadose soil and grounds,[389] (5) surfaces fluxes and emissions, and (6) surface water.[390]

Prior to actually undertaking remediation, detection is necessary, and that requires monitoring technologies that are installed and operational prior to the event.[391] The injection rates of the fluid CO_2 can be monitored; such technologies are in place in many EOR and AGI fields.[392] Injection rates can be affected by fluid density, temperatures, changes in pressure, and changes or differences in fluid composition.[393] Injection rates can be monitored at both the top of the well and the bottom near the perforations;[394] such measurements can also be made continuously and streamed live to a control room.[395] Automatic shutoff or shutdown controls can be programmed on a variety of injection rates problems based on those data streams.[396]

There are direct means of measuring migration. Monitoring subterranean movements of the injected fluids is possible, although improvements in the technology are desired.[397] There are a few limited means to directly track the migrating fluids, including isotopic sampling and sampling from minor wells.[398] One can also co-inject various tracing chemicals, called tracers, that are easier to detect than the CO_2 itself.[399] Most of these methods require historical baselines against which to examine later water or other samples, so testing must be done in advance of initial injections.[400]

Indirect means to test for migrations include seismic, nonseismic geophysical, nonseismic geochemical, and electrical measurements.[401] Seismic involves shaking the earth to get vibrational responses; doing this from multiple surface locations and across time frames can provide fairly accurate four-dimensional maps.[402] An interesting result of CO_2 density in subterranean structures is that it becomes more readily and discretely detectable as it approaches the ground surface, so as leaks become more likely, they also become much more detectable.[403] Tilt meters placed across the injection field can enable early detection of subterranean movements that result in minor surface displacements.[404] Changes to the underground rock layers will also result in changes in the patterns of gravity detectable at the surface.[405] Similarly, changes to those same rock layers would also affect their electrical conductivity and static electricity levels.[406]

If the incident is due to leaks from the injection well or previously abandoned wells, repairs and replacement of the well could address the problem.[407] Wells themselves can be monitored for weaknesses susceptible to facilitating an accident.[408] Engineers could monitor the cement bond logs, the well pressure stress tests, the injection pressure levels, and the so-called noise logs for clues to the integrity and function of the well structure.[409] The wells could be recompleted, the injection tubing could be replaced, or additional cementing could be added.[410] Injection wells could be plugged and abandoned if those remedies failed to take effect.[411] For more difficult wells, a "well kill" could provide cessation of the leaks prior to additional cementing, plugging, and abandoning efforts.[412] Such efforts would benefit from extensive field experience in similar oil and gas events.[413]

There are also known means to remediate CO_2 leaks and seepage from CCS storage sites: [414]

Leak mitigation approaches typically involve proposed short- term engineering interventions to reduce and/or stabilize the flow of CO2-rich solutions from the storage formation. These include: the manipulation of formation fluid pressure, physical extraction of CO2-rich fluids, the creation of *in situ* hydraulic barriers via intentional water injection and the intentional dilution of leaked fluids.

Remediation approaches may involve the physical blocking of the fluid leakage pathway via a number of physico-chemical processes and/or microbiological treatments. These latter options are intended to remediate fluid leakage by the permanent closure and sealing of fluid flow pathways.[415]

In summary, there are a variety of both direct and indirect means of monitoring subterranean events after the initial injection of CO_2 fluids. When problems are detected, there are multiple options for reducing or reversing risk factors. Finally, there are means to remediate certain types of CCS-related damages. In conclusion, operators have the means to prevent and detect accidents as their predictive factors emerge, and should injury occur, many forms of damages could be remediated or restored.

2.6 Summary

This chapter has provided a review and survey of the science and technology associated with the development of CCS storage systems. It has provided explanations of the geologic processes that enable the long-term storage of fluidized CO_2. In that explanation, the potential for the CO_2 to remain in motion for decades or even centuries after initial injection was explored. However, it was made clear that as long as the actual storage reservoir zone was sufficiently stable and not otherwise suffering from fractures or poorly abandoned wells or mines, then the vast majority of the stored gases should remain in storage for geological periods of time. Thus, there is a physical science understanding that the two most determinative issues are avoiding poor storage locations and the prevention of operational events that could convert good storage sites into poor ones.

The technology to achieve CCS-based injection and storage was reviewed. There is a century of experience of injecting, storing, and retrieving natural gas and decades of experience with helium and CO_2. The necessary technology to inject CO_2 into subterranean storage zones is currently in the field with EOR and AGI projects. The injection technology is ready, off-the-shelf, today. There are a variety of injection options based on both reservoir targets and based on subterranean geophysical differences. CCS injection and storage can be implemented both onshore and offshore; both formats have current operational projects. From experiences from earlier oil and gas EOR and AGI projects, building experience with currently operational CCS projects, it is becoming clear that CCS storage is a functional concept that can be implemented in the present time frame. Furthermore, those experiences suggest that as long as storage systems are well selected and well stewarded, such efforts can be undertaken with few to no reports of substantial injuries.

The technology to monitor and observe the injection and storage activities was reviewed. It was shown that there are diverse menus of options to monitor the movement and activity of the injected fluids both directly and indirectly. In many cases, it appears to be possible with existing technology to become aware of incipient harms before they occur. Such activity monitoring could enable feedback into decisions regarding both precautionary efforts and activity levels. Given the ability to observe the operational activities of the operator and the ability to observe the impact of the injected fluids as they move into long-term storage, the potential to minimize and mitigate at the earliest point in time could substantially reduce risks and concerns. Previous industrial experience shows that with sound stewardship, accidents can be kept rare. That modern technology would enable such responsiveness to incipient dangers prior to the event of injury should alleviate concerns some parties hold that CCS storage might present ungovernable risks and hazards and that it in fact presents a sense of "false hopes."[416]

The risks and hazards that could result from leaks, ventings, and subterranean aquifers intrusions were reviewed. Our historical review of similar accidents and events showed both natural and man-made CO_2 geysers and leaks; a body of scientific data on the mechanisms and impacts is available for CCS policymakers. It was shown that such events were either brought under control and remediated or were in fact of no substantial harm requiring restoration or repair. This is not to say that no injuries occurred, but very few fatalities appear to have been historically associated with surface-based venting events, and no reports of substantial fauna or flora damage were found. Furthermore, it was found that remedial actions that could be undertaken lessen the risks and repair what damages did in fact result. Even worst-case models of surface-based events appear to match certain historical examples; those cases were controlled and remedied. Subterranean damages remain a substantial concern; however, again, proper site selection and well-stewarded operational procedures would appear to minimize the potential risks.

But to be clear, there are risks and hazards to CCS storage activities; it is most assuredly not without them. The risks could manifest into harms at early periods and at periods much later. However, the geophysics and geochemistry of CCS storage systems appear to create a unique situation in that the activities that give rise to early injuries would appear to be the very same activities that give rise to later injuries, barring the introduction of late activities or energy sources. Thus, while normally long-latency periods might create a greater concern for the efficacy of regulations or civil liability rules to govern risks from dangerous activities, in this case the risks are the same. As long as the efforts to govern the near-term risks are effective, they would be likely effective in the governance of long-term risks.

The following three chapters address the potential to affect operators' decisions with civil liabilities, followed by a chapter on the function of public and private regulations in those decisions. But in conclusion, it would appear that the science and technology dimensions of CCS storage systems would be amenable to the needs of those governance mechanisms.

3 Clarifying Liability Rules for CCS

Scholars have called for clarifications to liability rules in order to further the application of CCS storage for greenhouse gases. This chapter provides a survey of the dominant rules of civil liability and how they might be applied to CCS liability concerns. It begins with a historical review of the literature and its mounting recognition of the complexity, confusion, and lack of certainty provided by existing, or previously existing, liability rules as they appeared to apply to CCS projects and their risks.

While the rules of civil liability can be relied on to provide efficient incentives to actors engaged in CCS projects, it is important to correctly choose which rule or rules might be applied to create those incentives. This chapter notes that there is currently confusion or uncertainty over how rules of civil liability might be applied to CCS projects. Beck et al. stated, "To ensure safe, permanent CO_2 storage, efforts to demonstrate the technical, safety and environmental viability of commercial-scale CCS projects must therefore be accompanied by parallel regulatory developments."[1]

Without additional clarity with regard to future financial responsibility for postclosure CCS-related liabilities, "it is unlikely that [CCS] projects in the United States will be developed at a level that, collectively, could have a measurable impact on climate change in the near future."[2]

This chapter examines the basic civil liability rule options that could be implemented to govern CCS operations. It reviews and surveys the scholastic literature that has identified a variety of liability questions and raised concerns for rule and standard setting to be developed to better enable the feasibility of CCS storage systems. It provides a review of the existing multiple, potentially overlapping, and simultaneous liability rules of strict liability, trespass, nuisance, and negligence that might be applicable to the CCS storage industry.

The chapter next evaluates the methodology of the law and economics school as a means to assist in the identification of robust liability rules and

standards for the CCS storage industry. It provides an overview of the literature's perspective on rule choice.

Once it is recognized that a rule of civil liability could be usefully
employed to govern the risks of an activity, it becomes necessary to choose
a rule or rules that might apply. The chapter then proceeds to examine the
two major rules of negligence and strict liability to determine which
one would prove more robust for the circumstances of CCS operations and
risks. The chapter concludes with a recommendation for the rule of strict
liability, and its related torts of trespass and nuisance, over the rule of
negligence.

Chapter 4 further explores policy issues related to the implementation of
civil liability rules. Chapter 5 is dedicated to the specific issue area of how
to most robustly address the liability concerns that might arise after the
closure of a CCS storage facility.

3.1 Scholastic Calls for Clarification of CCS Liability Rules

The need for clarity and stability for CCS-related activities has been a topic
of work for a decade. The literature early on identified the unique character
of CCS's long-term latent liabilities.[3] Although our study provides its own
list of policy concerns regarding CCS liability issues, an outstanding body
of literature preceded it. This section provides an introduction and survey
of that literature.[4]

Dana and Wiseman recently wrote that the main goal of policymakers
calling for the implementation of CCS technologies is the effort to provide
a sufficient carbon sink capacity to mitigate the impact of ongoing combustion of fossil fuels on anthropogenic climate change hazards. In order to
provide sufficient capacity, the nascent CCS industry would need to provide tremendous volumes of storage in a wide array of locations around the
world.[5] The scale is of a magnitude that likely no further discussion is
needed to justify that such global implementation of an novel technology
would benefit from the ex ante clarification of how to address some of the
novel liabilities likely to arise after its implementation. Thus, little discussion is found in the existing literature of a need to explicate the need for
clarity, but much discussion can be found in the literature dwelling
on which types of liabilities and injuries would be in benefit of such
clarification.

The lack of clarification on these postclosure liabilities prevents the
development of any substantial CCS projects within the United States;
there are insufficient incentives to develop even pilot projects.[6] "Ideally,

state or federal policy makers will design a CCS regulatory structure to diminish identified risks or to avoid them altogether through risk management," advocated Monast, Pearson, and Pratson.[7]

As early as 2005, McLaren and Fahey identified the concerns of long-term latency risks for CCS projects. They detailed how the continuance of the storage activity legally gives rise to a continuity of an act that could give rise to liability; barring legislative relief, there might be no time limits to the liability exposure for placing CO_2 into subterranean storage.[8]

In the normal case, where a tort is time limited to a defined period after the injury arises,[9] there is a safety point in time beyond which the tortfeasor would no longer be liable. Australian rules do provide limited examples of statutory provisions that enable certain latency torts to extend their liability to a period dating from the moment of the victim's discovery of injury.[10]

McLaren and Fahey offered an opinion that CCS storage systems were not likely to present such latency problems, yet they argued that the act of storage itself could give rise to fresh injuries at long temporal distances.[11] This is certainly an odd result, one arising from a set of circumstances not well aligned with the fundamental assumptions of more routine accidents. Clearly the circumstances of CCS storage activities would benefit from additional legal clarification on when liability arises, how long each injury remains within a limitation period, and if ever the liability from never-ending storage could be cut off or indemnified.

In 2007, Mace, Hendriks, and Coenraads found that linguistic confusion in international conventions governing environmental law and similar areas of concern could lead to confusion as to whether those conventions apply to regulating CCS operations.[12] They listed multiple terms, which were implemented in prior conventions without forethought to potential application to CCS concerns.[13] Clarity could be gained by a consistent approach to interpretation. In another context, the conventions sometimes relied on positive lists of assertively regulated materials, or negative lists of items not regulated, but often overlap of treaties or conventions could lead to interpretative frustration on CCS matters.[14]

Pollak and Wilson raised concerns as early as 2008 that the proposed rules in various states and at the federal level were developing inconsistently along three main lines.[15] They found that substantial differences existed in risk assessment requirements, the postclosure waiting periods, and the differing manners of addressing postclosure liability planning.[16] They also identified gaps in regulatory drafting.[17] They advocated a balancing of the advantages of the "policy laboratories" at the state level and the policy efficiencies attainable at the federal level.[18]

Also in 2008, Trabucchi and Patton identified the risk of a particular moral hazard: if the operator were to face little or no financial risk, this condition could lead to poor siting selection choices and management decisions.[19] They were therefore opposed to financial solutions whereby the risk would be transferred to the public since this would remove incentives for a proper site selection.[20] Trabucchi and Patton strongly rely on the deterrent effect of an exposure to liability, arguing that when CCS operators bear the costs of safely operating and closing their facilities, the companies will have a financial incentive to site, design, and operate facilities in a risk-reducing manner.[21]

The EU's CCS Directive was enacted in 2009.[22] In the preamble to the directive, several paragraphs address liability issues concerning CO_2 storage facilities. Paragraph 8 documents the historical call for the development of the "necessary technical, economic, and regulatory framework in order to remove existing legal barriers." Similarly, paragraph 12 reviews the amendments under the 1996 London Protocol and the OSPAR Convention to enable maritime CCS-related activities. Paragraph 30 provides guidance on how liability rules could interface with existing EU environmental and climate protection efforts.[23] The directive calls for member states to determine how liabilities arising from both the injection phase and the postclosure phase should be handled.[24] Thus, the liability issues are to be set at a national or domestic level.

In that same year, Flatt identified several central questions that needed resolution to clarify the liability surrounding CCS-related activities: (1) the legal jurisdiction of CCS rules, (2) the property rules to govern CCS activities, and (3) the assignment of liability to particular actors given the first two sets of rules.[25] He assigned several characteristics to the jurisdictional question, listing physical location, chronological position on a time line, notional ideas of the retention of CO_2 through those dimensions, and a notion of how to define failure.[26] Without clarification of these basic issues, Flatt argued that the decisions necessary for implementing CCS storage might not be properly resolvable.

One of the most formidable barriers facing potential CCS operations is the possible liability costs of these operations. If the costs of CCS outweigh the benefits, it should not go forward and, generally, we rely on complex common law liability to send that market signal. However, *liability is much more uncertain here, where economic signals do not operate efficiently.* ... Without this predictability, the threat of uncertain liability costs will likely deter a large number of potential operators.

Thus, *it is necessary to create and adopt a liability scheme* that encourages private industry to implement CCS operations, while protecting the public and the public interest [italics added].[27]

One need not agree with Flatt's position that the liability rules must be designed to foster CCS applications to see the logic behind his position that the existing situation provides poor or few economic signals to the market, which results in nearly certain inefficiency for decision makers on either side of the CCS policy debate. Poor-quality signals in the market lead to indeterminacy or inefficiency of results, neither optimal for policymakers regardless of intent to encourage or resist the development of CCS.

Flatt did convey a need for ensuring that public health, public safety, and property rights be protected regardless of the eventual result of rule setting: "Any liability scheme must also ensure that public health and private property rights will be protected and that any harms will be compensated."[28] Flatt found that EOR case law had evolved two distinct governance rules: one that favored unitization to align rights and protection and an alternative rule that held operators liable for resultant mineral rights infringements.[29] Subterranean injection of hazardous wastes is routine in the United States; Flatt noted that while torts cases have found liability based on nuisance, negligence, and trespass, courts have limited findings of liability to scenarios that interfered with the "'reasonable and foreseeable' use of the subsurface."[30]

Governance is routinely addressed within physically limited jurisdictions, but CCS-storage affects both local and global conditions; a policy decision needs to be made to guide which authority is responsible for which potential regulatory issues for CCS.[31] The very long-term time frame of CCS storage is another concern regarding which authority is likely to remain in place for the required duration, which authority should be concerned with which section of the life cycle of CCS storage, and which means should be used to determine such decisions; all must be addressed prior to the onset of CCS activities. Finally, assignment of liability to specific actors for specific consequences or legal acts should be established from the start to ensure that those actors can make reasonable decisions and plans to address their potential risk exposures.

Also in 2009, Rankin listed three options on how to address the liability problems facing in situ, or postclosure, CCS storage activities:[32] to provide an eventual transfer from the private operators to public authorities; to facilitate private contractual means of addressing potential liabilities, such as through insurance-based indemnifications; and a combination of the first two options.[33] Within the second option, Rankin advised on strategies such as self-insurance, performance bonding, private liability insurance, risk pools, and catastrophe bonds.[34]

Rankin observed that the United States, unlike the coordinated elements within the EU's CCS Directive, has not explicitly provided how CCS liabilities might interact with a host of federal laws and regulations, such as Comprehensive Environmental Response, Compensation, and Liability Act of 1980, commonly known as "SuperFund" (CERCLA), and the Resource Conservation and Recovery Act of 1976 (RCRA).[35] He advocates that the federal government should provide guidance to avoid confusion[36] in order to resolve the potential liability exposure of future operators and clarify and delimit the purview of overlapping regulatory agencies.

Rankin also called for a statutory clarification on the rule of civil liabilities to provide a rule of negligence for lesser accidents and injuries and for a rule of strict liability for cases of extraordinary circumstances;[37] however, he is unclear as to how such an instrument might function.

In 2010, Ingelson, Kleffner, and Neilson advanced the earlier discussions on the extent of developer and operator long-term liability for CO_2 leaks; in particular, they addressed leaks from petroleum reservoirs as one of the critical issues to be addressed prior to the onset of private investment in CCS.[38] They claimed that while other stages of CCS development liability were previously well addressed, the risks associated with long-term storage lacked comprehensive analysis in the literature.[39] They developed a long-term liability model for North America predicated on the comparative analysis methods of John Reitz. Their model targeted the promotion of private CCS investment and the minimization of long-term liability costs to taxpayers.[40] They further identified four basic challenges for the "very long-term risks" of CCS storage facilities.[41]

First, Ingelson, Kleffner, and Neilson were concerned with balancing the principle of the polluter pays with potential needs for limited liability.[42] Second, they were concerned with the lack of knowledge on the distribution of risk in terms of both probability curves and potential instances.[43] Third, they were concerned that as CCS storage facilities went into postclosure mode, knowledge might be obtained as experiences were gained and that previous negotiations on risk allocations might be discovered to have been misforecast.[44] Furthermore, it appears that they were worried that the evolving knowledge base might not be smooth or consistent, at least in the early years, compounding the confusion on the allocation of risks.[45] Fourth, they were concerned that there might be substantial difficulties in properly pricing private insurance for CCS storage facilities, particularly for their postclosure liabilities.[46]

For these four reasons, Ingelson, Kleffner, and Neilson first found that it was unlikely that CCS storage operators would be held liable for long-term storage risks;[47] they also implied that this uncertainty itself should be resolved by efforts to clarify the liability conjectures for operators, authorities, and the general public.[48] They called for taxpayers to bear a minimum of liability, so they advocated that the operators of CO_2 storage facilities bear all of the associated liabilities of carbon storage until a time close to the postclosure certification of the storage facility.[49]

In 2011 new perspectives emerged that certain subterranean risks might be more manageable and preventable than previously assumed by earlier research literature. Abend highlighted that it is improbable that subterranean migrations can be completely prevented or controlled but that many kinds of migration might also be harmless.[50] She raises an argument that standards will need to be set in order to balance the risks of migrations versus the risks of additional risks to climate change.[51]

Adelman and Duncan suggested that common law liability is likely to play a modest role in promoting safe sequestration of CO_2.[52] The reason they argue so is that CCS potentially creates a mix of short-term and long-term risks that are difficult to be addressed in a liability regime. They refer to studies (e.g., that of Trabucchi and Patton) that were strongly opposed to transferring the long-term stewardship to government since this would de facto remove liability from operators. That could potentially create a risk of moral hazard—for example, in location choice if operators knew that the government would pay the costs of long-term liability.[53]

Swayne and Phillips offered a terse assessment of Australia's state of liability rules for CCS in 2012, reporting that "differing liability rules are presented across the jurisdictions, creating unnecessary legal uncertainty and higher transaction costs, acting as a barrier to the commercial development of CCS technology."[54] This reflects that Pollak and Wilson's earlier concerns remained problems.

In order to determine what was preventing the necessary development of CCS storage capacity, Davies, Uchitel, and Ruple confidentially polled potential CCS operators, current CCS researchers, CO_2 emitters, potential regulators, and other groups engaged in the CCS industry.[55] In 2013, they reported that in contrast to their expectations that the primary hindrances to CCS development were insufficient technological capacity and a lack of historical experiences or full-scale test projects, their survey had demonstrated that parties close to CCS development felt that the three of the four largest factors holding back CCS development were the unresolved uncertainties in CCS liability risks, climate change liability risks, and the general lack of clear CCS regulatory schemes.[56] Thus, eight years after McLaren and Fahey, uncertainty over a variety of civil liability exposures were the predominant factors holding back CCS development.

In 2014, Jacobs reviewed which liability questions were most critically in need of clarity and found several key concerns that need addressing prior to the onset of CCS storage activities.[57] Jacobs reports that industry has two primary legal concerns: the unknown extent of potential long-term liability exposure and the financial reporting impacts of carrying such risks.[58] She also listed needs for clarity on property rights to land, water, and pore space; on ownership of fugitive gas plumes; and on the protection of subterranean waters.[59]

Jacobs countered those corporate concerns with a reminder that federal laws in the United States fail to provide clarity on long-term liability for hazardous wastes and hazardous substances, yet those chemicals are stored nevertheless.[60] Jacobs did recognize that the long-term time frames of CCS storage were of a potentially differentiable character from that of the previous hazardous storage operations and could merit stronger justification for clarifications on liability questions.[61] Jacobs concluded that there were "three major legal obstacles" to address before the "widespread commercial development of CCS": (1) resolution of long-term liabilities, (2) clarity on the ownership of both pore space and injected gas volumes, and (3) ownership and identification of injury to subterranean waters.[62]

Thus, it would appear that in the literature, the question progresses immediately from, "Is clarification needed?" to, "Which clarifications are most needed?" According the literature we have surveyed, the civil liability rules need both major and minor points clarified to facilitate CCS development. Although the list of critical issues has been corralled to a smaller set, it is useful to review the list of concerns in the review of over a decade of CCS civil liability literature:

1. Long-term latency of unknown injuries
2. Temporal delay in onset of fresh injuries from long-ago acts
3. Mismatch of federal or national policies and state or local policies on CCS liabilities
4. Linguistic lacks of clarity within international environmental conventions as to whether CO_2 injection is waste disposal, a resource storage project, or some third unknown category of activity
5. Moral hazard of operators facing limited or no liabilities
6. Uncertain jurisdiction of geographically large storage sites or zones
7. Lack of clarity for certain property law issues

8. Choice of actors to be held liable
9. Determining who holds liability over the long run of CCS storage after closure
10. Determining how compensation might be provided after closure
11. Query on whether negligence and strict liability might be better suited to different types of accidents at CCS sites: negligence to lesser events and strict liability to more severe events
12. Questions on balancing the polluter-pays principle with encouraging the implementation of CCS carbon sinks
13. Questions on the actual probability curves for risks associated with CCS projects
14. Questions on the impacts of rapidly advancing practical knowledge on CCS risks and the potential hazards of overly specific rules

While questions remain on which liability issues are more critical to resolve, scholars generally seem to agree, particularly as far as site selection, injection, operation, closure of the site , and postclosure monitoring is concerned, that liability rules can play an important role in providing incentives.[63]

However, the literature equally argues that the role of tort liability is much more limited as far as long-term stewardship (the long-tail risk) is concerned.[64] There seem, however, to be overstated expectations with respect to the role that can be played by the liability system with both the operators and the public at large. As already noted,[65] many consider the creation of a proper and clear liability regime important to provide certainty to industry. Concerns about potential liability would hence be an important barrier to develop CCS projects.[66]

There is apparently a demand for a clear liability regime not only from industry but also from the public at large. Wilson, Klass, and Bergan stress that a clear and transparent liability regime may help the public understand and have confidence that risks will be actively managed and, in the event of an accident, effectively remediated and compensated.[67] The way in which (also long-term) liability is addressed may therefore have an impact on public perceptions of CCS.[68] This shows that the demand for a liability regime is not always based on objective valuations of the risk but may have a symbolic function—on the one hand, toward the involved industry as reinsurance concerning the limits on their long-term liability exposure and, on the other hand, toward the public at large. The latter is, however, a double-edged sword; for example, transferring liability for long-term stewardship to government could signal to the public that industry may not be liable for (remaining) long-tail risks, but could provide reinsurance that would be available (by government) in the event of damage compensation.

Adelman and Duncan correctly pleaded for a realistic approach concerning the possibilities and limits of liability rules for CCS. They distinguished between the different functions of liability rules (prevention/deterrence and compensation) and addressed how these functions can play a different role in the various phases of the CCS life cycle and point at alternative regulatory instruments to liability rules that could fulfill this role. As a result, they argue that liability rules could provide meaningful, if albeit limited, deterrence to poor site selection and operation.[69] Nevertheless, civil liability rules would likely efficiently address a range of circumstances facing the activities and risks of CCS operations.[70] Pendergrass has called Adelman and Duncan's efforts to focus primarily on regulatory efforts to govern the risks of CCS operations "too narrow" and argued that a broader view encompassing rules of civil liabilities would be needed.[71]

Moreover, to the extent that liability rules may not provide perfect incentives for deterrence, incentives could equally be provided through regulatory standards aiming at rigorous site selection, diligent project management, and monitoring.[72]

In conclusion, dozens of scholars have called for clarity on issues related to planning for liabilities related to CSS storage activities. A variety of theoretical assumptions have been employed in supporting those calls. Our study attempts to address many of those questions with answers and policies drawn from and supported by law and economic research.[73] From here, this study will review the legal options available to address claims of civil liability. Four primary rules could be brought to bear: (1) strict liability, (2) trespass, (3) nuisance, and (4) negligence.[74] Those rules have been well examined within the literature of law and economics, and certain circumstances are known to be efficient to certain rules; thus, the circumstances of CCS storage activities can be evaluated in light of those findings to determine which rule might be more robust than the other rules of civil liability.[75] From those analyses, we produce a policy recommendation for the rule of strict liability.[76] In the following chapter, we also provide additional analyses on several policy options on how best to implement and operate that rule of strict liability.[77]

3.2 Multiple Rules of Overlapping Civil Liability

Another point of potential confusion is determining which rule of civil liability might be applied in the case of particular injuries resulting from CCS storage accidents. There are four basic rules that could be applied, and in certain cases they be applied for legally distinguishable injuries arising from a singular accident. As such, it would be useful to determine which rules might best be applied to the circumstances of CCS operations and set the appropriate standards for choice of liability rule(s).

Haan-Kamminga provided a definition of a liability system: "A liability system is outlined by defining the damages that are to be compensated, how they can be measured and proven, the persons liable for that damage, the type of liability and the period for which liability exists (liability horizon)."[78]

She described the liability options as basically of two camps: that of "tortious or fault-based liability" and that of "strict or risked-based liability."[79] She describes fault-based liability as any rule that satisfies five elements: (1) unlawful acts or omissions, (2) that impact the plaintiff, (3) are imputable to the respondent, (4) which result in legally recognizable damages, and (5) that damage is suffered by the plaintiff.[80] Haan-Kamminga contrasted that with the rule of strict liability, wherein a "legislator has determined" that an activity, even if otherwise lawful, is "so uncertain or carry such high risk" that those who undertake such activity should be held liable for any resultant damages.[81]

Thus, Haan-Kamminga would likely bundle negligence, nuisance, and trespass together, predicated primarily on her first, fourth, and fifth elements from her fault-based liability test. The common law perspective divides the rules differently: negligence requires a failure to undertake a due level of care, whereas there are no prescribed standard levels of care for strict liability, trespass, or nuisance.[82]

Within common law countries, the rules of civil liability can be applied in parallel to the same accident or injury; thus, it is important to be aware of the range of potentially coincidental grounds of civil liability. It is clear that the contamination of land or its associated air and water could readily fall within trespass, nuisance, negligence, or strict liability. Furthermore, the finding of trespass, nuisance, or strict liability or negligence would present no bar to the finding of the other members of this list; thus, it is quite possible that several of these torts could be present at the same time under the same facts.[83]

This is an interesting result that traditional rules of civil liability did not, and in many current jurisdictions do not, operate in exclusive but in parallel use. Thus, a single injury from a CCS accident might be acted on from multiple different civil liability rules simultaneously. Clarity of which rules or rule might be applied would certainly add certainty, and the provision of clarity in the liability standards is of great importance in ensuring efficient decision making on the behalf of the CCS operator. We next endeavor to resolve which rules of civil liability might enable CCS operators to make efficient risk decisions on storage activities.

3.2.1 Strict Liability

The rule of strict liability is one of the primary rules of civil liability, rivaled only by the negligence rule.[84] The literature at large appears to broadly support a call for a rule of strict liability for CCS storage activities. We therefore provide a survey of the various calls made in support of various versions of potential implementations.[85]

Strict liability basically means that the operator has to compensate damage regardless of any due-care efforts. The only requirement under strict liability is that a particular act under the control of the operator caused the damage. Strict liability can be established by both case law and regulatory rules.[86] In practice, strict liability can be found under several conditions—for example, if the activity of the first party is abnormally hazardous and if there is a causal connection of the first party's acts and the resultant harm; no intent to harm or duty of care needs to be established in such cases. In such cases, a rule of strict liability might be found to apply to CCS-related activities.[87] Furthermore, the Restatement (Second) on Torts clarified that the application of the strict liability rule could be applied to both private property and public property, thus enabling a broader range of application for the rule in CCS storage sites.[88]

To some extent, the analysis of liability for CCS-related damage reassembles that for environmental risks.[89] Those arguments have justified the introduction of strict liability for environmental harm.[90] Many legal systems have followed those arguments and introduced strict liability regimes for environmental harm.[91]

Ingelson, Kleffner, and Neilson called for the choice of strict liability to "avoid complicated and costly litigation arising from the difficulty in proving damage and causation from CCS facilities and the appropriate technology that may well evolve."[92]

Cypser and Davis appear to have been the first to argue for the application of strict liability for accidents resulting from injuries due to anthropogenically induced seismic activity.[93] Their arguments were predicated on the scientific potential that artificial injection of fluids into the earth could trigger seismic activity that could cause injuries.[94] They first extract anthropogenic seismic events from consideration as "acts of God" or as "superseding causes" due to the explicit role of human decision making to undertake acts that result in novel seismic events, enabling them to address causation.[95] They then find basis for the application of strict liability from (1) vibrational trespass,[96] (2) *Rylands*-based analysis for blasting damage and fluid injection damages to drinking water supplies,[97] (3) and the first Restatement's "ultra-hazardous activity" tests[98] and the second Restatement's "abnormally dangerous activity" six-factor test.[99] They also presented an argument suggesting a rationale based on the incentive to include the cost of the potential damages within the price of the product created by the risky activity, albeit no reference to law and economics literature was provided.[100] They did not reject application of a negligence rule or a nuisance tort rule, but found their results on accident management and in failing to provide sufficient incentives for safety improvements less attractive than the rule of strict liability.[101]

An alternative method of identifying the application of strict liability is given by Haan-Kamminga: when policymakers seek to channel liability to an operator, especially when the activity under governance is a highly technical one or engages a dangerous substance.[102] She adds that strict liability is generally applied (1) in certain circumstance predefined by legislators, (2) for particular types of dangers, (3) when the identification of the tortfeasor's fault is not a policy goal or need, and (4) when policymakers might set standards to allocate risk for specific activities.[103]

Bidlack evaluated the potential precedence for applying strict liability within the United States and found product liability rules and implied warranty rule potentially applicable.[104] He also suggested that a plaint in trespass could be used to reach a strict liability result within certain jurisdictions.[105]

Several authors have analyzed the way in which damage related to CCS would be treated in particular legal systems and have found that in different systems, different legal bases could be found to constitute CCS-related liability, but that in fact the strict liability rule is the most common standard.[106] This seems to be the standard now not only in the United States[107] but in many European countries as well.[108] In the EU, the operation of storage sites pursuant to the CCS directive has now, further to article 34 of the CCS directive, been brought under the framework of the Environmental

Liability Directive. The CCS storage has been brought under annex III of this directive, and as a result, a strict liability regime applies as well.

Adelman and Duncan mention that under US law, strict liability is favored when an activity is deemed to be "abnormally dangerous."[109] They argue that since CCS in fact is not an abnormally hazardous activity, it should be precluded from the application of a rule of strict liability.[110]

Following Adelman and Duncan's lead, Schremmer argued against the application of strict liability with particular application to potential injuries to underground waters from hydraulic fracturing activities.[111] He found a rule of negligence with a burden of proof shifting assumption of res ipsa loquitur would be preferable.[112] His arguments basically follow the six-factor tests of the Restatement (Second) of Torts and finds that there is no (1) great harm presented,[113] (2) no "fair" high degree of risk,[114] (3) reasonable care could eliminate the risk of fracking,[115] (4) fracking was potentially of "common usage," [116] and (5) the social value of fracking to the community was net positive;[117] yet he did find that the location of fracking could be inappropriate to local populations.[118] Given that Schremmer found five of the six factors not applicable to the Restatement's "abnormally dangerous activity" definition, he found a call for strict liability to be inequitable to the operators.[119]

It will be the conclusion of our analysis that the rule of strict liability would be more robust than other liability options;[120] in that conclusion, we join with many of the authors we have mentioned. However, there are important jurisdictions with limited access to the rule of strict liability, such as Australia,[121] and we review other means of civil liability in the subsequent sections.

3.2.2 Trespass

At common law, landowners are entitled to enjoy the "exclusive possession and physical condition of land"; trespass of the CO_2 plume onto or into other lands or subterranean zones could establish a trespass.[122]

The elements of trespass are:[123]

1. Intentional entry[124] or causes "direct and tangible" entry[125] on the land possessed by another person
2. Unless privileged to do so[126]
3. No physical harm or injury is required[127]

Trespass has been referred to as an interference with the land.[128] The landowner need not be on-site or actively engaged with the land in question, so long as he or she is an owner in fact; even a reversionary interest would substantiate the necessary degree of ownership.[129] Thus, a simple finding of fact could establish liability for trespass. Some commentators have seen trespass as a form of strict liability,[130] in that no level of exercise of due of care will excuse the tortious performance of trespass.[131]

Given that trespass is often well established in jurisdictions where strict liability might be of limited application, the literature on trespass is notable for providing analogous remedies in those areas. We offer a brief review of that literature.

Trespass of subterranean gases has litigation history in Australia, Canada, and the United States.[132] Subterranean trespass by a gas occurs when an "invisible gas and microscopic particles" enter into or onto the property of a third party and result in injury to the property or substantial interference with the enjoyment of that property.[133] No proof of actual damage is necessary to establish a claim under nuisance,[134] just a demonstration of trespass itself. Ingelson, Kleffner, and Neilson reported on a range of historical subterranean trespass cases involving injected gases from Texas,[135] Michigan,[136] and Ohio.[137] They note that shale oil developments could lead to additional case law based on injuries from injected gases that could affect CCS liability concerns.[138]

Underground invasions by liquids, which an injected plume of CO_2 would be, have been found to be trespasses.[139] Subsurface invasive acts of pollution of water or oil reserves have been found to have been torts of trespass,[140] although the more indirect the act of trespass is, the less likely courts have been willing to find trespass.[141] The key to trespass for CCS would appear to be any entry of a thing (e.g., a plume of CO_2 or displaced brines) onto the subterranean property of a third party that conflicts with that party's exclusive enjoyment of that property. The act of invasion need not be underground, as common law recognizes trespass even when aerially performed.[142]

Arguments have been raised that operators of CCS facilities should not be liable for climate change liabilities.[143] Flatt presented an argument that it could be counterproductive to assign climate change liability to actors taking financial risks to prevent damage to the climate by undertaking investment in CCS storage sites.[144]

In conclusion, where more overt forms of the strict liability rule are difficult to bring forward (see section 3.2.4), trespass is often well established and available in most cases; thus, trespass is a useful alternative in those cases.[145]

3.2.3 Nuisance

Nuisance law resolves "competing land use claims of private individuals."[146] It can occur when the acts of a party create "substantial and unreasonable" interference with another party's use and enjoyment of his or her own land and property.[147] The focus with nuisance is on the "type of harm rather than the motivation for the intrusion"; thus, the underlying quality of the moral intent is of little relevance.[148]

The elements of nuisance are (1) an act (or omission) that affects the reasonable enjoyment of property, including freedom from its injury; (2) the tortfeasor initiated, continued, or adopted the causation of the nuisance; and (3) no legitimate defense or legal privilege exists that would excuse the tortfeasor.[149] The inappropriate presence of smoke, fumes, water, smells, gas, noise, and vibrations are all well-established causes of nuisance.[150] It is foreseeable that several of these categories connect well with the potential migration and mischiefs associated with CCS injection and storage.

There is no accounting for whatever benefits might accrue to numerous people beyond the location of the conflict;[151] unlike negligence, nuisance makes no provision for welfare analysis. However, nuisance does require continuing or recurrent interference to use or enjoyment; thus, when the interference from a CCS operation either is noncontinual or of a minor nature to the victim, a finding of nuisance would be unlikely.[152]

Health and property damage can qualify as torts in nuisance.[153,154] Dobbs defines *private nuisance* as "a condition or activity that interferes with the possessor's use and enjoyment of her land by incorporeal or non-trespassory invasions."[155] The Restatement (Second) of Torts also provided for public nuisance, in which a similar interference affects the general public with regard to public lands.[156]

The nuisance-derived injuries could be tangible or nontangible "enjoyment," but the interference must be to the normal use and not be of an unusual or sensitive character.[157] While a nuisance claim could be traditionally grounded on either a negligence or a strict liability rule, the most traditional avenue would be that of either intentional or negligent nuisance.[158] For example, while the invasion of a plume of CO_2 onto a rancher's property could qualify as a trespass, the venting of that CO_2 to ground level and the resultant ruin to her grasses and the impact on her cattle could qualify as a nuisance. Thus, trespass and nuisance can arise simultaneously yet remain separate legal injuries under the common law.

Nuisance is intriguing, in that many jurisdictions recognize both private and public versions of it.[159] A public nuisance is "a nuisance that is 'so widespread or indiscriminate in its effect that it is not reasonable to expect one person to take proceedings on his own responsibility or put a stop to it, but it should be taken on the responsibility to the community at large.'"[160] For a private citizen as a member of the public, it is unnecessary to establish any particular interest in the public lands affected, enabling a broader range of parties to bring action.[161] But that same private citizen must be able to demonstrate a claim in "special damage" resulting from the nuisance to the public lands.[162]

3.2.4 Negligence

The literature on the application of the negligence rule to CCS storage activities is not unilaterally against the implementation of the rule. Some research advocates for the rule, and some discusses its potential implementation given the likelihood that strict liability might not be available or be very limited in certain jurisdictions. In this spirit, we provide a quick review of those two threads of previous literature.

The negligence rule is defined as a rule according to which the injurer will have to bear the loss only if he or she uses less than a legally required level of care, referred to as the due care level.[163] The negligence rule as defined here means that the injurer will be held liable if he or she spends less than the due care level required by the legal system—in other words, if this person's acts to avoid foreseeable harm were insufficient (i.e., negligent).

Negligence to property could result when the acts of a first party cause injury to the property of a third party if (1) a duty of care was owed to the victim, (2) a breach of duty of care occurred by the tortfeasor's unreasonably risky behavior, (3) an injury in fact resulted, (4) the injury was a result of proximate causation from the acts of the first party, and (5) legally recognized injuries and damages resulted.[164] The injuries or harms must have been foreseeable given the breach of duty.[165] A key concern would be the establishment of a reliable duty of care against which a breach of duty could be asserted.[166] There are no known CCS negligence cases, so there is also no known existing stare decisis, or preexisting bench guidance, on what constitutes such duty or the breach thereof.[167]

Some scholars who have written on CCS liabilities have called for applying the rule of negligence, at least for cases not bearing particular evidence of malfeasance. Flatt called for a negligence rule to be employed for CCS storage projects, although he said that the negligence rule should be based in statutory enactment and not derived from court decisions.[168] He also advocated for a statutory plan requiring "'reasonable' behavior" on the part of operators.[169] He then raised concerns that "a strict liability regime should not be adopted for CCS operations, in that it would greatly increase the risk of liability costs for CCS operators without commensurate benefits to the public."[170]

Flatt found that a rule of negligence was the most likely tort rule to get invoked in the United States.[171] Bidlack noted that negligence was more likely to be found applicable within the United States due to the limits of strict liability in general in the country.[172] Because of the "inherently broad nature of negligence," because of the impact of the res ipsa loquitur rule on leaks from sequestered CO_2 volumes, and because a lack of prior clarifying case law that CCS activities would be subject to strict liability, Bidlack found it more likely that negligence could be applied.[173] Similarly, Canadian courts have found negligence for subsurface leaks that led to environmental injuries.[174]

It has been reported that in New Zealand, strict liability is not available for injuries related to CCS activities; only claims raised in negligence, nuisance, and trespass would stand.[175] New Zealand has precedence on how to address the setting of a "duty of care" for cases arising from novel situations; the rule requires that (1) the loss was reasonably foreseeable in nature, (2) the novel accident was proximate in causation, and (3) there were no "external factors negating a duty of care."[176] It appears that the appropriate duty of care would be that of a "skill expected from a skillful and experienced person in the profession, judged at the time the work was done." Enabling an increasing level in the duty of care as knowledge is gained from practice in the realm of CCS storage, if not also being comparable to a "best practices" test.[177]

A logic thus exists that the provision of CCS injection and storage is a public good, generative of many positive externalities, and that the general welfare benefit to society outweighs most harms to private victims.[178] Therefore, only in cases in which "best practices" were not undertaken to avoid foreseeable injuries should liability be found—and thus a logic for favoring a rule of negligence.[179]

But that requirement for best practices raises its own alert. There are both academic and case law concerns that the rule of negligence would face challenges to its assumed efficiency, even in common law jurisdictions otherwise broadly favorable to the rule. Havercroft and Macrory have raised a concern that the lack of a preestablished duty of care for CCS operations could prevent negligence from being effective.[180] They present an argument that for such novel hazards as CCS storage, courts might be faced with calls to apply "especially high standards" in duty setting and might need to invoke the precautionary principle.[181] Perhaps in this light, American courts have held that when tortfeasors encounter circumstances of heightened danger, they owe a duty of "utmost" or "extraordinary" care;[182] in such cases, the duty of care might be so high that it approaches results similar to those of strict liability.[183]

Again, in conclusion, the policy recommendation of this study is that the rule of strict liability would be more robust than the rule of negligence.[184] However, it is recognized, as detailed in the literature, that the implementation of strict liability might be hampered or limited in certain jurisdictions. In that light, it is important to analyze why we affirm the role of strict liability and encourage its adoption by regulation where not already provided for by common law. It is also important to learn in what circumstances a rule of negligence might be superior for CCS storage operations. The sections that follow provide those analyses on the comparative efficiencies of strict liability and negligence for the circumstances of CCS-storage activities.[185] Those arguments necessarily dive into theoretical models and issues and provide the foundations for our policy arguments.

3.3 Law and Economics on the Role of Civil Liability Rules

The literature of law and economics assumes that actors within legal contexts are subject to similar rational conditions as actors within economic contexts. That is to suggest that given sufficient information, sufficient levels of rationality, and sufficient freedom of choice, an actor would choose the optimum result; in other words, the actor would select the efficient choice among his or her options.[186] This section provides a brief introduction to the policy perspectives of the law and economics school of literature and to its insights into both liability and compensation policy concerns.

It is important to be aware from the beginning of this analysis that the literature of law and economics prioritizes the goal of efficiently reducing the incidence of accidents over the alternative goal for liability rules to provide compensation and relief to the victims of those accidents; much of the argument has been grounded on welfare economics. Thus, Benjamin Franklin's old adage has been found in favor in this vein of research: "An ounce of prevention is worth a pound of cure."[187]

At the end of the twentieth century, economists and lawyers engaged in fierce debates on the goals of tort law, and some attempts were made to reconcile the legal (corrective-justice-based) approach with the economic (deterrence)-based approach.[188] Over time, the legal community and policy-makers have become increasingly convinced on the importance of liability rules as instruments of prevention, especially in the environmental arena.[189] One of the reasons for this change is the increasing availability of empirical evidence demonstrating that at least industrial operators, as often found in the environmental issues, respond to financial incentives provided through the liability regime.[190]

Another reason is that economic research has demonstrated that it is inefficient to attempt to both create incentives to guide actors to efficient levels of risk undertaking and provide for compensation goals; the overlap could lead to failure to attain either goal.[191] While civil liabilities could be seen as a system of justice, in that they could attempt to provide compensation to victims in an effort to give them a sense of restitution,[192] the law and economics literature views this use of liability rules as less efficient than other means of compensation, such as by insurance.[193]

Ultimately the literature has not attempted to shift the character of traditional liability notions[194] but to evaluate when each particular tort rule might be efficiently applied and if there are underlying frameworks across the set of tort rules that policymakers might gainfully employ. Perhaps the most important findings for policymakers are the central roles of information and who possesses which bits of information and of initial allocations of rights.[195]

From an economic perspective, a liability system has an important social function in remedying market failures.[196] When economic actors lack information on how their activities affect other parties, whether positively or negatively, those impacts to the third parties are called externalities;[197] they are external in the sense that pricing data are not transferred between the parties, and thus the pricing data are external from the decision-making process of the first actor.[198] Thus, liability rules can play an important role in curing market failures caused by externalities.[199]

The literature has examined the potential of civil liability rules to use a variety of tools to influence the decisions of actors engaged in risky activities. The literature broadly sees the use of civil liability rules as a means to set standards to guide actors in undertaking efficient levels of risky activities.[200] Economists state that the internalization of externality costs, consisting of the costs of prevention and of expected damage caused by the accident risk, should enable the optimal level of social costs from accidents to be attained.[201]

One of the most effective ways to achieve this result is to shift the ex post costs of damages to victims back to the ex ante deliberations of the potential tortfeasor. Liability rules can be used as a kind of a time machine to transport future potential costs of hazards and harms back to the period when the actor makes a decision on how to undertake a risky activity; the transposition of future ex post costs to serve as ex ante data bits enables policymakers to provide incentives to decision makers to optimize their risky behaviors.[202]

In the economic analysis of law, liability rules are therefore seen as instruments to deter activities worth being avoided on efficiency grounds through liability rules.[203] The goal of liability rules is, however, from an economic perspective not to necessarily prevent all accidents at all costs, for the simple reason that that would be too costly. Calabresi formulated this as follows:

> Our society is not committed to preserving life at any cost. In its broadest sense, the rather unpleasant notion that we are willing to destroy lives should be obvious. Wars are fought. ... Ventures are undertaken that, statistically at least, are certain to cost lives. We take planes and cars rather than safer, slower means of travel. And perhaps most telling, we use relatively safe equipment rather than the safest imaginable, because—and it is not a bad reason—the safest costs too much.[204]

Calabresi provided an alternative goal for tort law: he displaced the provision of compensation for injuries received with his new goal to minimize the total social costs of a given tort.[205] For Calabresi, compensation for injuries was just one of several costs that policymakers could minimize as a set; for example, he included the administrative costs of the judicial system, the costs of insurance, and the costs to avoid the accident by both the lead actor and those who might become victims.[206] Thus, the goal is to reduce the overall burden of accidents to society and not merely provide justice for those injured. This provided the law and economics movement with momentum toward their research on efficient levels of accidents.

Finsinger and Pauly extended Calabresi's arguments by mapping his ideas into social welfare theory by arguing that tort law should optimize the net social welfare benefit of the risky activity:[207] "The main goal of tort law is to internalise the externalities in order to enhance optimal decisions on the level of precaution."[208] This could be achieved by using liability rules to affect the decision making of both lead and secondary parties to a risky activity, the actor's decisions on activity levels, and Coasian negotiations.[209]

In summary, the literature has provided insight into when which types of civil liability rules might be more efficient for certain circumstances. The literature holds that by careful drafting and selection of civil liability rules, incentives can be provided to the actors as they decide on risky activities that would result in actors' choosing efficient levels of activity and precaution; by so doing, policymakers could attain their goal of minimizing the cost of accidents.[210] This chapter explores these matters in particular relation to the unique circumstances of CCS storage activities.

3.4 Evaluating Strict Liability for CCS

Since Shavell developed his model on efficient liability rules,[211] a standard model of civil liability rules has emerged in the law and economics literature.[212] The model generally follows Shavell's lead that standards should be set to attain optimal levels of accident avoidance by providing ex ante incentives, but not necessarily to prevent accidents in line with Calabresi's insights.[213] The standard model generally finds that a rule of strict liability would be more efficient in a broader array of circumstances than a rule of negligence.[214] This section provides a review of those circumstances and evaluates the fitness of the circumstances of CCS operations to the rule of strict liability.

Strict liability is preferable in five circumstances: (1) the accident is based in abnormally hazardous activities,[215] (2) the accident is unilateral in character,[216] (3) the accident is based on novel risks and harms,[217] (4) decentralization is a policy goal,[218] and (5) the justice system would be challenged by the transaction costs of applying the civil liability rule.[219]However, there is a paradigmatic encapsulating manner from which to join these five cases.[220] Strict liability will likely be more efficient than a negligence rule when the risk-taking actor has superior information on the risks and their potential abatements and remedies than other parties, especially in the courtroom. This is doubly so for operators who might want to effectively control their financial exposure to risk; strict liability leaves the penultimate decision on risk and liability with the risk taker, not a courtroom and its parties.

CCS itself is a highly complex undertaking from scientific and engineering perspectives; thus, the basic decision-making calculus necessarily requires a decision maker with sufficient technical and scientific training and the data from the situational circumstances of the CCS activities. It would be surprising to find a judge or jury with better access to such data as contrasted against the technical teams associated with the CCS operators. A rule of negligence would leave the decision of duty of care to the courtroom, whereas strict liability places that decision with the CCS operator.

Even when the court might become fully informed in some sense of equivalence to the operator's state of knowledge, it would be after the event of the accident, whereas the operator would have those data at or prior to the accident. Strict liability enables the actor with the best information, at the right time, to make the decision on setting efficient levels of care and activity. Implementing a negligence rule is hampered by both data-gathering problems and the timeliness of that data gathering and is thus less likely to provide an efficient level of care and activity.

3.4.1 Abnormally Hazardous Activities

Most students of law and members of the general public commonly associate strict liability with the idea of abnormally hazardous activities . While there are multiple other theoretical foundations for the application of strict liability, the concerns of abnormally or ultrahazardous activities would appear to be front and center for many policymakers in determining the appropriate rule of civil liability for CCS activities.

3.4.1.1 Theoretical Framework Dobbs begins his discussion on strict liability in this way: "Strict liability is liability without fault. … An even older form was found in early tort law, which, according to traditional views, imposed strict liability for all direct and forcible harms to person or property and perhaps for the spread of fire as well."[221] The element of dangerous undertakings thus appears to have always played a role in distinguishing the rule of strict liability from other forms of rules of liability. Earlier forms of strict liability included the dangers presented by wild animals (*ferae naturae*) or abnormally dangerous domestic animals.[222]

Modern strict liability cases predicated on "abnormally hazardous" conditions can be found in the nineteenth century, with *Rylands v. Fletcher*[223] serving as an instrumental case across common law jurisdictions. *Rylands* held that parties who accumulate a foreseeable hazard on their land that then escapes are liable for any injuries that result. In a manner, the case follows the earlier intuitions of animal liability, both domestic animals in trespass and of animals *ferae naturae*, that the escape of a dangerous animal would result in strict liability for the owner.[224]

Lord Cransworth held in his concurrence,

If a person brings, or accumulates, on his land anything which, if it should escape, may cause damage to his neighbour, he does so at his peril. If it does escape, and cause damage, he is responsible, however careful he may have been, and whatever precautions he may have taken to prevent the damage.[225]

Key to his logic was a flooded coal mine case that featured an operator that had, with all due professionalism and prudence, introduced additional water volumes in his own coal mine that then injured a nearby coal mine belonging to a separate party. The operator had been found liable for those resultant damages.[226] Cranworth summarized:

The Defendants, in order to effect an object of their own, brought on to their land, or on to land which for this purpose may be treated as being theirs, a large accumulated mass of water, and stored it up in a reservoir. The consequence of this was damage to the Plaintiff, and for that damage, however skillfully and carefully the accumulation was made, the Defendants, according to the principles and authorities to which I have adverted, were certainly responsible.[227]

The passage of time has resulted in the development of the concept as summarized within the Restatement (Second) of Torts:

In determining whether an activity is abnormally dangerous, the following factors are to be considered:

a)   existence of a high degree of risk of some harm to the person, land, or chattels of others;
b)   likelihood that the harm that results from it will be great;
c)   inability to eliminate the risk by the exercise of reasonable care;
d)   extent to which the activity is not a matter of common usage;
e)   inappropriateness of the activity to the place where it is carried on; and
f)   extent to which its value to the community is outweighed by its dangerous attributes.[228]

Coman noted that American courts have relied more on factors a, b, and c as instrumental in determining the applicability of strict liability to a particular set of circumstances but rely on factors, d, e, and f more as instruments to provide judicial discretion over social policy concerns.[229] She also notes that American courts have found the existence of public regulations for a specific abnormally dangerous activity to provide judicial discretion to not find need for the application of strict liability; this was particularly true for cases involving natural gas accidents.[230] Despite the generally limiting tendencies of factors d, e, and f, it was found that certain activities and injuries are generally found in want of application of strict liability.[231] Coman found that abnormally hazardous activities that resulted in permanent damage to clean water supplies generally resulted in the finding for application of a strict liability rule; activities engaging oil and gas more often than not found strict liability applied to their cases.[232] These applications of strict liability are reportedly due to the extensive harm that could fall onto innocent bystanders with little to no control over the activities in question.[233]

This perspective can be found in Shavell's analysis as well, which basically supports the rights of victims to continue in their mode of life mostly unaffected by the tortfeasor's assumption of ultrahazardous activities. In other words, Shavell found that in contrast to negligence in which both parties had a duty to avoid, the assumption of ultrahazardous activities by one party should not require the second party to adjust his otherwise ordinary life habits.

Shavell had suggested that ultrahazardous activities could be characterized by two fundamental traits that drove the policy to apply strict liability in lieu of a negligence rule.[234] First, the activities are readily identifiable from routine activity and introduce nonnegligible risks onto nonparticipant victims, which "make[s] the activity worthwhile controlling."[235] Second, the victim's choice of activity lies within the domain of normal life and thus is "activity that cannot and *ought not be controlled*_[italics added]."[236] Shavell thus identifies a clash of the tortfeasor's unique and risk-imposing activity onto unwilling and nonparticipatory others: that abnormally hazardous activities inflict nonnegligible risks on nonparticipant victims, who were otherwise engaged in activities than cannot and should not be controlled. Therefore, when the risky activities are ultrahazardous, it becomes efficient to govern them with a rule of strict liability.

Theoretically, Shavell placed abnormally hazardous activities among his bilateral accident models in which tortfeasor and victim are strangers, specifically within his model's propositions 4 and 6.[237] Shavell found that the choice of civil liability rule determined which party would have his or her activity level reduced. A policy choice for a rule of strict liability with defense of contributory negligence left the tortfeasor with an expected reduced activity level while leaving the victim unaffected in activity level. A policy choice for a negligence rule went the other way: it was expected to reduce the activity level of the victim while not affecting the activity level of the tortfeasor. Thus, policymakers are forced to ask, "Which party do we want to control: the tortfeasor or the victim?"[238] The answer depends on Calabresi's cheapest-cost-avoider rule: the party who could have prevented the accident with the least cost of taking care should be deemed the liable party.[239] In such ultrahazardous activity models, the best sets of incentives were provided by the rule of strict liability with defense of contributory negligence because it placed both the activity decisions and the resulting costs of damages with the actor engaged in the ultrahazardous activity; therefore, the tortfeasor is given incentives to attain the efficient level of accidents.[240] While a policymaker might hope to attain perfectly efficient results, Shavell found that in bilateral ultrahazardous accidents, that was not attainable;[241] setting the rule to strict liability, with defense of contributory negligence, would be the most efficient of the available policy options.[242]

Schäfer and Schönenberger reported that Polinsky had found that under a rule of negligence,[243] an actor might excessively undertake risky activities; they contrasted that with Shavell's finding that a rule of strict liability could lead an actor to underproduce from a risky activity even if that activity was of social benefit.[244] This leads to a result that adopting a rule of strict liability tends to create net social welfare gains, if less than maximal, whereas adopting a rule of negligence might lead to a net social loss.[245] Given this difference in social welfare impact for ultrahazardous activities, the choice should clearly be made for a rule of strict liability.

It has been argued (e.g., historically with the application of strict liability to CERCLA-related hazardous wastes) that the application of strict liability to certain hazardous activities could result in difficulties in obtaining insurance or otherwise providing market-based risk-spreading options.[246] However, the concerns raised in those circumstances of difficult establishment of unique causation (e.g., that too many polluters contributed to a common accident[247] or that the chemicals might be unique to some polluters and not others[248]) are not likely to be present within the circumstances of CCS and are readily distinguishable.[249]

3.4.1.2 In Application to CCS That the original facts from the case of *Rylands* involved such closely related vocabulary, such as the linked ideas of artificial accumulated reservoirs that seep and migrate into abandoned mineshafts, would make it difficult to distinguish the circumstances of most CCS storage accidents from the case in *Rylands*. Under a closer *Rylands* analysis, it is quite reasonable to draw an inference that the case might well fit as precedent for CCS accidents. Under a *Rylands* perspective, CCS would involve the bringing of CO_2 and its accumulation. Furthermore, the storage of the CO_2 would be brought on to the land or licensed area of the operator to accumulate in an unnatural way with respect to the prior nature of the geological storage area. The CO_2 would be imported with the goal of accumulating a large mass and storing it in a reservoir. If the CO_2 were to escape, it could cause a wide variety of climate, environmental, personal, and property damages. If a jurisdiction holds that a valid *Rylands* analysis leads to a finding for a rule of strict liability, then that jurisdiction would likely apply strict liability to CCS-related accidents.

Under current Restatement of Torts analysis,[250] there are six elements to review. First, given the particular range of hazards from CCS accidents,[251] one might find a high degree of risk to some harm to people, lands, or chattels involved in such an accident. While some experts have found that risks are most likely limited to freshwater contamination, that injury itself could seen as sufficient; other scientists are not necessarily in agreement due to certain shortfalls in the historical experience of actually operating CCS storage systems for industrially captured CO_2 fluids at the pressures, volumes, and time durations required. Perhaps one day the evidence will accumulate to more strongly suggest that there is a lack of a "high degree of risk of some harm"; in the meantime, legal scholars have generally preferred to implement the precautionary principle when the facts are not yet in evidence. Given the government's role in approving, permitting, and regulating the risks of CCS storage, it is probably reasonable to conclude that until more evidence to the contrary is presented, the more reasonable path forward is to presume that the needs of this first element are met.

The second element reviews whether there is a likelihood that the resultant harms would be great. The scientific evidence points to the potential to lose massive accumulations of stored CO_2 to the atmosphere and enable great risks of climate change impacts and endanger the lives of nearby humans and animals, or the potential to lose subterranean control of the CO_2 that results in harm to drinking and agriculturally useful water resources, or the potential to cause a variety of other harms. It therefore is reasonable to conclude that the likelihood of great harm is certainly positive in value and probably of substantial magnitude, given the very active academic debate on how to regulate the expected risks of CCS storage.

The third element requires an analysis of whether the risks of CCs storage could be eliminated by the exercise of reasonable care. While the answer of that very question might more properly require its own book, it is reasonable, given the technical uncertainties and the unknowable geological, geophysical, and geochemical issues of the subterranean storage reservoirs, to assume that a certain amount of the risk would remain present even under the most diligent care and best efforts to minimize any risk of injury from the operations of the storage systems. So the first three elements would appear to qualify CCS storage activities, while injecting fluids and during postclosure periods, as matching the requirements for the application of strict liability and, as Coman presented, would normally suffice in most cases.

An examination of the second set of elements, those of the last three elements, is often used to strengthen or weaken the findings of the first three elements, often in the development of public policy by the courts.

The fourth element asks if the activity of CCS storage is a matter of common use. While there are a variety of metaphors to address the vague notion of "common usage," the plain facts are that there are extremely few existing CCS projects and they are complicated in technology and require highly trained scientists and engineers to operate them. When more CCS storage systems are built and placed in operation, one would likely expect them to be operated exclusively by either corporations with very large capitalizations or by governmental authorities. CCS storage facilities are not likely to ever become anything other than highly regulated and extremely limited in ownership. Thus, it is reasonable to conclude that CCS storage activities are not likely ever to be seen as being in "common usage."

The fifth element requires an analysis of whether the CCS storage activity is undertaken in an appropriate location. With this test in mind, site approval procedures, if well executed, should reduce the application of this element for the application of strict liability. However, both economic and political motives might exist to place CCS storage sites adjacent to industrial areas with nearby residential areas. As such, this element is likely weaker than the other elements but not removed from supporting the application of a rule of strict liability.

The sixth element requires balancing the value to the community from CCS storage versus its dangerous attributes to the same community. Arguments can be made on both sides. The successful storage of harmful greenhouse gases is widely held to be one way to reduce the risks of anthropogenic climate change; thus, this is the primary motivation to bring CCS operations into practice. However, others have argued that the development of CCS technologies only furthers the development of fossil-fuel-based energy systems and misdirects investments from alternative carbon-free or carbon-neutral forms of energy production. In addition, the onset of CCS storage systems would certainly introduce new risks into the world that previously had not been in place. Thus, this sixth element is difficult to reasonably resolve on either side of the question.

In sum, the argument has been presented that four of the six elements would support the application of a strict liability rule, one element appeared to benefit the application of strict liability, and one element was found unresolvable on the merits. Since the rule under the Restatement (Second) of Torts does not require the support of all six elements, it can be inferred that the application of a rule of strict liability could be supported under the Restatement's rule on abnormally hazardous activities.

Shavellian analysis of ultrahazardous activities can be developed similarly in support of a rule of strict liability for CCS operations. First, CCS activities can readily be distinguished from routine activities, and the activities are expected to introduce substantial risks (i.e., nonnegligible risk) to nonparticipant citizens and residents. Second, citizens and residents should have the right to continue in their private lives and private activities unaffected by the operation of any CCS facilities or injections over which they would likely have no means of control. It is unclear from where would arrive a duty on the part of the citizen or resident to somehow avoid affecting the risk profile of a CCS project, and thus it is very difficult to imagine how a negligence rule might result in higher levels of safety based on citizen or resident behavioral choices. Thus, Shavell's suggested analysis for ultrahazardous activities would support the application of strict liability to CCS activities.

3.4.2 Dominant Actor Accidents (Unilateral Accidents)
In addition to ultrahazardous activities, another major reason for the application of a rule of strict liability is when there is fundamentally only one actor with sufficient information about the hazards and benefits of a risky activity. In such a case, that dominant actor would be the best informed and best placed to make the optimal decision on attaining an efficient level of engagement in that activity. Strict liability appears to be dominant against a rule of negligence in such circumstances.[252]

3.4.2.1 Theoretical Framework
In Shavell's models, there are both unilateral and bilateral accidents. In unilateral accidents, only the tortfeasor's actions affect the probability and extent of harms. In bilateral accidents, both the tortfeasor's actions and the victim's actions affect the probability and extent of harms. Within both unilateral and bilateral accidents, two main factors determine the extent of accidents: (1) the level of preventive care and (2) the activity level (how often the risky activity occurs).

The injection of CCS would closely approximate the idea of a unilateral accident, in that only the operator would have access to both information and control over events that could lead to an accident. Victims in nearby locations would have very limited impact on any ultimate accidents. Therefore, the modeling results from unilateral accidents are potentially applicable to CCS operations.

The impact of the two rules can be contrasted. Under a rule of strict liability for unilateral accidents, both efficient preventive care and efficient activity levels are attained.[253] This is basically the result of the tortfeasor's optimal strategy being in alignment with that of public welfare[254] because the tortfeasor is bearing the complete social cost of the injuries and damages.[255]

However, under a rule of negligence for unilateral accidents, the tortfeasor behaves at the due care level of preventive care,[256] but because due care shields liability for any level of activity,[257] the tortfeasor pursues a higher-than-efficient level of activity. Therefore, strict liability correctly sets goals for both preventive actions and activity levels, but a rule of negligence sets only the preventive actions correctly. Negligence would be expected to lead to excessive risks,[258] and resultant excessive damages to victims,[259] due to overconsumption of the risky activity. However, Shavell found that both rules were superior to having no rule of civil liability.[260]

In conclusion, a rule of strict liability forces the complete set of costs for injuries and damages onto the tortfeasor, effecting a Kaldor-Hicks-type welfare efficiency.[261] However, under a rule of negligence, so long as the tortfeasor maintains the appropriate duty of care, she or he would bear no liability for any injuries that occur despite the preventive measures undertaken; thus, the tortfeasor bears some costs of preventive measures but none for damages.[262] So in the case of unilateral accidents or dominant actor accidents, the literature commonly finds that a rule of strict liability is robust over the rule of negligence.[263]

3.4.2.2 In Application to CCS In the circumstances of CCS operations, only one recognized actor would have the scope of decision making with regard to the issues that determine the risks and hazards of the CCS operations, the operator. The operator would likely not only be delegated by shareholders, coventurers, and public authorities as the singular decision maker and activity undertaker, but would also likely be expected to provide reasonable efforts to keep nonessential personnel at a safe distance from the CCS facilities to prevent any accidental third-party interference with the safe operations of the CCS facilities. The operator would likely be the best informed with regard to the risk of the CCS operations and would likely have exclusive access to the decisions and control processes to affect those risks. Whether the CCS operations are properly characterized as unilateral or bilateral, the conclusion is the same: the operator would be a dominant actor, and thus the rule of strict liability would be more robust than a rule of negligence.

3.4.3 Novel and Uncertain Risks

When risky activities are novel and lack sufficient historical events from which to develop probabilistic models, policymakers might find it difficult to set efficient standards before initial operations. In the absence of a well-established duty of care, a rule of negligence is not likely to function well.[264] But because a rule of strict liability has no such standard, it would not be expected to be less efficient when data are too scarce to assemble an efficient duty of care standard.[265]

3.4.3.1 Theoretical Framework Policymakers would want to consider which rule of civil liability is more likely to produce the necessary incentives to reduce the overall level of risks over the long term. Strict liability requires the tortfeasor to bear all costs of harms, so it also provides strong incentives for safety innovation.[266] Negligence is less effective, because it provides incentives only to the level of established and prospective due care, not for the full range of potential harms.[267] By selecting a rule of strict liability, the policymaker provides the strongest incentives to the tortfeasor-cum-operator to develop and provide improvement in safety and preventive technologies.

Nussim and Tabbach found that in certain complex circumstances, preventive care decisions and activity-level decisions could create nonlinear impacts on expected injuries. For example, in a variety of settings, additional levels of preventive care and activity could mix to create fewer harms, but in some settings, additional levels of preventive care and activity could mix to create more harms.[268] In such cases, where the tortfeasor's decision engages complex results, a rule of strict liability remains efficient. However, not only would a rule of negligence be inefficient, but it would create standard-setting problems that would result in the due care levels being set higher than they properly should be set to be efficient[269] because the complexity of the interactions of preventive care and activity-level decisions is too murky to plumb.[270] The costs of solving the standard-setting problem could be too expensive for a rule of negligence; as a rule of strict liability need not solve that problem, it can remain efficient in the face of such complexities.

Strict liability is found to be more robust for novel risks and hazards. The logic is that the tortfeasor-cum-operator would need to bear the costs of any resultant damages; the costs of the externalized hazards, though novel, are internalized. A rule of negligence flounders in its inability to efficiently formulate the efficient level of due care.

3.4.3.2 In Application to CCS Are CCS technology, activity, and its risks novel from a theoretical perspective? If so, then a rule of strict liability would be more robust than a rule of negligence. The earlier part of this book set out in detail that the technologies and activities of injecting gases and of injecting CO_2 are not new but rather benefit from decades of practical experience within the oil and gas industries. Gas storage facilities for CO_2, methane, and helium have functioned, and occasionally had accidents, for a very long time. However, the critical issue here is whether the relevant duty of care has been set by prior court precedent or if the duty-of-care standard remains yet to be set, and, if so, how novel are the circumstances that would affect that standard setting.

The setting and purpose of CCS storage are novel in that it is intended to accomplish a scale of injection and storage never before achieved for durations never before attempted, for a project never before thought necessary: the protection of the global climate from anthropogenic greenhouse gas emissions. A jury composed of ordinary citizens usually performs the determination of a duty of care within common law settings. How that jury might decide to set the standard for duty of care in operating such unprecedented storage programs might reasonably range from high to low standards, depending on their information, their fears of CO_2 accidents, and their fears of climate change and the benefits of preventing it. It would be difficult to ascertain in advance if such a determination would reach an efficient setting of the duty-of-care standard. Thus, it is reasonable to present the argument that the risks of CCS, as far as a setting of the duty of care goes, could be considered sufficiently novel and be more robustly governed under a rule of strict liability.

3.4.4 Efficiency of Bespoke Safety Plans (Policy of Decentralization)

It has been shown that when activities are abnormally hazardous, in the control of a dominant actor, or of a novel character of risks and hazards, strict liability is expected to be more robust in attaining efficient levels of accidents. Here, a related argument is presented: that when the risks and the technologies to manage them are sufficiently unique to each tortfeasor or operator, then, assuming the ultimate goal is to achieve the efficient level of accidents across the jurisdiction, it would be more efficient to set safety standards unique to each risky activity and its tortfeasor.

3.4.4.1 Theoretical Framework Decentralization is the policy that allows each tortfeasor to determine his level of preventive care based on his unique cost structures. The technology and experience of each tortfeasor is different, so the technique and costs to attain similar levels of safety might be different. Decentralization enables each tortfeasor to set his or her own internal standards to attain efficient levels of preventive care and activity level.

Because a rule of strict liability allows each tortfeasor-cum-operator to uniquely determine his or her risk exposure, because this person bears full liability for all damages, strict liability directly provides decentralization.[271] If decentralization is a goal for policymakers, employing a rule of strict liability guarantees it.

Negligence works by examining the behavior of the tortfeasor against judicially determined norms; thus, attaining decentralization is a challenge but not impossible under negligence.[272] Partial liability can provide decentralization under negligence.[273] "Highest degree of care" can similarly provide decentralization.[274] Extremely high settings of the duty-of-care norms result in de facto strict liability, as the tortfeasor is almost always liable for harms.[275]

3.4.4.2 In Application to CCS Several factors argue for decentralization in the governance of CCS risks and thus in support of a rule of strict liability for those risks. It is clear that CO_2 could and likely would be injected into a variety of storage conditions within singular jurisdictional areas. Even when the basic geology might be classified into types, such as depleted oil reservoirs or coal-bed seams, each oil reservoir or coal-bed system would have a unique fracture and fault system, unique saline and freshwater concerns, and local-to-the-site surface-level conditions and issues. To efficiently govern the risks of each field would require specific and unique understandings of the circumstances of each injection field and the development of safety and preventive measures to accommodate the unique challenges present at each site. Decentralization is the only functional answer given the broad variety of potential geological and injection technology profiles. Thus, a rule of strict liability would be more robust than a rule of negligence for CCS.

3.4.5 Overwhelming Transaction Costs of Justice
While concerns of institutional capacity for the administration of judicial processes might not be of concern in countries expected to lead in CCS development, such as the United States or the Netherlands, such concerns might be part of the consideration in rule choice in many developing countries that might also need to implement CCS storage systems.

3.4.5.1 Theoretical Framework Schäfer et al. found that some court systems face limited budgets, limited court personnel, or other constraints in their ability to render due process; they might be challenged to meet the transaction costs of court proceedings, for example. The arguments are not as well resolved in this area, but it would appear that a rule of strict liability has the advantage over a rule of negligence.

Shavell stated that under negligence, tortfeasors could meet their duty of care and thus evade liability, so no suits result and this avenue is less costly for courts.[276] However, Schäfer et al. found several reasons that a rule of strict liability would probably be cheaper and easier to implement than a negligence rule.[277] Finding of facts and determination of duty-of-care level is expensive; negligence requires such findings, but a rule of strict liability would not.[278] Thus, the informational costs of litigating under strict liability would be less expensive.[279] Because litigation would be cheaper under strict liability, the expected strategic payoff for a victim would be better, so a victim would be more likely to bring suit and be heard. Also, because victims have good incentives to bring these simpler cases under strict liability,[280] they would also be likely to benefit from more pretrial settlements,[281] reducing court costs further.

Within a hydraulic fracturing context, Posner's concept that strict liability should be applied when a rule of negligence cannot make plaintiffs whole, especially due to complexities of establishing causation, could be well applied given the difficulties of tracking subterranean chains of causation.[282] Similarly for fracking lawsuits, it was argued that based on the need to be responsive to the decision-making factors of risk allocation, strict liability placed liability "justly" on the parties controlling that decision.[283] Both of these arguments could resonate within the circumstances of CCS storage accidents: that causation might be difficult to clarify and that the allocation of risk principle would find one party in control of most of the critical risk decisions.

3.4.5.2 In Application to CCS One of the problems of CCS accidents will be the presentation of causation-related evidence at trial. It is clear that it might be difficult to determine what the operator might have done, what happens deep within the subterranean geological layers, and how those efforts, energies, and migrations might lead to tortious injuries. The simpler the burden of proof is made, the more likely it is that victims of such torts could bring suit and receive damages. A rule of strict liability alleviates investigations into whether operators met duty-of-care standards. It also facilitates litigation in general by providing a simpler means of foreseeing the likely outcome of litigation and thus facilitating prelitigation settlements to the victims.

Will the locations of CCS efforts place litigation in locations with courts facing institutional challenges or transactional costs challenges? Yes. It is foreseeable that wherever there is substantial industry with greenhouse gas emissions, if CCS technologies were to become viable, then CCS storage projects, and their resultant litigation as well, would soon become present. And that implies most of the world today, as greenhouse gas–emitting industries include crude oil- and coal-burning power plants, coal-burning iron and steel forges, and many more categories of heavy industry, which are increasingly located in developing countries. While developed economies would have already had reason to apply a rule of strict liability for reasons given, it would also help to support alignment with jurisdictions in developing economies that could also benefit from lessening the transaction costs of providing judicial relief for CCS-related accidents.

3.5 Evaluating Negligence for CCS

Although the previous section explored the multiple settings in which a rule of strict liability would be preferable for policymakers, there are also circumstances in which the rule of negligence might be more effective for policymakers.[284] This section provides a survey of those settings.

There are broadly two main sets of circumstances in which a rule of negligence would be expected to be more effective than a rule of strict liability: (1) when decision makers deviate from "perfect rationality" and (2) when courts render inaccurate damages. Imperfect decision makers could be separated further into three subsets; (1) when risk aversion, risk allocation, and incomplete insurance are part of the decision process of tortfeasors;[285] (2) when tortfeasors might be insolvent and unresponsive to costs of damages; and (3) when tortfeasors undertake strategic avoidance measures to shrug liabilities.[286] When court decisions result, or would be expected to result, in inaccurate judgments on damages, negligence has been found to be more robust than a rule of strict liability.

Thus, the particular circumstances of CCS operations would need to be evaluated to determine if policymakers might efficiently apply a rule of negligence.

3.5.1 Risk Aversion and Imperfect Insurance
Shavell had assumed, as neoclassical microeconomics does, that the actor is risk neutral; the actor can accurately assess a variety of risk-based scenarios

with contemplation of the probability-weighted outcomes. However, it is reasonable to wonder what might occur if those rational skills are missing or defective, or otherwise reduced due to various missing markets in information. This section explores the potential reactions of actors who demonstrate risk aversion or are unable to rely on the market for efficient insurance products.

3.5.1.1 Theoretical Framework Nell and Richter updated Shavell's and other earlier modelers' reliance on risk-neutral actors in accident scenarios;[287] they examined how Shavell's seminal results could be affected by including risk aversion.[288] When they did so, they found that Shavell's earlier findings on unilateral accidents would not hold for models with risk aversion and other "defects."[289] They also found that when tortfeasors are risk-averse actors, a rule of negligence was more robust than a rule of strict liability.[290] A rule of strict liability was efficient only if the tortfeasor was risk neutral,[291] as classical economic models commonly assume.[292]

They also found a similar result when insurance is costly to purchase,[293] with "costly" meaning to them that loading fees would prevent full insurance.[294] Again strict liability would be efficient only when full, or "perfect," insurance was available on the market.[295]

Also intriguingl, Nell and Richter found that a rule of negligence performed better, and strict liability worse, as the numbers of potential victims increased.[296,297] For a rule of negligence, as the level of expected victims increased, the efficient level of due care increases so that eventually, the maximum level of due care would become optimal.[298] In such a case, the extremely high setting of the due care standard would emulate a rule of strict liability to the tortfeasor's perspective.

Overall, Nell and Richter concluded that in those circumstances, a rule of strict liability would lead to insufficient levels of engagement and thus attain inefficient activity levels.[299] In particular, they found that when tortfeasors are risk averse and face incomplete insurance products[300] and when the number of potential victims is very large, a rule of negligence would be more robust than a rule of strict liability.[301] However, even under those conditions, the rule of negligence that they find robust would in fact operate close to the incentives more traditionally associated with a rule of strict liability.

3.5.1.2 In Application to CCS CCS projects are not generally operated by one person but by sophisticated commercial operators that are able to both divest and plan for risk while also being controlled to attain specific outputs and revenue goals; risk aversion is not likely to cause them substantial cognitive bias. Additionally, it is foreseeable that the early providers of CCS technologies and operational strategies would be parties closely affiliated to the existing gas injection and gas storage facilities within the oil and gas industry, providing field experience with both the technical risks and hazards and the means of addressing the financial risks of damages. Thus, it is reasonable to expect that risk aversion is not likely to be a substantial concern within CCS policy for operators.

Given that neither risk aversion nor imperfect supplies of insurance are likely to be insurmountable problems for CCS operators, the arguments in support of a rule of negligence lack support.

To the extent that financial assurances should be made mandatory or that the government might facilitate the provision of insurance, regulatory efforts might be more robust to address these particular conditions.

3.5.2 Insolvent Tortfeasors

When one discusses insolvent tortfeasors, the concern is that if tortfeasors are out of cash, then liability rules could fail to provide incentives by transference of ex post damages to ex ante decision making. Furthermore, if the tortfeasor had an expected or probabilistic likelihood of future insolvency, then the idea itself that a civil liability rule could transform the ex post damages to ex ante considerations in decision making would be substantially imperiled.

3.5.2.1 Theoretical Framework Shavell found that a rule of negligence appeared to remain efficient in most instances of insolvent tortfeasors.[302] However, he found that a rule of strict liability led to the tortfeasor's facing effectively lower-than-accurate damages and thus would lead to an inefficient result. Both insufficient preventive measures and overengagement in the risky activity would result under strict liability.[303]

Nussim and Tabbach employed a "durable precaution" model and found that while Shavell's results remained robust, extra nuance could be discerned. They opted for a three-tier model, although in effect there were only two tiers: (1) when assets exceed expected damages, both rules remain efficient in most cases,[304] but (2) when assets match or are lower than expected damages, then tortfeasors are likely to engage in the maximum level of activity and underperform preventive activities. For a negligence rule, this depended on the increasing amount of shortfall posed by the damages against the assets.[305]

3.5.2.2 In Application to CCS It is not likely that CCS operators would go insolvent during the operational phases, given the existing appearance of requirements for financial assurance and performance bonds. In addition, almost every existing permitting process appears to require the payment of fees concurrent with the injection operations to cover a substantial part of the financial needs of postclosure activities. Should an operator go insolvent, the authority could either replace the operator with a new solvent operator or close the facility and operate from collected funds. Thus, there is little risk of insolvency with respect to the operational period.

It is clearly foreseeable that many CCS operators could be insolvent when the storage facilities incur accidents at some future point, perhaps centuries after the cessation of injection activities. However, given the likelihood of complete insolvency and, frankly, likely legal dissolution given the time frame, it is unclear how either strict liability or a rule of negligence would bestow any effective incentive for events that occur that far out in time. A secondary consideration would be the impact of discounted cash flow accounting on the perceived impact of those future damages. The effective damages might be negligible given sufficient years and a positive rate of interest, rendering no incentives to affect operational period decision making on risks.

Thus, it is unclear if there would be any benefits from choosing a rule of negligence over a rule of strict liability under these circumstances. Regulatory efforts might be more robust to address these particular conditions.

3.5.3 Strategic Avoidance of Liability by Tortfeasors

Avoidance strategies limit liability. When these strategies are effective, both strict liability and negligence result in insufficient precaution levels. This is primarily due to a reduction in preventive care levels.[306]

3.5.3.1 Theoretical Framework Negligence appears to be more robust under more circumstances, particularly for unilateral accidents,[307] akin to its ability to retain limited effectiveness under insolvency. See the previous discussion in this chapter for the theoretical framework.

3.5.3.2 In Application to CCS It is reasonable to foresee that operators would undertake all the legally allowed measures to limit their risk exposure and thus engage in a certain number of liability-avoidance strategies. However, it would appear that in the context of CCS operations, with the circumstances of extensive governmental permitting and oversight, regulatory measures might be more effective to ensure that liability-avoiding strategies are limited or ineffective. For example, regulatory measures already in place in multiple locations require performance bonds that provide the necessary capital for environmental remediation, monitoring, or damages payments in advance of receipt of operational permits. Given the delivery of the performance bonds, liability-avoidance strategies are rendered effectively neutral because the capital remains available to the authority even in the absence of the operator. Thus, while a negligence rule might be more robust than a rule of strict liability in this particular theoretical concern, regulatory measures may well be even more robust.

3.5.4 Inaccurately Determined Damages

The incentives created by civil liabilities transport ex post damages to ex ante decision-making processes, so if the damages rendered by courts are not accurate, the derived incentives will not be accurate either.[308] Yet such problems do exist,[309] many due to various transaction cost problems.[310] Policymakers must thus determine which rule of civil liability would respond more robustly in those settings.

3.5.4.1 Theoretical Framework It appears from the literature that negligence is more robust than a rule of strict liability in the event of errant judgments of damages;[311] in that case, it will not need the damages to equal the injuries to remain efficient in attaining efficient levels of accidents.[312] When damages are overestimated, the tortfeasor remains at the due care level of precaution at maximum activity levels.[313] When damages are slightly underestimated, the tortfeasor retains the due care level of precaution at maximum activity levels.[314] But when damages are very underestimated and substantially below the due care level, the tortfeasor finds himself under a de facto strict liability rule.[315] Negligence is more robust, in that the duty of care is exogenously determined, but as seen above, its efficacy can be damaged when the error is sufficiently large.[316,317]

A rule of strict liability needs the judgment damages to equal the injuries to retain its policy effectiveness.[318] Without that strict connection, strict liability becomes increasingly inaccurate,[319] in proportion to the inaccuracy of the damages' forecast model. When expected judgments are errantly assumed too large, strict liability led to overprecaution.[320] When expected judgments are errantly assumed too small, strict liability led to underprecaution and resulted in a surplus of accidental harms.[321] Because strict liability directly depends on the assessed damages for the effectiveness of its incentives, when the damages are incorrect, there are no efficient results from a rule of strict liability.[322] Thus, for strict liability, the problem is a rational response to error-laden information.

3.5.4.2 In Application to CCS It is difficult to state precisely what types of accidents will actually occur, but the scientific consensus is that given properly sited and carefully constructed and operated injection facilities, only limited injuries to water resources would occur. In such cases, the extent of water volumes no longer available and the market costs of those damages should be readily determinable. To that extent, one would not foresee substantial problems with inaccurately determined damages, and, thus, no evidence of a need to prefer the application of a rule of negligence.

In the case of reckless operators, resulting in inappropriately sited, poorly constructed, and unsafely operated CCS facilities, potentially other types of accidents would be more likely to occur. Eruptions of CO_2 volumes into the atmosphere could occur, resulting in injuries to personal health; property, including agricultural assets; and the climate from the massive volumes of released greenhouse gases. However, it is likely that the injuries to personal health and personal assets, including real estate, ought not to provide particular frustration in determining damages, as such injuries routinely occur from other sources of accidents and pollution. The impact on damages from climate change–related injuries would indeed be more likely to be unforeseeable in the amount ultimately decided; however, it would appear that technology is readily in place to address all but the most extreme well blowouts or subsurface leaks so that the extent of the injuries should remain controllable and not open ended. Thus, even in the face of a broader range of potential accidents, it is not likely that the foreseeable types of accidents would render particular trouble in accurately determining damages.

3.6   Conclusion and Results

From a review of the literature, it is clear that the majority of scholars who have inquired into the liability issues facing CCS storage systems have found that the potential operators, landowners, and other stakeholders would face confusing and unclear liability rules if the rules were not improved prior to the onset of the CCS storage industry. For CCS storage system operators to make efficient decisions. they will need clarification of the rules they would face should damages result from their activities. Thus, an effort should be made to determine how to best clarify the existing rules so that liability rule could provide efficient and effective incentives to best enable the goals of CCS operators and of society to align harmoniously.

This chapter has examined that research by engaging the tools from the literature of law and economics. That school of literature has developed a platform to ascertain when various rules of civil liability might be more robustly applied. In an examination of the two major rules of strict liability and negligence, the unique circumstances of CCS storage systems were reviewed.

Strict liability was found to be a more robust solution for the circumstances facing CCS operations than a rule of negligence. Strict liability provides more robust solutions under the circumstances of CCS, rules of abnormal hazards, and guidance for dominant actors; in the face of novel risks; for the diverse settings of CCS; and to assist areas with transaction cost challenges facing their judicial systems.

It is important to draw a line under that conclusion. It is not only due to the hazards of CCS storage that we find a rule of strict liability more robust. CCS storage activities might well qualify under the notions of ultrahazardous or abnormally dangerous, but neither qualification is required for the rule of strict liability to be found more robust than a rule of negligence within the circumstances of CCS storage. It is the theoretical framework that reveals that when one party has an overwhelming informational advantage in decisions regarding risky activities, the rule of strict liability is found to be more robust. The major decisions to be made regarding CCS storage safety are primarily those of site selection and operational oversight of the injection strategies. Both activities are normally primarily within the stewardship of the operator and of its joint venturers, and by the public or public authorities that might otherwise have regulatory oversight. Furthermore, strict liability would better provide for the operators to more legally and efficiently adopt their own safety solutions, enabled by a policy of effective decentralization. The transactional costs of litigation under a rule of strict liability are also thought to better enable ready resolution and settlement of damages, again enabling economic efficiency for both the operator and the community.

Of the four potential foundations to apply a rule of negligence, none of the scenarios, once examined under the facts and circumstances of CCS operations, strongly supported the application of a negligence rule. When the circumstances did suggest that strict liability might not be resilient or robust in addressing certain problems, it was identified that regulatory solutions already in place might be more robust than the application of a rule of negligence. The advantages that might be available under a rule of negligence might be more robustly provided by the adequate implementation of regulations, a topic explored in a later chapter.

Therefore, the rule of strict liability has been found to be a better match and more robust for the facts and circumstances of CCS operations.

To the extent that a jurisdiction cannot avail itself of a strict liability rule, as might be the case in certain common law jurisdictions, it should implement a version of negligence that closely approximates the economic results of a strict liability rule to minimize welfare losses. For example, a negligence rule could set a very high duty-of-care standard to mimic the levels of care achieved by "faultless" liability rules required under a rule of strict liability. Also, a rule of negligence could minimize the opportunity for contributory negligence to better ensure that the costs of damages remain with the operator-tortfeasor. Finally, high standards could be set within a parallel set of regulations to better enable the support of a high duty of care using a high standard for arguments based on a per se negligence rule. But this discussion invokes a need for a systematic review of policy options within implementations of civil liability rules, a task to which the next chapter is fully dedicated. For the separate and unique issue of how best to address civil liability for time frames beyond the expected longevity of the operators, that of long latency periods, the subsequent chapter covers multiple rationales for the potential transfer of liability from the operator to a public authority.

4 Policy Options for CCS Liability Rules

The previous chapter addressed which particular rule of civil liability might be most robust for the circumstances of CCS storage systems. This chapter follows on that study by evaluating various subpolicies that attend the implementation of such civil liability rules.

These matters include questions on how best to determine the relevant actors, what role for the notion of force majeure, potential interactions with public regulations, and rules on establishing causality, among other considerations. This chapter provides analysis, review, and reflection on these themes and concludes with recommendations.

4.1 Refocusing on the Goals and Limits of Liability Rules

Before discussing various aspects of how a liability regime for CCS-related damage could be shaped, we return to the discussion from the previous chapter concerning the goals of a liability system for CCS and the importance of liability rules generally for CCS.[1] The reason to come back to this is that the particular way in which one wishes to shape a liability regime will of course strongly depend on the expectations concerning such a regime.

It is striking that the literature concerning CCS-related damage strongly stresses the need to expose operators to liability, arguing that otherwise a moral hazard risk would emerge.[2] Authors therefore hold that operators should be held financially responsible according to the individual site characteristics and operational methods, since otherwise a risk of moral hazard could emerge.[3] Granting operators broad-scale indemnity would, it is held, create moral hazard.[4]

Adelman and Duncan demonstrated that many of the earlier studies viewed the potential damage resulting from CCS in a rather unbalanced manner, without sufficiently distinguishing among the various phases in the life cycle of the CCS project. They first generally point to the importance of regulation in the prevention of harm[5] and also argue that as far as long-term stewardship is concerned, there is in fact no moral hazard for the simple reason that these risks are so remote in time that they can hardly provide any meaningful deterrence on the operator's behavior. That is obviously not an argument to totally exclude CCS operators from liability, but rather to limit the exposure to liability to the moment when the long-term stewardship starts and liability is transferred to the state. For all damage that could emerge to that moment, operators would still be held fully liable and would of course be able to choose appropriate insurance or other coverage mechanisms.

We have added to this argument that beyond the governance of the risks that lead to accidents that arise while the operator is actively involved at the injection site and up to the closure of the storage site, the governance of the risks of events beyond that time period might also be well managed by the application of civil liability rules. The vast majority of events that would lead to potential long-term events of harms are identical to the events that would lead to shorter-terms liabilities. It is perhaps, in colloquial terms, a matter of two birds with one application of civil liability, because both imminent and long-term risks arise from the same near-term acts and behaviors of the operator: primarily those of site selection and operational prudence.

This relates to the more general point mentioned in the introduction to the role of liability rules:[6] liability rules have to balance both positive and negative externalities. It is an issue that has been raised by Gilead.[7] Those who stress only the potential negative externalities that could follow from a CCS project tend to forget the positive externalities (in terms of mitigating climate change) that CCS may generate as well. Hence, a liability regime should be shaped in such a way that positive and negative externalities are appropriately balanced.

If one were, for example, to stress only potentially negative externalities and hence argue that operators should be held liable for losses that could emerge at an undefined moment in time, this could lead to a crushing liability, and socially desirable CCS activities would not be undertaken at all. This dilemma created by a so-called crushing strict liability has also been described by Trebilcock, who warns that a combination of strict liability with joint and several liability and retrospective liability, especially in case of long-tail risks, may lead to defensive practices, with the result that socially beneficial activities would no longer be undertaken.[8]

Finally, from an economic perspective, the primary goal of a liability regime is to provide incentives for deterrence. Given the primary role of regulation to prevent risks related to CCS, this role of liability rules will, moreover, be only a supplementary one. From an economic perspective, liability rules do not have compensating victims as a primary goal. The effect of a strict liability rule can, of course, be that it leads to victim compensation as well.

However, it may be dangerous to (broadly) shape liability rules with the aim of victim compensation since that may, for example, endanger the insurability of liability risks. The compensatory function of liability rules should therefore not be leading in determining the shape of liability rules; other mechanisms[9] can better fulfill the compensation function than liability rules.[10]

4.2   Determination of Responsible Actors

In determining liability, a party or parties must be selected, but in the case of CCS, that could be less obvious than in more traditional cases of environmental or personal injuries. The issue of responsible actors can be viewed as a confluence of two distinguishable areas of law: stakeholder identification and property law.

4.2.1   Stakeholders

The gas stream could belong to various parties. One assumes that industrial generators of the CO_2 would want to transfer ownership and liabilities to the transporter so as to reduce their liability exposure once the CO_2 is no longer within their control. We could assume repeated transfers of liability as the CO_2 moves downstream and into subterranean storage. Once it is stored, property law concepts and the passage of time could enable or require the transfer of liability from party to party. Thus, it might be challenging to determine responsible actors without establishing standards. For the sake of early investors, the preference would be to gain that clarity ex ante and not ex post.[11]

A first challenge is that the literature indicates that various stakeholders can be involved in CCS projects. Indeed, multiple contemporary stakeholders of potentially adverse interests may be a complicating factor in determining liable parties. One might consider the direct actors operating the project to be a stakeholder, but government agencies, local landowners, various oversight groups, and potentially local communities and their residents as well might function as stakeholders to various degrees.

The energy industry is expected to be an important player in the development and operation of CCS projects. The literature stresses that given the large amount of coal reserves (e.g., in North America), the continued use of coal may be critical to energy security.[12] For the energy industry in states that are largely dependent on coal, the large-commercial-scale development of CCS technology can be considered critical.[13] The question therefore arises how incentives can be provided to induce (potentially risk-averse) fuel extractors and power producers to invest in CCS, which, given existing uncertainties, may be a risky adventure. An adequate matching between these various business sectors with a view to providing adequate incentives to develop storage facilities will be of crucial importance.[14]

Another important stakeholder in the CCS process is government. This could be plural actors in multiple government agencies and across both local and national levels of governance. In order to provide incentives to industry to invest in CCS, which could require long-term stewardship, the government may become an important player, as well in case responsibility for the storage site is transferred to the government.[15]

A second challenge in determining responsible actors is one of multiple time periods. One problem is the uncertainty of the extent of the operator when the injury is discovered in some future time period, perhaps long after the operator has been legally extinguished or gone insolvent.[16] Victims might not become aware of their injuries until long after the tortious act; especially given the long-term storage goal of CCS storage, there are potentially extensive latency concerns to be addressed.[17] Governmental units are proposed as being potentially available for victims seeking redress in later time frames, given governmental ability to endure potentially centuries.[18] Key among the resources enabling governmental endurance would be the ability to draw on public resources such as taxes to support future capital needs.[19]

A third challenge in determining responsible actors is how to address first movers in the nascent CCS industry. Precisely in order to deal with the uncertainties and problems in providing adequate incentives to stakeholders, some have argued that an incentive structure should be provided in order to stimulate first movers. Klass and Wilson have argued that given the inherent uncertainties of the technology and the necessary involvement of government, the first CCS projects should be encouraged under a shared public-private liability regime.[20] The advantage would be that through these types of pilot projects, an experiment could take place, with the result that additional information could be collected that could ultimately benefit the entire sector. Klass and Wilson therefore hold that industry should be helped to gain confidence and experience by providing first movers a financial cap on liability.[21] The advantage of a cap is that for those first movers, liability would be restricted and information would be collected that could create positive externalities to the entire market and from which all could benefit.[22] The result of this strategy is that the mantle of responsible actor would be shared between industry and the government; such liability sharing might be challenging to implement without a clear setting of standards and expectations.

4.2.2 Property Law Clarifications

To whom does injected gas belong: the injector, the landowner, or a third party?[23] The answer to ownership could clarify who is liable at what points in time for CCS projects. The literature is clearly concerned with potential confusion over who owns what at different stages of the storage period.

Generally state property law in the United States has held that injected gases remain the property of the injector even in circumstances in which the gas migrated into other reservoirs and even if the original injector lacked all forms of property rights to the final reservoir location.[24] Even traditional answers, pre-CCS concerns, diverge in ways that complicate developing a broader common international approach. In most states in the United States, surface rights and mineral rights for oil and gas are bifurcated, usually with the surface owner taking command of the pore space once the hydrocarbons are removed.[25] Canadian and English rules appear to provide that the pore space remains with the mineral rights owner even after the depletion of the hydrocarbons.[26]

However, there is some divergence on who owns the pore space within the subterranean storage zones, and how; in essence, it asks if the pore space is part of the mineral rights or part of the surface rights.[27] In this question as to which property owner should receive the default allocation of property rights to the emptied pore space, there is much in analog to Coase's original analysis of spectrum allocation and how best to resolve the problem of two adjacent wavelength owners that experience frequency overlap:[28]

What this analysis demonstrates, so far as the radio industry is concerned, is that there is no analytical difference between the problem of interference between operators on a single frequency and that of interference between operators on adjacent frequencies. The latter problem, like the former, can be solved by delimiting the rights of operators to transmit signals which interfere, or might potentially interfere, with those of others. Once this is done, it can be left to market transactions to bring an optimum utilization of rights.[29]

It was this analysis that underlay the more abstract treatment of the same model that is more popularly known as the seminal paper on Coasian allocation of rights.[30]

With regard to pore space allocation rules, there are two leading legal theories: the American Rule and the English Rule.[31] The American Rule provides that if gaseous minerals are removed from a subterranean zone, the emptied space remains the property of the surface owner and not of the mineral extractor who once owned rights to the minerals in situ.[32] The English Rule holds, to the contrary, that the mineral owner and extractor would retain its property rights to the emptied pore space.[33] Most American states follow the American Rule, whereas Canadian provinces and Kentucky follow the English Rule.[34] In Australia, most jurisdictions place ownership of the pore space with the Crown, regardless of surface rights or mineral lease holders.[35]

To identify basic torts and injuries to subterranean waters, there will need to be clarifications to the rules on water ownership. There is a division of theory in riparian rights in the United States.[36] There are loosely three main paradigms that allocate property rights to subterranean water volumes.[37] The first paradigm, absolute dominion, provides that the surface owner will own all ex ante rights to waters contained within their lands; the surface owner has the rights to use these waters without risk of liability to third parties such as adjacent land owners.[38] The second paradigm, reasonable use, provides a limited sense of accession of the water to the surface owners, limited to uses that are "reasonable and beneficial to the land itself."[39] Under the reasonable use rule, "unreasonable uses" are recognized as torts and give rise to liability to other parties reliant on the water volumes.[40] The third paradigm, the appropriation rule, provides water rights to the first party to use those waters; the third paradigm is common in western states.[41] In a sense, the third rule is the riparian rights version of *Pierson v. Post* on first capture and assession.[42]

Havercroft and Macrory raise a separate concern of when the injected fluids might become recognized as a fixture within the land and no longer as chattel.[43] In logic based on *Holland v. Hodgson*,[44] they provide that if the CO_2 fluids are bound only by their own mass or buoyancy, they remain chattel; if they are bound to the land by other means, such as fingering, the fluids might properly become characterized as fixtures.[45] In such a case, the liability could become transferred to the landowner or pore space owner as permanence in the storage is attained.

It is clear that the rules guiding property ownership will need to be clarified in advance of early development and planning for CCS storage sites, as fundamental control issues would become invoked by determinations as to who owns the pore space, the injected CO_2, and the migratory CO_2 in escape from containment. This need for ex ante clarification of property, and thus of resultant liability, derives immediately from previous discussions on the need for the operator, as risk taker, to have sufficient information on the liability rules to make efficient determinations on levels of precautionary efforts and overall activity levels in the risky activities.

4.3 Force Majeure

One question that has often been asked in case of ultrahazardous risks on which a strict liability regime applies (as in the case of nuclear liability, for example) is whether operators should still be held liable if the damage is caused by an act that was beyond the control of the operator, often qualified as force majeure or an act of God.

The basic premise of the economic analysis of tort law is that an exposure to liability should provide incentives for a behavioral change by the operator (i.e., for additional preventive measures). If force majeure is defined as an act that is beyond the control of the operator, in principle the operator should not be held liable for damage resulting from it since in that case, liability could not have a positive effect on his incentives for prevention. However, this requires a qualification in case of damage that would be caused by natural hazards such as flooding or earthquakes. These may be hazards to which CCS sites could be exposed and could lead to substantial damage.

Natural events might be foreseeable and thus not quite acts of God from a liability perspective. The mere fact that damage resulting from a CCS site is caused by a natural disaster should of course not automatically exclude the liability from the operator to the extent that those events are reasonably foreseeable. A proper location choice, which is crucial for the prevention of risks emerging from CCS, will take the possibility of flooding or seismic events into account and hence avoid a CCS storage site in flooding zones or earthquake-prone areas. Only when expert opinion holds that a particular natural hazard in the location of the site was reasonably unpredictable for the operator, liability should be excluded.[46]

Note that in the nuclear liability conventions, operator liability is excluded only in case of a natural disaster of an "exceptional character."[47] Similar language is used in the Japanese Act on Nuclear Liability.[48] What might be a surprise to some readers is that the tsunami of March 2011 was not considered a natural disaster of an exceptional character in that region of Japan.[49] As a result of that characteristic, the event of the tsunami did not exclude the operator, TEPCO, from liability for the subsequent Fukushima nuclear disaster.[50]

The relevance of the force majeure exception is also discussed in the literature related to CCS. Since in Europe CCS storage sites were brought under the Environmental Liability Directive through the CCS Directive, it is held that the liability of operators can be excluded on the basis of force majeure.[51] However, the same authors hold that liability would still be possible in case of earthquakes, which could be a significant type of risk in relation to long-term CO_2 storage.[52] This shows that a natural disaster does not automatically constitute a form of force majeure that excludes operator liability.

The question will have to be asked whether the damage resulting from a natural disaster was reasonably foreseeable, for example, when the CCS storage site was constructed in a flood-prone area or an area exposed to earthquakes. In those cases, the site selection could be considered wrongful and thus lead to operator liability. Only when the damage resulting from a natural disaster was to be considered totally unforeseeable and not preventable by reasonable measures to be taken by the operator would force majeure exclude liability.[53]

4.4   Attribution of Liability

The question of attribution of liability has different angles. It amounts basically to the question of who, in case many parties are involved, should be held liable for risks related to CCS. As was explained when addressing the project life cycle of CCS[54] and the stakeholders involved,[55] there may be different parties in the various phases of the CCS chain, differing particularly in the CO_2 capture, transportation, and storage.

In this case, we mostly focus on storage, whereby liability will mostly be allocated to the licensee of the storage site, that is, the operator. A question that could be asked is whether, following the example of the nuclear liability conventions and the conventions with respect to marine oil pollution, liability should be channeled to the licensee. Channeling of liability basically means that only one party can be held liable, thus excluding the liability of other parties involved. These types of channeling regimes have been defended as facilitating the litigation efforts of the victims who only had to address the licensee;[56] channeling was also defended in cases where, through complicated contractual arrangements, more parties could potentially be involved. It is also held that for operators, channeling facilitates insurability since only one party would have to carry insurance coverage.

However, in the law and economics literature, channeling is usually considered inefficient. It has negative effects on the incentives to take care more particularly of all other parties who could equally have influenced the accident risk; it enables nonchanneled parties to avoid costs that otherwise might have provided risk-reducing incentives. Channeling of liability to the licensee is therefore generally rejected[57] and as a consequence should not be recommended in a future liability regime concerning risks related to CCS.

There is, however, one aspect concerning the attribution of liability related to CCS that should be mentioned: the need to transfer the long-term liability to the government, whereby operators remain liable only up to the moment of transfer of liability.[58] This is to a large extent the approach followed in the EU CCS Directive. According to this directive, provided that specific conditions are met, a storage site shall be transferred to the member states, which equally includes a transfer of the liability.[59]

4.5   Effect of Regulation

This section reflects on the role of regulations when civil liability rules are implemented. The discussion focuses on potential liability limits afforded by regulatory compliance.[60]

Later in this book, we suggest that regulations will be a primary tool aiming at the reduction of CCS-related risks.[61] The question arises whether regulation compliance (e.g., meeting the conditions of a permit) would excuse an operator from liability in tort. Although this is a much-debated issue,[62] most legal systems reject such a "regulatory compliance defense."

One can find a clear economic rationale for this result. If compliance with a regulatory standard or license would automatically result in a release from liability, the potential injurer would have no incentive to invest more in care than the regulation asks, even if additional care could still reduce the expected accident costs beneficially.[63] The regulations would effectively set a minimum but sufficient level of prevention under such a defense.

A first reason to hold an injurer liable, given that other conditions for liability are met, although he has followed the regulatory standard, is that this standard is often a minimum. The complete compliance defense prevents any incentive to take precaution in excess of the regulated standard.[64] Exposure to liability will give the potential injurer incentives to take all efficient precautions, even if this requires more than just following the license.[65] Since the regulatory standard cannot always take into account all of the efficient precautionary measures an injurer could take, testing the measures that the injurer took will provide additional incentives, even though the regulatory standard was followed.

Allowing a regulatory compliance defense would also largely remove the beneficial incentive effects of strict liability. As explained in the previous chapter, strict liability without this defense has the advantage that it provides the injurer with the incentives to take all efficient measures to reduce the risk,[66] even if less demanding regulatory conditions were followed. This outcome has been shown formally by Kolstad, Ulen, and Johnson[67] and by Burrows.[68] They argue that the exercise of a complete compliance defense prevents any precaution in excess of the regulated standard. If there is serious underenforcement of standards, the role of liability as an incentive to take precautions remains important.

A second reason is that exposure to liability might be a good remedy for the unavoidable capturing and public choice effects that play a role when permits are granted.[69] If a permit would always release from liability, all a plant operator would have to do is get a good permit with easy conditions from a friendly civil servant. That would then exclude any lawsuit for damages from a potential victim. Obviously the capturing and public choice effects should also be addressed with direct tools. In this respect, one can think about the liability, even under criminal law, of the licensor.[70] Liability of the licensor (and appropriate sanctions within administrative law) can provide incentives to civil servants to act efficiently when granting licenses.[71] This, however, still requires tort law to take into account the fact that regulatory standards are not always set efficiently. If the optimal level of care is higher than the regulatory standard, liability will efficiently provide additional incentives.

Third, tort law can also be seen as a stopgap for situations not dealt with by statutes.[72] This makes clear that the exposure to liability notwithstanding the permit is an important guarantee that the plant operator will take efficient care. However, all scholars follow this position. Bergkamp, for example, powerfully argued that a polluter should "only pay once," that is, pay only for compliance with regulatory standards as a result of which he could call on a compliant defense to be excused from liability.[73] That position, however, has not been followed in the literature or in case law. It is mostly held that compliance with regulation (e.g., obtaining a license) merely removes an administrative obstacle to pursuing an activity and hence does not exclude liability toward third parties.[74]

The reasons for not accepting a compliance defense also seem compelling for the case of CCS activities. If compliance with regulation were to free an operator from liability, liability rules could not play their desirable supplementary effect.[75] Moreover, compliance with regulation defense would de facto make a strict liability rule potentially meaningless. Especially since safety requirements may develop rapidly in this potentially very volatile technological environment, if compliance with potentially outdated regulation were to free an operator from liability, this would potentially seriously restrict the incentives provided by a strict liability rule.[76]

That is not to say that safety regulations do not have an influence on the exposure of operators to liability. To the extent that safety regulation is effective and adequate, it would force operators to choose optimal locations, monitor risks, and implement safety standards that could contribute to a prevention of the risk exposure. Compliance with efficient safety standards could, in effect, reduce the liability exposure of operators without formally excluding liability in case damage nevertheless occurred. Regulations have the potential to effect control on the behavior of the operators and thus protect them from themselves.

Even when arguments have been presented for a regulatory compliance defense, the arguments reach only a limited defense, not a complete defense. Flatt has encouraged the use of regulatory compliance as justification for a complete liability shield for operators of CCS storage sites.[77] He advocated that regulatory rules could approximate a duty of care and achieve an "acting reasonably" result from the operators.[78] However, he also suggested that such treatment would present problems with the polluter-pays principle and would likely require the government to provide compensation for damages, both of which he saw as reasons to support only limited liabilities for regulatory compliance.[79]

Thus, the arguments against allowing compliance with regulations to provide a justification for liability limits or caps are persuasive. However, there are other arguments for providing forms of liability protections or indemnifications to the operators of CCS sites; those arguments are reviewed in the next chapter.

4.6 Causation

The way in which the law deals with uncertainty over causation has an important bearing on the potential scope of the overall liability of CCS site operators. Causal uncertainty is especially important with regard to so-called toxic torts whereby, for example, a part of the population has been exposed to hazardous substances or radiation and subsequently a certain disease (e.g., cancer) is discovered.[80] In those cases, it is often not known which of the victims (among a larger population) was a victim of a tort and which developed the disease as a result of the so-called background risk.[81] In that case, uncertainty concerning the causal relationship often prevails, whereby experts can, for example, establish only that there is a probability of, say, 40 percent that a particular event "caused" a particular damage.[82]

The traditional approach taken to causal uncertainty in the legal system was often a so-called threshold liability. That meant that the probability that the event caused the damage had to pass a certain threshold. It was often referred to that it had to be "more probable than not" that the event caused the damage, and hence the threshold was 50 percent. This type of threshold liability was considered highly problematic since injurers could get systematically off the hook if, for example, they were to expose victims with a 40 percent probability.

The proportionate liability rule is an alternative that increasingly is gaining ground with scholars. From the operator's perspective, proportional liability is desirable since it limits liability exposure exactly to the proportion in which the operator contributed to the risk. Proportional liability is therefore desirable compared to alternatives, such as a reversal of the burden of proof in case of causal uncertainty.[83] A proportional liability rule may therefore also help keep operator's liability within reasonable limits.[84]

Causation may also be an issue as far as the potential damage resulting from CCS is concerned. The question to what extent causal uncertainty will be an issue will of course depend on the nature of the damage.[85] In the rare case that very high concentrations of CO_2 would result in the sudden death of humans, for example, it may not be difficult to lay a causal link with the CO_2 release. Causation may, however, already become more difficult in case of geological impacts, particularly for damages related to water resources. The more remote these impacts are from the CCS site in time or distance, the more difficult proving the causal link will become. Causation will be especially hard to prove in case of atmospheric releases giving rise to climate change liability.

In all of those cases of causal uncertainty, the most effective remedy is to rely on expert evidence concerning the likelihood that a particular damage was related to the CO_2 storage site and to subsequently translate that likelihood proportionally into an amount of compensation to be awarded to the victim. Only in that way could a crushing liability (and thus overdeterrence) be avoided by making the operator liable for only the damage that was actually caused by the CO_2 storage site.

The effectiveness of this system will of course depend to a large extent on the question of whether experts are indeed able to reasonably assess the likelihood that the damage can be allocated to the CO_2 storage site. In practice, this may obviously give rise to debates.

4.7 Joint and Several Liability

Joint and several liability basically means that a victim can choose which operator to sue and claim full compensation from one operator even when more than one operator could have contributed to the loss.[86] To some extent, joint and several liability may be the reverse of proportional liability or channeling of liability already discussed.[87] Under joint and several liability, joint tortfeasors are held liable for all the damage to which their behavior might have contributed.

Joint and several liability regimes have often been introduced to relieve the burden of proof from victims. Victims could collect the entire damage from one of the contributing tortfeasors, whereas the tortfeasor who was compensated could exercise an action in recourse against the other liable tortfeasors. An argument in favor of joint and several liability is that it gives incentives for mutual monitoring by potential injurers.[88] However, in case of the insolvency of one of the actors, inefficiencies may arise since recourse may become impossible.[89] Joint and several liability is debated since an injurer could in principle also be held liable for a part of the damage not caused by his activity and could thus potentially increase liability exposure.[90]

Questions of joint and several liability obviously do not arise when there is only a single operator, in which case attribution of liability is not problematic.[91] The question of the desirability of joint and several liability arises only when multiple (potential) tortfeasors are involved in different phases of the project life cycle or multiply operate the storage site.

Ingelson, Kleffner, and Neilson called for operators of CCS storage facilities to bear joint and several liability.[92] They advised that this would mirror the existing treatment of oil and gas operators in their non-CCS injection activities.[93] This is because those firms often invest through joint venture entities and retain their independent legal liabilities.

Joint and several liability will increase the potential scope of liability for CCS operators.[94] Therefore, Adelman and Duncan argue that the case for joint and several liability in case of CCS-related damage is weak. Given the scale of operations, the number of potential defendants should be trace-able, and joint and several liability could potentially lead to overdeter-rence.[95] With a view of the positive externalities equally generated by CCS, the policymaker may be cautious with the introduction of joint and several liability for CCS operations.[96]

4.8   Remedies: Limited Financial Amounts?

The remedies that could be applied in case of a finding of liability are strongly linked to the types of damage that could result from CCS activi-ties.[97] The most important distinction to be made is between CO_2 emissions that could lead to a climate change liability and other damage that could result from CCS. The climate change liability would result from the fact that in legal systems where duties to mitigate greenhouse gases exist (such as in the EU), a release of CO_2 from a storage site could potentially give rise to emissions for which operators have no allowances. Potentially this would lead to a duty for the operator to provide allowances to cover the emissions and potentially it could equally lead to a penalty (of 100 euros per tonne) for excess emissions.[98] Hawkins, Peridas, and Steelman indicate that this can potentially create price risks, related to the fact that there may be a need to buy allowances in the future at a higher price if CO_2 would be released.[99]

More traditional damage could be created as well, such as the contami-nation of groundwater [100] or, in extreme cases, personal injury resulting from exposure to extreme concentrations of CO_2. There is a lot of experi-ence with the valuation of these types of more traditional, yet environ-mental, damages. It is possible to identify and monetize the different parts of the damage that could result from a CCS site differently from the release of CO_2.[101]

More important, some literature discusses the desirability of putting a financial limit, a so-called cap, on the amount of liability. The arguments in favor of a cap are advanced in some literature either generally or to stimu-late first movers.[102] For example, De Figueirodo, Reiner, and Herzog argue, with the example of nuclear liability in mind, that a liability cap may be desirable.[103] However, they also argue that a liability cap could be detrimen-tal to carbon storage from a public perception point of view. This may give a wrong signal to the public that CCS (like nuclear energy) is a catastrophic type of risk and can lead to a stigmatization of CCS.[104] Trabucchi and Patton seem to be more generally in favor of liability caps. Referring to additional enabling legislation, they note the need to introduce damage thresholds,[105] but they provide little motivation. They refer to the example of the Price-Anderson Act Nuclear Industries Indemnity Act of 1957,[106] regulating nuclear liability, and the Oil Pollution Act of 1990,[107] which has limitations on liability.[108] Also, Wilson, Klass, and Bergan argue in favor of liability caps but only during the postclosure period.[109] In another study, Klass and Wilson provide a more balanced picture. They ultimately conclude that generalized damage caps (e.g., those that exist in the Price-Anderson Act for nuclear liability) would not be appropriate for CCS as a general matter but would in the early years to encourage pilot projects.[110] A general liability cap would undermine the credibility of CCS in the eyes of the public,[111] but a more limited cap may help first movers manage the financial risk of the new CCS technologies.[112]

Adelman and Duncan are strongly opposed to liability caps, even to promote first movers. They argue that caps are unnecessary because the projected magnitude of the potential damage for pilot projects would be limited. Moreover, it would not be the potential damage that would constitute the most important barrier but technological uncertainties. Those would not be removed with a financial cap.[113] In that sense liability caps would have only a symbolic value.[114]

Economic analysis strongly supports the arguments against financial caps. From an economic perspective, it is important for the potential injurer to be fully exposed to the social costs of his or her activities. Otherwise the desirable internalization of the negative externality would not take place.[115]

The literature has indicated that there may be good reason to favor a strict liability rule for environmental risks, the main reason being that only a strict liability rule would lead to a full internalization of those highly risky activities.[116] This strict liability rule is especially put forward in a so-called unilateral accident situation; this is where only one party influences the accident risk. Only with strict liability would the potential injurer have an incentive to adopt an optimal activity level.[117] This full internalization is obviously possible only if the injurer is effectively exposed to the full costs of the activity she engages in and is therefore in principle held to provide full compensation to a victim. An obvious disadvantage of a system of financial caps is that this will seriously impair the victim's rights to full compensation.[118] But if the cap is indeed set at a much lower amount than the expected damage, this would not only violate the victim's right to compensation; the full internalization of the externality would not take place either. From an economic point of view, a limitation of compensation therefore poses a serious problem since there will be no internalization of the risky activity.

Indeed, if one believes that the exposure to liability has a deterrent effect, a limitation of the amount of compensation due to victims poses another problem. There is a direct linear relationship between the magnitude of the accident risk and the amount spent on care by the potential wrongdoer. If the liability is limited to a certain amount, the potential injurer will consider the accident as one with a magnitude of the limited amount. Hence, he will spend on care to avoid an accident caused with a magnitude equal to the limited amount and will not spend the care necessary to reduce the total accident costs. Obviously the amount of care spent by the potential injurer will be lower, and a problem of underdeterrence arises. The amount of optimal care, reflected in the optimal standard, being the care necessary to reduce the total accident costs[119] efficiently, will be higher than the amount the potential injurer will spend to avoid an accident equal to the statutory limited amount.[120] Thus, as a result of the cap, too little care is taken.[121]

Another effect of a financial limit on liability (in addition to providing under-compensation to victims and under-deterrence of operators) is that it would constitute an indirect subsidization of the industry enjoying a particular limit on liability.[122] This point has been proven with respect to the liability limit in the US Price-Anderson Act [123] and with respect to the international nuclear liability regime as well.[124]

It should also be mentioned that in the areas where financial caps on liability exist, such as for marine oil pollution and nuclear liability, those are also heavily criticized in the literature.[125] These economic arguments therefore support Adelman and Duncan, who argue against financial caps in general and for pilot projects.[126] Financial caps may, as they argue, be unnecessary and ineffective. They could also lead to undercompensation and underdeterrence, as the economic analysis showed. Moreover, they could have the undesirable effect of signaling that CCS is in fact highly risky activity (like nuclear power), which would thus have the precise opposite effect of reducing public support for CCS.[127] As a result, the liability of the operator should in principle be unlimited in amount.

4.9   Conclusion and Results

This chapter has evaluated a diverse range of policy options for the implementation of civil liability systems. While the previous chapter concluded that a rule of strict liability would be the more robust option, many subsidiary considerations needed to be addressed. This chapter, plus the discussion on long-term latency concerns in the next chapter, aims to supplement the conclusions of the previous chapter with the necessary policies to enable proper implementation of civil liability systems that could efficiently address the needs of CCS storage systems.

Stakeholding and the determination of who should be liable for what and when were found to be fundamental challenges facing civil liability policy. Multiple parties from potentially adverse perspectives might simultaneously be potentially liable without additional ex ante clarity. There are actors and victims, multiple time frame issues, and concerns to incentivize certain early adopters of the CCS technologies. The overall conclusion was squarely centered on the need to make clear ex ante choices, so that the operators, or other to-be-deemed liable parties, can make efficient decisions. In a separate but affiliated discussion, the chapter found arguments against channeling of liability. Joint and several liability might be seen as the reverse case of channeling, when every potential stakeholder is allowed to be held liable for the damages. It is not a substantial concern if there is a simple or singular entity held to be liable, such as a single operator. Traditionally, it has been applied within the oil field context, but there are arguments that it could lead to overdeterrence and reduce the availability of CCS storage systems. In short, policymakers need to provide ex ante clarity from the start if they hope to help operators and investors move efficiently.

"Who owns the minerals after injection at the time of potential injury?" is the next policy needing clarification. Is the postinjection CO_2 a private, movable asset that remains the property of whoever had it prior to injection? Or does the gas become united, unsevered, to the land and thus become an immovable asset belonging to whoever owns the land into which the CO_2 was injected? Or what if the CO_2 drifts in subterranean transit? Does it follow the property rules of fugitive minerals, as if an animal ferae naturae, or would it become aligned with flowing water rights? And if under the rules of flowing waters, would it reflect the American Rule or the English Rule, or perhaps the First Appropriation Rule found in the American West? Again, the most important aspect of the policy process is to ensure that these questions are resolved prior to the onset of exploration and development for a CCS storage system.

Our review of the policy option of regulatory defense found multiple reasons against it. First, regulations often set minimum rather than optimal standards, and thus a regulatory compliance defense would enable a CCS operator to behave at insufficient standards. Second, civil liability rules act to protect regulatory authorities from capture and other related problems; the regulatory compliance defense would prevent that function. Third, regulations are by their very nature explicit and limited, whereas civil liability rules are open-formatted and could address concerns not contemplated in the drafting of the regulations. Thus, a regulatory compliance defense would trap in time the standards to be applied and miss more efficient opportunities.

Causation can be complex in environmental and health torts due to the complexity of the chemical and physical mechanisms, the time spans that are sometimes required to give rise to detectable injuries, and the potential for multiple causes to join in the creation of an injury. Both threshold and proportionate liability rules were reviewed. Both have merits to the application of CCS circumstances. It was concluded that causation itself benefits highly from the accumulation of expert evidence; therefore, data observation and collection will be critical for the resolution of CCS-related causation questions.

Finally, the CCS literature is split on limits to liability exposure. So-called caps are seen as potentially inducing for early investors, but they are assuredly not efficient at optimizing the risks associated with that induced activity. Caps would certainly create information problems for the efficacy of a strict liability rule. This analysis is tempered somewhat by the affiliate question of how to address long-term latency concerns for operator liabilities. That issue is addressed in the next chapter. But in brief, we find that caps should not be offered to operators for the periods when the operators are present; thus, for the exploration, development, injecting, and early closure periods, the operators should remain liable without protection of caps. Thereafter, in the later stages of postclosure, we advocate the transfer of liability to a public authority; again, more detailed comments are found in the next chapter.

5 Postclosure Liability Transfers and Indemnifications

In order for CCS to work, it not only needs to remove CO_2 from emission points, transport the CO_2 to an injection site, and then secure the CO_2 into subterranean storage; it also needs to keep that CO_2 in place for centuries, millennia, or longer to protect the atmosphere from anthropogenic climate change. While it is in that century or those millennia of storage, events could arise that result in injuries and damages. For civil liability rules to work effectively, there needs to be a tortfeasor to be held liable. But what if the passage of centuries or millennia has eliminated the once-operator? Who could be held liable? Who might address the resultant damages?

In the prior chapters, a central research concern has been whether "there should be a departure from the ordinary position in order to accommodate the special character of CCS."[1] In this chapter, a more particular question is raised: Should the liability from CCS activities be bifurcated into periods of before the cessation of CO_2 injection and after the cessation of injections in order to be responsive to the factual needs of the very long run?[2]

The risks in the permanent sequestration phase of the CO_2 activities are very interesting. While there are several phases to that sequestration,[3] and whereas other phases are well bounded temporally, the final phase of long-term stewardship could potentially take hundreds of years and is in other terms indefinite.[4] That is why, especially as far as this long-term stewardship is concerned, many point to the potential role of the state and at the fact that liability for this long-term stewardship should be transferred to the state. This is a model that is already incorporated in EU Directive 2009/31/EC on the geologic storage of carbon dioxide. Article 17 of the directive addresses the closure and aftercare of the storage site.

Uncertainties are especially higher in the postclosure phase of a CCS project since this may occur ten, twenty, or fifty years from now.[5] Although the nature of the risks may largely be the same as during the period of CO_2 injection and operation of the site, what changes in the long term is the added risk of the changing science. Also societies' understanding of what it means to store CO_2 in perpetuity may evolve, which adds uncertainty in the postclosure phase of a CCS project as compared to the previous phases.[6]

Thus, this chapter engages this question of risks with long-term latencies from several perspectives. It first engages the question, "What are the long-term liability issues?" It then reviews several rationales for why policymakers might want to consider bifurcating the liability assignments into pre- and postclosure frameworks. It then considers the economic matters of externalities under long-term latency perspectives and provides a review of existing policy enactments.

The chapter concludes that the frameworks for assigning liability should be bifurcated and that planning should be undertaken to ensure that some form of public authority is capable of assuming the management, liabilities, and responsibilities once a threshold stage of closure and abandonment has been secured. The recommendation includes the potential for operators to retain their costs by providing capital to the respondent public authority to provide for future capital needs related to the long-term needs of the closed storage facility.

5.1   Long-Term Liability Issues

The long-term liability issues of CCS are held to be the "most challenging and complex aspects of regulating CO_2 storage activities."[7] The majority of the debates or discussions on long-term liability have centered on who should bear those liabilities: private actors or public actors.[8]

This section begins with a survey and review of discussions on CCS's long-term liability issues within earlier research literature; after this review, arguments for a variety of long-term liability policy options are reviewed, primarily on the impact of the polluter-pays principle, on the means to address potential operator extinction, and of means to support nascent CCS operations.[9] Then a survey of policies enacted in different jurisdictions is provided.[10] Finally, we present our policy recommendations for a transfer of these liabilities after certain conditions and events have been successfully met.[11]

The phrase *long-term liability* has a specific meaning within the CCS context: liabilities that arise after the complete cessation of injection activities and after the active monitoring of the site has been reduced or eliminated.[12] As Barton, Jordan, and Severinsen stated, a balancing is needed to provide recognition that "operators need to have security of

investment, but not at the cost of imposing financial burdens on land owners and the general public."[13]

Broadly speaking, there is an option to limit the application of civil
liability rules, or even regulatory liability rules, for operators and owners
of carbon storage facilities. There are four main versions of this option:

1. Transfer of liability from the operator to a governmental authority or
   other institution, such as a specific-purpose fund[14]
2. Offers of indemnification from governmental authority to operators
   for some or all of the long-term liabilities derived from CCS
   activities[15]
3. Retention of long-term liabilities by operators without financial
   responsibility requirements[16]
4. The creation of industry-funded pooled trust funds[17]

Monast, Pearson, and Pratson stated that "the majority of academic
writing advocates a transfer of liability to a government entity at some
point during the post-injection period coupled with an industry financed
trust fund."[18] Under such rules, "the operator is liable when the risk is the
greatest, during injection and immediately afterward, which promotes
appropriate safety precautions and protects the state from excessive
liability."[19]

Haan-Kamminga supported long-term relief for operators by providing
them with liability caps or liability exemptions (indemnifications).[20]
However, she was concerned that such reliefs could create moral hazards
for the operator; in response to those concerns, she advocated that
public regulations be drafted to guard against lapses in precaution and
prevention.[21]

Ingelson, Kleffner, and Neilson identified two basic pathways currently
undertaken by states that have enacted CCS regulations.[22] They identify
the first as a path that strictly retains the polluter-pays principle and provides no clear means for the authorities to assume or indemnify the liabilities of the operators of CO_2 storage facilities.[23] The second pathway is for
the authorities to assume either all or a significant subset of the postclosure liabilities from the operators of the storage sites.[24] They suggest that
a key issue in the choice of liability pathway is the social perspective
on CCS storage facilities as primarily a private good[25] or a public good,[26]
respectively.

Jacobs reports that "many legal commentators," including herself, have
called for operators to bear liability for CCS storage operations because "the
site owner/operator has the ability to control risk by means of careful and a

comprehensive pre-sequestration site characterization, careful site operation and prompt corrective action."[27] The paradigm that Jacobs supports is that an operator should retain a broad portfolio of liability exposure from the beginning of the CCS project until some time period after closure.[28] Given a decade or a score of years and evidence that the CO_2 presents no expected risks of venting or migrating, the operator could either receive indemnity for its CCS liabilities or become able to transfer the CCS properties and their associated liabilities to an authority.[29] Given the possibility of such an enactment, Jacobs advocates for a federal enactment over potential state enactments to provide broader common alignment on the critical CCS issues within the United States.[30]

Thus, as far as the operation and injecting phase is concerned, all agree that the CCS facility remains financially responsible for harm that would occur.[31] There should equally be financial responsibility of the operator for the second phase of the postclosure monitoring.[32]

Where the literature seems to disagree (or at least is confusing or unclear) is to what extent operators should still be held liable for damage that may occur during the long-term stewardship, that is, the phase that in principle is indefinite in time. Some mention that operators still must "pay into a central fund, as prepayment for long-term stewardship,"[33] thus holding that operators still need to "cover the costs of long-term stewardship."[34]

Only Adelman and Duncan take the radical position that financial responsibility for those long-tail risks (which the long-term stewardship would necessarily imply) cannot generate any ex ante incentives and from that perspective would not make any sense to impose. For that reason, they argue that after the postclosure period, monitoring liability could be transferred to the state, hence limiting the liability exposure of operators in time.[35]

We agree that given the fact that liability for long-tail risks cannot efficiently generate ex ante incentives[36] and, moreover, creates uninsurability, operator liability for damage during the phase of long-term stewardship should be excluded.

5.2   Latency Period Planning and Civil Liability Rules

An issue that is extensively discussed in the CCS-related literature is the way in which liability law should deal with long-tail risks, given the potential long latency period of risks related to CCS. There seems to be a consensus in the literature that the exposure to liability of operators should depend on the different phases in the CCS project life cycle. For the entire CCS sequestration, that is, from design and operation (injection) until closure and postclosure, all seem to agree that operators should in principle be held liable.

It is only as far as the long-term stewardship is concerned, which can in principle take an indefinite time, that operators should no longer be exposed to liability. This is especially strongly argued by Adelman and Duncan,[37] to some extent going against earlier literature that in a more unbalanced manner seemed opposed to any type of exclusion of the liability of CCS operators.[38] Adelman and Duncan's argument is that long-term liability, given long latency periods, in fact offers only a nominal deterrence.[39] Liability for the long-term stewardship after the closure of the site should, they hold, be transferred to the government.[40] They provide convincing economic arguments for such an exclusion of liability: the mere fact for an operator to be potentially held liable over, say, more than fifty years in the future will be discounted to a present value and thus will have a very limited deterrent effect.[41] They even argue that these long latency periods could create perverse incentives to operators to organize their insolvency in order to avoid long-term liability.[42]

Although rules of civil liability are adequate instruments for targeting short-term risks, significant latency periods substantially reduce the deterrent value of those liability rules.[43] It is therefore more useful to introduce a rigorous regulatory framework to address measures aiming at the prevention of long-term liability risk, but at the same time limiting the liability of operators to the moment that the site is transferred to the government. Moreover, in many legal systems, such as in the United States, a statute of limitations will also put a limit on the temporal liability of operators.[44]

Others stress the importance of keeping operators liable during the entire operational phase,[45] which can include not only the phase of operation (CO_2 injection) but the period of postclosure monitoring as well.[46] Given that the period of operation could take several decades[47] and that the period of postclosure monitoring could take ten to thirty years,[48] operators would potentially still be exposed to liability for cumulative periods of a couple of decades for the operational years plus an additional ten to thirty years of postclosure monitoring, yielding a total of between eleven and sixty years of liability exposure. It is only for the long-term stewardship that liability in this model would shift to the state.[49]

Others seem to agree with that model but argue that if a transition of CCS projects to the state were to occur, operators would have to provide sufficient funds to cover the costs of long-term stewardship.[50] This seems to be the approach followed to some extent at the policy level.[51]

These solutions suggested in the literature and in policy documents seem to comply with the economic insights already presented: holding operators liable out of principle reasons even in the period of long-term stewardship may have no additional value as far as providing incentives for deterrence are concerned.[52]

5.2.1 Scope of Polluter Pays

The polluter-pays principle is a long-standing rule within environmental law; it appears to have been first enunciated in a 1972 OECD recommendation on the International Economic Aspects of Environmental Policies.[53] It stated;

The principle to be used for allocating costs of pollution prevention and control measures to encourage rational use of scarce environmental resources and to avoid distortions in international trade and investment is the so-called Polluter-Pays Principle.[54]

This principle should also guide policies on the prevention and control of pollution. It requires the polluter to bear the full cost of his or her liabilities, so subsidies should be avoided when possible.[55] The polluter-pays principle has been adopted in a variety of legal contexts,[56] including for the unconventional production technology of fracturing.[57]

However, despite the legal norms set out by the principle, many governments have drafted or enacted legislation to limit the liabilities of carbon storage facility operators, particularly with regard to long-term liabilities after the closure of the injection facilities.[58] These transfers of liability or indemnification have managed to dovetail with the polluter-pays principle by requiring the operators to provide various means of financial support for the state's expected future costs.[59] Thus, the operation of liability transfer or indemnification need not be in stark opposition to the polluter-pays principle in practice.

Ingelson, Kleffner, and Neilson see nuclear waste as an important analog for CCS's "extremely long-term risks."[60] In Canada, its federal government has authority over nuclear waste concerns, and it enforces the polluter-pays principle for nuclear waste.[61] The waste owners are required to cover the costs of development, licensing, construction, and operation of nuclear waste storage facilities.[62] However, Canada's federal government has provided a carve-out to the general rule: it will assume liability for nuclear wastes when "'the original producer cannot be easily be held responsible,' or when the producer no longer exists."[63] The United States has combined the liability coverage for both waste and nuclear energy incidents, with the federal government stepping in once private insurance capacity is exceeded.[64]

Historically, the influence of the polluter-pays principle was the spirit behind the resistance to the continuation of CERCLA's Post Closure Liability Fund, which, prior to its repeal, had supported the public assumption of private liabilities associated with certain hazardous waste sites.[65] Environmental groups protested that the fund enabled a minimalist strategy that reduced operator costs and left hazards extant at the time of transfer.[66] However, the purpose of CERCLA and CCS proposals can be readily distinguished, in that CCS attempts to provide for a public good: it generates positive externalities for the public by reducing risks of climate change, and thus it is arguably more appropriate to consider balancing the polluter-pays principle in favor of providing this public good.[67] It was in this light that the US Interstate Oil and Gas Compact Commission (IOGCC) and the EU have supported such liability limits for CCS postclosure storage activities.[68]

Thus, there are recognized precedents for balancing pure polluter-pays implementations against the practical problems of fiscal capacity and long-term insolvency concerns.

5.2.2 Risk of Operator Extinction

Another rationale in support of a transfer to a public authority of the CCS operations and their correspondent liabilities is that the operators-cum-actors-held-liable might no longer be present when injuries eventually occur. A parallel aspect of permanence is that for CCS storage to be consider successful, it needs to sequester the CO_2 for a very long time—longer than the life span of the vast bulk of corporations and, indeed, of most governmental bodies.

Can corporations survive long enough to address the long-term storage needs of CCS operations, and will they be present if injuries arise later? There are substantial doubts that most companies could survive long enough to address the long-term risks of CCS operations. In such cases, it is extinction of the operator, rather than insolvency, that might limit the application of rules of civil liability.

Sweden's Stora Kopparberg Bergslags Aktiebolag is commonly held to be the world's oldest public corporation; it began as a copper mining operation in 1288 A.D., making it shy of eight centuries in age.[69] But very few corporations have survived that long; barely any last even a century.[70] Business studies have found that routine corporations face half-lives of under a decade, whereas very large corporations may face half-lives of seventy-five years or so; some argue that such firms face a much shorter half-life today: fifteen years.[71] Those numbers imply that very few firms would last one hundred years, and far less than 1 percent of the very largest firms might survive in the time frame of secure CCS storage.[72]

The simple fact is that the vast bulk of corporations, even very large ones, do not have a long life. And that is a problem: many operators might become insolvent or dissolve during the periods of permanent storage after closure of the injection activities. One reasonably assumes that public authorities would continue to exist in the long run in one form or another and that they would be able to steward both responsibilities and capital funds associated with the storage sites long after their original operators have gone extinct.

5.2.3 Support of Nascent Industry

Another potential rationale for limiting the liability exposure of early CCS storage site operators is to reduce their costs and better enable their early entry and viability to support the nascent CCS industry. There are also arguments that the support of nascent CCS operators could be coordinated with other older industries and thus provide support to both. This section reviews and discusses these arguments.

The Price-Anderson Nuclear Industries Indemnity Act of 1957 (Price-Anderson Act)[73] provided a form of governmental indemnification for operators of nonmilitary fissile nuclear energy plants.[74] The Price-Anderson Act enabled two regulatory goals: it encouraged private investment in an industry with worrisome liability exposures and provided a methodology to compensate victims and otherwise clarify certain liability issues in advance of those accidents.[75] The provision of federal indemnity for future liabilities and the financial commitment to cover certain damages offered the legal clarity that enabled the development of the nuclear industry. Such relief could similarly encourage the development of CCS storage capacity.[76] The Price-Anderson Act's indemnifications have been in place for decades and could provide a strong precedent for ensuring that the offered indemnities for CCS operators could survive constitutional reviews.[77]

Similarly, the Uranium Mill Tailings Radiation Control Act of 1978 (UMTRCA) addressed the transfer of operator liabilities to the US federal government.[78] It provides that an operator of a thorium or uranium by-product materials storage facility may transfer liability to the Department of Energy (DOE) after establishing a long-term surveillance plan and paying a one-time fee.[79] While a specific time period is not enunciated within the

UMTRCA, the rules require that the site is no longer in need of active maintenance prior to the transfer;[80] thus, its rules are analogous to those found in CCS regulations that require closure plus an observation period to be completed prior to transfer. The United States also established forms of liability protection and transfer for the vaccine industry and for the technologies developed to address threats from terrorism.[81] As Ingelson, Kleffner, and Neilson stated, "In each of these industries, as well as the CCS models described earlier, in the case of transfer of liability to the state, a social judgment is being made that there are risks that society should encourage companies to take."[82]

Another argument for indemnification has arisen: it allows the operator to remain engaged in the deliberation of the determination of damages and thus provides incentives for ongoing efforts to minimize injuries and damages because its name and reputation would remain on the line and also the operator would remain exposed to periods of liquidity calls when damages would need to be paid prior to reimbursement from the indemnifying government.[83] Thus, indemnification retains certain costs for the operator that a transfer of liability might not. These costs would provide incentives for the operator to remain engaged in the development of standards as the industry developed.[84]

A countervailing view can be found in the US Safe Water Drinking Act (SWDA); the operator remains financially responsible for all postclosure liabilities but without any requirement for advance payment into a fund or otherwise.[85] Currently the SWDA does regulate CO_2 injection wells to the extent that they interface with reservoirs of drinking water.[86] To the extent that an operator meets the regulatory requirements for ensuring the protection of drinking water supplies and receives approval for the operations to begin, the operator would be free of liability, albeit with a limit to the domain of the SWDA.[87] However, postclosure liabilities remain with the operator, who would remain liable for any calls from the EPA for additional remediation or maintenance.[88] However, given the long-term risk exposure of centuries or longer, this approach might not be merited if a substantial number of the operators could disappear with the passage of time.[89]

Flatt has argued that the ex ante setting of civil liability rules could facilitate planning by operators for the long-term risks of their storage facilities after closure.[90] He advocated for a federal enactment, within the United States, to address the needs of liability planning for the postclosure concerns.[91]

Eames and Anderson argued that one of the reasons that the United States continues to lack a functional CCS facility is that the current regulatory framework relying on the EPA's injection well regulations remains too onerous and leaves too many liabilities either unassigned or with the operators.[92]

Thus, it has been shown that the support of an older industry can be dovetailed to the support of a nascent industry; the longevity of the coal mining industry might be lengthened if CCS storage options are better provided for.[93] For this reason, it is expected that within the United States, the coal industry would be a strong supporter of limiting the long-term liabilities for CCS operators.[94]

5.3 Addressing Externalities and Beneficiaries

While it is simpler to see liability as a single-factor operator issue, there are multiple beneficiaries of CCS storage beyond the initial operator. It is reasonable to consider if some or all of the other beneficiaries could or should bear a share of liability at some point. Here, we provide a survey of arguments raised in previous literature.

Flynn and Marriott framed the need to provie for a transfer of the liabilities and needs to respond to potential damages from CCS storage operations from a different perspective: that of the multiple beneficiaries of such storage:[95]

To incentivize development of this key technology and encourage commercial use, limits on liability must be provided to allocate and distribute risk between those parties that benefit the most from GS: the CO_2 generator and the public. Accordingly, at some definite and predictable point, liability should be transferred from the facility owner/operator to the state or federal government.[96]

Their argument is that CCS operators benefit while actively engaged, whereas the general public is the primary beneficiary once the storage has transitioned to a postclosure status due to the provisioning of a lack of ambient climate-changing emissions.[97] Thus, it appears from their logic that the liabilities should accrue to the primary beneficiary of the risk activity. Operators would bear liability while performing the injection and early storage periods, whereas the public, through the state, could be liable for those later periods when they are the primary beneficiary of the storage activity.

Separately, the authority itself as a regulatory body and a potential respondent to liabilities has a complex decision to balance. Should it prefer strict enforcement (or drafting) of its regulations to minimize the levels of transferred risks, or should it prefer less stringent controls on the closure approvals to provide fewer costs to operators and thus provide greater incentives to undertake CCS operations, availing the public of greater climate security?[98] An additional counterposition is the amount of funds required to be transferred prior to the receipt of the closure certificate and the transfer of liabilities or the granting of indemnifications; similar balancing issues exist.[99]

Haan-Kamminga reviewed three related arguments for limiting the long-term liability exposure for operators.[100] First, without those limitations and corresponding transfers to a responsible authority, victims would lack respondents, as many operators would likely cease to exist by events decades or centuries after the final injections.[101] Second, the benefits of energy, and thus of the storage of its greenhouse gas emissions, is a public good and should be supported by the public benefiting from its use.[102] Third, given the likely low risk to society from CO_2 leaks and the potential to mitigate or eliminate climate change risks, encouragement to invest in CCS would be facilitated by the clarification of eventual long-term liabilities.[103]

In contrast, Monast, Pearson, and Pratson raised both the upside of a liability cap as to enable "increase[d] market penetration" of CCS storage options and the downside as creating a moral hazard for operators to shift consequences of risky behaviors onto later taxpayers.[104]

Thus, in developing policies for CCS storage activities, there are strong arguments that the operator is not the only party that benefits from CCS storage operations. Given the potential global impact of reduced carbon emissions, the list of potential beneficiaries is large in scope. As is, these arguments are not in contrast to our recommended policies to enable a transfer from the operators to some form of public oversight after the postinjection closure activities have been completed.[105]

5.4 Policies in Place

The questions studied in this chapter have not gone ignored by policymakers. Indeed, many states and nations have taken up these issues and delivered various legislative and regulative answers to these concerns. But those policies remain divided and differentiated across jurisdictions.

In the next three sections, we provide a survey of some of the key legislative and regulatory efforts that have been undertaken in Europe, North America, and Australia.[106] The theoretical arguments presented earlier can be examined against these enactments. It can also be informative to view which efforts have been repealed or reduced in effect. After this multiple jurisdictional survey, we explain a final set of policy recommendations.[107]

5.4.1 In European Countries

In Europe, for those areas that have endorsed transfers of long-term liabilities from private actors to public actors, generally three requirements need to be met prior to those transfers: "evidence that there is no significant risk of physical leakage" of the injected fluids, a completed period of observation with no leakage or other problems, and delivery of capital funds from the private actor to the public actors to provide for future capital needs in addressing remediation or compensation needs.[108] Stability of the injection site and its underlying geology are key to the effectiveness of these liability transfers.[109] That the injection site is "behaving in a consistent and predictable manner" is important to establish within a reasonable time period after the cessation of the fluid injections.[110] The required waiting period after that cessation of injection can range from two to five decades; in the United States, certain provisions place the liability time frame much longer, especially in the context of protecting drinking water.[111] While the idea of capital transfers is common, the means can vary, with different states preferring a variety of financial tools drawn from royalties, fees, trust fund, and insurance.[112]

The EU CCS Directive,[113] adopted in 2009, explicitly requires member states to provide for the assumption of liabilities after a waiting period beyond the closing of the storage site.[114] The directive provides a four-part test for the operator to satisfy before liability for the storage site could be transferred to a competent authority:[115] (1) abundant evidence that the sequestered CO_2 will remain "completely and permanently contained";[116] (2) some time period, of no less than twenty years in duration, has passed since the closure of the storage site;[117] (3) the operator has met certain article 20 financial obligations;[118] and (4) the injection site has been sealed (i.e., properly plugged and abandoned) and the injection facilities have been removed from the surface.[119]

After the postclosure transfer of liability to the competent authority, there are to be no additional measures to seek capital or other funding from the operators.[120] However, in the time prior to the transfer, the CCS Directive provides several opportunities to obtain such funds during both the injection period and the closing period prior to the issuance of the reports required under its article 18.[121]

France provides for the transfer of postclosure liabilities after a thirty-year waiting period.[122] It requires the operator to pay a "stewardship contribution" as part of the transfer of liabilities; the contribution is designed to cover thirty years of monitoring costs for the site.[123]

It has been proposed that the rules of Norway's Pollution Control Act, Mining Act, and Petroleum Act could all be used as a model for the transfer of liabilities resulting from hydrocarbon extraction activities, as they are substantially similar to the activities of CCS storage.[124] Norway's rules provide for strict liability and several liability prior to the transfer of liability.[125] Its Mining Act and Petroleum Act also provide unlimited time horizons of liability to restore and remediate any resultant damage.[126]

Spain provides for the transfer of postclosure liabilities after a thirty-year waiting period.[127] The operator must meet a list of measures prior to transfer of the liabilities.[128] Spain's legislation did provide a carve-out for when operators demonstrate misconduct; in such cases the liabilities would not be transferable.[129]

The United Kingdom provides for a transfer of long-term liabilities from the operator to the national government after a twenty-year waiting period following the closure of the carbon injection activities.[130] It adopted those regulations in 2011.[131] The transfer of liabilities respected the date of transfer; the operator would remain responsible for any leaks or resultant injuries that occurred prior to the transfer.[132]

5.4.2 In North America

In Alberta,[133] the government has established that it will assume the long-term liabilities for all CCS projects.[134] But it does not assume any liabilities that are present at the time of the transfer, only those that arise after transfer.[135]

The US federal government is barred from offering open-ended indemnification under the Anti-Deficiency Act unless otherwise enabled by specific congressional legislation.[136] Just such a specific bill of legislation was provided for experimental CCS projects under the proposed Department of Energy Carbon Capture and Sequestration Program Amendment Act, but it did not pass.[137] To date, the federal government has not enacted the specific grant of power to cover indemnifications for CCS-related projects, experimental or otherwise.[138]

The US EPA did evaluate several options on how long after closure might be sufficiently long to reduce the risks transferred to an appropriate authority.[139] For the Underground Injection Control (UIC) well program, the EPA chose a fifty-year waiting period in its initial proposal.[140] It discounted literature that suggested that thirty-year periods might be sufficient due to the variety and variability of geologic substructures.[141] However, even with the EPA's preference for a minimum fifty-year period, it reserved powers to extend that waiting period to up to one hundred years "if monitoring and modeling information suggest that the plume may till endanger USDWs throughout this period."[142]

The IOGCC recommended that US postclosure policies should be implemented at a state, not federal, level, and include an initial step of closing and stabilization with liability attached to the operator and a latter step with monitoring and liability transferred to a state trust.[143] The initial step would begin once the injection well was plugged and abandoned.[144] The IOGCC recommended a ten-year waiting period before transfer of the operator's liabilities to a state trust; coincident with that time period, the operator would need to secure a performance bond to cover any postclosure leakage problems.[145] Before that transfer, the operator would be required to ensure that the sequestered CO_2 had become stabilized in its storage.[146] Once the stabilization of the CO_2 is evidenced and after ten years, the liabilities could be transferred and the performance bond released.[147] The trust fund was to be funded by a per ton fee charged against injected volumes accrued at the time of injection.[148]

Monast, Pearson, and Pratson found substantial inconsistency with state law approaches to postclosure liability guidance: seven states had enacted liability rules for CCS operations, but four chose to not provide any guidance on postclosure liability concerns, though one assumes that the operator would retain liability for CCS-related events.[149] They raised concerns that without federal guidance, the uncoordinated state rules would lead to an inefficient regulatory environment.[150]

Lepore and Turner found that indecisiveness at the state level to resolve the liability transfer issues was predicated on several overlapping concerns: that other states would fail to assume liability,[151] that eventually the federal government might assume such liability transfers or indemnifications,[152] and that some voiced opinions that long-term liability rules should not be specially crafted just for CCS operators and that other industries might make similar claims of need.[153]

Kansas has resisted enacting rules that would enable a transfer of liability to state authorities.[154]

Illinois has enacted legislation for experimental CCS projects that could enable the transfer of certain liabilities for CCS projects if third-party insurance for long-term storage concerns were to be available. Within that legislation, the title to the CO_2 transfers to the state at the same time that the operator receives indemnification.[155]

Louisiana provided indemnification to operators if they contributed to both the CO_2 Geologic Storage Trust Fund while in operation and thereafter complete and obtain a certificate of completion at some point after closure of the storage facility.[156] The certificate of completion would be available ten years after the actual closure of the facility.[157] If the operator or owner were subject to performance bonds, those obligations would also be lifted under section 661.[158] However, Louisiana specifically did not assume the liabilities for the storage facilities, even after the state took title to them; the storage trust fund itself would bear the liabilities associated with the postclosure storage site.[159] Other unique features set Louisiana's solution apart: the lack of a general indemnification holds should the fund be depleted and the operator's civil liabilities for noneconomics are capped at $250,000.[160] Louisiana requires only a ten-year waiting period prior to a potential transfer of liabilities.[161]

Montana has enacted legislation to provide operator indemnification provided that the operator has contributed to a long-term management fund.[162] Operators are required to post a performance bond that cannot be released until both fifteen years have passed since cessation of injections and a certificate of completion is issued to certify that the plume is stable and all other regulatory remediation efforts have been met.[163] Montana Code 82–11–183(4) provides:

(4) Subject to subsection (5), the certificate may be issued only if the geologic storage operator:

(a) is in full compliance with regulations governing the geologic storage reservoir pursuant to this part;

(b) shows that the geologic storage reservoir will retain the carbon dioxide stored in it;

(c) shows that all wells, equipment, and facilities to be used in the post-closure period are in good condition and retain mechanical integrity;

(d) shows that it has plugged wells, removed equipment and facilities, and completed reclamation work as required by the board;

(e) shows that the carbon dioxide in the geologic storage reservoir has become stable, which means that it is essentially stationary or chemically combined or, if it is migrating or may migrate, that any migration will not cross the geologic storage reservoir boundary; and

(f) except as provided in subsection (11), shows that the geologic storage operator will continue to provide adequate bond or other surety after receiving the certificate of completion for at least 25 years following issuance of the certificate of completion and that the operator continues to accept liability for the geologic storage reservoir and the stored carbon dioxide.

After the required waiting period, Montana can assume all the long-term liabilities and provide the former operator with indemnification;[164] however those indemnities might be constrained to "regulatory requirements and liabilities" and not liabilities from common law torts.[165] Similar to other states, Montana would also gain complete assession to the mineral rights and the storage facility.[166]

North Dakota enacted legislation to provide indemnification for operational CCS facilities.[167] The state provides indemnification and assumes all long-term liabilities at the moment it acquires the complete and affiliated set of mineral rights to the stored CO_2; it also requires sufficient funding from the operator into a fund for long-term site management.[168] North Dakota requires only a ten-year waiting period prior to a potential transfer of liabilities.[169] However, the state's assumption of liability is conditional on a lack of federal action. Should the federal government provide a means to assume those same liabilities, the North Dakota laws would require transfer of those liabilities to the federal government.[170]

Texas has enacted similar legislation to that of Illinois for experimental CCS projects.[171] However, the state did not address third-party insurance in their enactment, and thus at the transition to postclosure, the mineral rights to the carbon transfer to the state and the operator is granted indemnification.[172]

Wyoming's Geologic Sequestration Special Revenue Account is intended to provide for the government's costs to monitor and administer the closed facility, but it is not a legislative recognition of state liability or a grant of a governmental indemnification to the operators; such terms were explicitly rejected within the bills.[173] Hayano found Wyoming's performance bonding requirements to reflect the underlying trend in the state to reject a transfer of liabilities or to provide indemnification.[174] Performance bonds are required to be maintained not only during active operations but also during the postclosure period; no clear indication of when the performance bonds might be releasable has been provided.[175] This deviation from the IOGCC's recommendations was one of the few enacted by Wyoming.[176]

Texas was considering the potential to provide some type of indemnification or a governmental assumption of liabilities for carbon storage projects, but such determinations were not clear as of 2012.[177]

5.4.3 In Australia

Australia enacted the Offshore Petroleum and Greenhouse Gas Storage Act in 2006.[178] Once a certification process is accomplished after closure of the storage facility, a fifteen-year waiting and observation period would begin.[179] Once the waiting period is completed, the operator would receive specific indemnities against specific liabilities.[180]

The Australian national government, however, is limited in the type of liabilities it can indemnify.[181] Primarily, the only liabilities it can indemnify are those arising from licensed activities for damages arising or incurring after the end of the "closure assurance period."[182] Thus, the operator would remain liable for damages resulting from an event prior to the completion of the closure assurance period or from activities not explicitly covered by license under the Offshore Petroleum and Greenhouse gas Storage Act (OPGGSA).[183]

The Australian State of Victoria opted to avoid the assumption of liability for postclosure long-term liability.[184] Its enactment, the Victoria Greenhouse Gas Geological Sequestration Act,[185] does call for sustainable practices.[186] Queensland has followed a similar path.[187]

5.5 Conclusion

This chapter has provided a review of the rationales for bifurcating the liabilities from CCS storage sites and found that such provisions and rules should be put in place. The polluter-pays principle was reviewed; it was concluded that there are means to ensure that both the polluter pays, or at least heavily contributes,[188] for the future event of long-term latent risks.

The risks of operator extinction were reviewed; it was seen as a very likely outcome that few companies or operators could last long enough to be viable to bear liability. It is, in a sense, a foreseeable outcome that most operators would eventually become insolvent or completely dissolved prior to events in the very long run.

In what might seem the reverse case, the potential to encourage early adopters of CCS technologies by limiting their long-term risk exposure was seen as beneficial. Not in the least was the argument that the provision of CCS itself carries certain aspects of a public good and thus merits a certain risk sharing with the public. The setting of the bifurcation point at sufficiently advanced postclosure efficiently enables that risk sharing.

Some jurisdictions have adopted such guidance, whereas others have either wholly prevented the adoption of such a rule, or, having once adopted such a rule, later amended their rules to remove it. Much of this activity occurred against the backdrop of a global recession, and local governments might have become concerned with near-in-time financial obligations. However, areas that did not adopt these liability bifurcation plans also would appear to be less likely to see near-term implementation of CCS storage capacity.

How should one read the reasoning presented earlier in this chapter—that from an economic perspective, there should be no liability for long-tail risks, in combination with the reasoning in the previous chapters, that there should be in principle no financial limits on liability? This conclusion, at first blush perhaps seemingly contradictory, has a perfectly logical foundation for the economic starting point that liability makes sense only if that liability could provide incentives for the prevention of damages. If exposure to liability affects decisions on risk, then that assignment of liability is beneficial; otherwise, if the imposition of liability provides no functional incentives, that assignment is not merited.

For that reason, we concluded that a liability for long-term stewardship does not make sense since operators today will already discount this long-term future liability. The foresight of being held liable for damage many years in the future would have no deterrent effect on incentives for prevention today. It takes just a few decades before the time value of money erodes financial incentives. That is an argument to limit the exposure to liability in time to the moment that the management of the CCS storage site has been transferred to the state.

However, as far as the remaining liability is concerned, and it is assumed that an operator's exposure to liability can have a positive effect on his or her incentives for care, it was equally argued in section 5.4.8 that this liability should be unlimited in order to fully provide correct incentives for prevention.

The conclusion is that liability should be limited in time but not in amount.

6 Publicly and Privately Regulating CCS Activities

While the call for public regulation of CCS activities is a constant presence within the broader research literature, less focus has been paid on exactly what roles such regulations should play and where regulations might be most robust in governing the potential risks and hazards of CCS activities.

This chapter endeavors to provide answers to those questions. It is important to delve into the theoretical details in order to reveal the origins and logical structures of our policy recommendations. In anticipation of the results of that effort, this chapter presents arguments that regulations would provide robust means of addressing the risks of CCS storage activities, that both public and private forms of regulations should be brought to bear, and in which circumstances different forms of regulation would be most robust.[1] The chapter also reviews the harmonic results from the availability of both civil liability and regulatory systems.[2]

The chapter examines two distinctly different forms of regulation: public regulation and private regulation. Public regulation is that more traditional sense of regulations: rules and standards as administered by a public authority. Private regulation encompasses the potential of private bodies to provide and monitor rules and standards, often within a market-based context. Private regulations can include both third-party organizations, such as by public interest groups or insurance agencies, or forms of self-regulation, such as by industrial groups.

The methodology again follows the law and economics literature. This provides a ready framework in which to draw comparisons with and contrasts to the previous chapters on civil liability rule systems.

In summary, we present the argument that both public and private regulatory efforts have robust roles to play in governing the risks and hazards of CCS activities. Furthermore, the joint implementation of public and private regulation would enable better and more efficient formulation of the optimal rules and standards while also enabling efficient enforcement of those same regulatory efforts. We also present a complementary argument: that such a regulatory response would also improve the function and efficiency of the liability systems advocated in previous chapters. Thus, the optimal governance strategy would be to design a coordinated system that integrates public regulations, private regulatory efforts, and civil liability systems that could efficiently set and enforce optimal standards for CCS-related activities and risks.

6.1   Evaluating Public Regulations for CCS

Because many public authorities have already enacted some form of public regulation or drafted preliminary versions of it and because many scholars, legal and otherwise, have called for the regulation of CCS activities, one might wonder why a review of the theoretic literature on regulation might be called for. First, a solid review of the theoretical foundations for the implementation of that idea should be evaluated to ensure where theory and scholastic opinions concur. Second, given that the idea of regulations for CCS were to be found valid, the theoretical literature could further enlighten where those regulations might be most efficiently applied,[3] for example, where regulations might be more effective than rules of civil liability and where not. Finally, there is more than one type of regulation, and a theoretical review might assist policymakers to ascertain which type of regulation might be most effective for each policy issue. Thus, we provide a review of the literature drawn from law and economics.

Regulations can be enabled as public regulation and private regulation; in turn, private regulation can be either self-regulation or private but not self-regulation. Either way, public and private regulations set standards to enable certainty and decision processes; civil liability rules can set standards only after accidents. In other terms, regulation enjoys a broader toolset than rules of civil liability and thus has been identified as "the preferred approach."[4]

Regulations can be drafted both before accidents occur and before norms and standards are settled by private parties. When done with due process and with the participation of the public, the regulations might render an enhanced sense of democratic participation.[5] By setting those standards prior to undertaking risky activities, public and private regulation can respond not only to the injuries to which rules of civil liability respond, but they can also address the quality and character of the risky behavior before those harms occur. Behavioral incentives can be provided to the actors before an activity begins, [6] improving the likelihood of efficient policy. Regulation therefore can act earlier in time; when potential injuries might be severe or guided by concerns of irreversibility, those injuries can be addressed before they occur. [7] In addition, the authority accompanying public regulations suggests a higher likelihood of enforcement of the standards: "public enforcement appears attractive whenever the probability of punishment under a private regime appears to be low."[8]

Public regulation enables the efficiency of a public-financed authority to acquire the necessary information from a broad range of involved parties to a risky activity. Regulations can thus include input from the broader public community, whereas rules of civil liability might lack a mechanism to do so.[9]

Private regulation enables the use of information that is likely more recent in innovation and from those most experienced with the risk profile of the activity.[10] Perhaps that information might otherwise be unlikely to be revealed by the dominant actor in a unilateral accident event.

The choice between public regulatory efforts and liability rule systems was examined by Steven Shavell in 1984 in a paper in which he advanced several criteria that influence the choice between safety regulation and liability rules.[11] Shavell advances six circumstances as efficiently benefiting from safety regulation: (1) when the conditions that support implementation of civil liabilities do not avail, (2) when information is distributed in a manner better addressed by a public authority, (3) when privately held information needs incentives to be revealed, (4) when there is a substantial risk of insolvency on the part of the tortfeasors, (5) when lawsuits are unlikely to be brought by the victims themselves, and (6) when the institutional capacity to support the necessary litigation is lacking.[12]

There are also certain circumstances when regulations would be expected to fare poorly in attaining efficiency. These issues are evaluated in the following sections.

6.1.1 Regulations Work When Civil Liability Rules Would Fail

Regulations can be effective when rules of civil liability would fail. Regulations can overcome Shavell's four weak spots of civil liability: (1) lack of sufficient knowledge, (2) insolvency risk, (3) effective absence of lawsuit threat, and (4) lack of institutional capacity for court processes.[13] We address each of these separately, but each leads to a failure of incentives to be generated that could guide the tortfeasor to efficiency.

Shavell's analysis for these threats assumed a fully responsive lawsuit capacity. Regulations could address three potential weaknesses that might become evident if that capacity was dysfunctional:[14] (1) the probability of detection of violation, (2) the probability of prosecution, and (3) the probability of punishment.[15] The argument runs that there is a probability that a tort is not detected, but if it is, then there is a probability that a lawsuit would not be brought, and even then, there is a probability of limited or missing enforcement of the judgment for damages. So the ex ante incentives, which are required for the efficacy of civil liability rules, could be missing; therefore, public regulations could leverage state resources to ensure that torts are detected, prosecuted, and punished. In the circumstances of CCS operations, there is a real concern that the duration of the time frames involved (e.g., centuries) could greatly reduce the ability of victims to be able to gather enough information and bring lawsuits against the operators if they still exist at the future time. The presence of a regulator ameliorates those concerns by being able to (1) consistently track the operator and its activities, (2) gather and collect a variety of data, and (3) retain, store, and provide to future citizens the information needed to ensure detection, prosecution, and punishment.

There are also concerns during the operational period of the injection facilities that CCS technologies are very sophisticated and would require advance means of monitoring, not generally viewed as sustainable by ordinary citizens. A regulatory authority would be better placed to have a sufficient budget, to be able to hire and retain adequate experts, and to be able generally to integrate the various sources of information to better support detection, prosecution, and punishment needs.

The importance of regulation as a primary tool to prevent risks emerging from CCS is also stressed in specific literature dealing with liability related to CCS. An important reason that the literature mentions that traditional liability rules will not provide efficient incentives for prevention is related to the potentially long-term nature of liability. Liability for risks that may occur over, say, fifty years may not have a strong deterrent effect on decisions that have to be taken ex ante.[16] This has important consequences: little would be lost as far as deterrence is concerned if operators would be relieved from this long-term liability, and the fact that liability rules are weak, especially as far as this long-tail damage is concerned, is a strong argument in favor of safety regulation.[17]

Especially the risk of spinning off parts of the business into small, low-capital companies and thus creating a judgment-proof problem is considered as a serious issue in Shavell's criteria.[18] The same goes for the likelihood that a lawsuit would be brought as a result of damage caused by CCS. Latency may substantially reduce the effectiveness of a liability suit, and, moreover, damage may be spread over a large number of victims.[19] It is well known that in those types of situations, the incentives with the individual victim to file a liability suit will be very small:[20] their personal benefits from a lawsuit would be low, and given free-rider problems, individuals would have no incentives to bring suits that could benefit the collective.[21] The only argument that might favor a liability framework would be information asymmetries. But that would favor liability rules only when it could be argued that operators would be better placed than government to assess the potential risks emerging from CCS.[22]

Also other studies clearly argue that liability rules alone are not able to sufficiently control the risks posed by CCS.[23] Hence, they argue, a tailored regulatory structure is a crucial component of risk management.[24] Direct regulation controlling the conditions under which the various phases in the CCS project life cycle are executed is therefore crucial in order to reduce the residual risk tail for a particular CCS site.[25] Voices hence plead in favor of the installation of a hybrid private/public board that controls the elements influencing the risk such as proper siting, immediate corrective action, and early shutdown of highly risky facilities.[26] All of the literature therefore agrees that notwithstanding the virtues of a liability system, it cannot be the primary system of prevention. Ex ante safety regulation will necessarily be the preferred instrument to draft and enforce tailor-made safety regulations aiming at the reduction of risks related to CCS.[27] Moreover, given that the literature equally indicates that the specific preventive measures to be taken largely depend on the specific location and are hence site specific, regulation should not be of a general nature but site specific. The administrative instrument to enable such an adaptation of regulatory conditions to the specific site is obviously the permit or license in which regulatory authorities can lay down specific safety standards for the site.

6.1.2 Addressing Asymmetric Information

Regulation can address informational asymmetries better than the rules of civil liability could. Regulations can be more effective than rules of civil liability when the information on the preventive measures and potential harms of a risky activity are scattered across the actors associated with that activity.[28] Rules of civil liability generally assume that such information is fully available to both the tortfeasor and the victim and that both could take rational action based on that knowledge.[29] When that knowledge is asymmetrically possessed across the parties, then a regulatory authority could canvass the involved parties and informed experts to collect a public repository of the necessary information.[30] That information could then be used to guide standard setting and rule enforcement efforts.

Information deficiencies have often been advanced as a cause of market failure and as the justification for government intervention through regulation.[31] Also, for the proper operation of a liability system, information on, for example, existing legal rules, accident risk, and efficient measures to prevent accidents is a precondition for efficient deterrence. According to Shavell, the parties in an accident setting generally have much better information on the accident risk than the regulatory body possesses.[32] The parties themselves have, in principle, the best information on the costs and benefits of the activity that they undertake and the optimal way to prevent accidents. This "assumption of information" will, however, be reversed if it becomes clear that the parties in an accident setting do not readily appreciate some risks. This may more particularly be a problem if costs are external. These cannot always be easily assessed by the parties involved. Therefore, for every activity, the question that will have to be asked is whether the government or the parties involved can acquire the information at the lowest cost.

Public regulation enables the efficiency of a publicly financed authority to acquire the necessary information from a broad range of involved parties to a risky activity.[33] Regulations can thus include input from the broader public community, whereas rules of civil liability might lack a mechanism to do so.[34]

Whereas routine victims or citizens might lack the education, the time, or the potential to completely monitor CCS sites, and the access to interview or audit the operators of CCS facilities, a regulatory authority would normally be assumed to have access to those types of resources. A regulatory authority would be more efficient in gathering disparate sources of information on the performance and risks of operational and postclosure CCS projects.

The availability of private regulatory groups could greatly facilitate the collection and distribution of necessary information. From shareholder groups, to private insurers, to CCS industry groups, each could play a role in ensuring that all of the necessary data were collected and made available.

The public regulatory authority could also enable full-time employees to liaison with private regulatory groups, something not viable for ordinary citizens. Once the data have been assembled, the public regulatory authority could ensure that they were accessible to the general public and other interest groups. Thus, the role of regulations could greatly facilitate the reduction of data asymmetries and better ensure efficient governance of the risks of CCS operations.

If one looks at the first criterion, that of information costs, it must be stressed that an assessment of the risks of a certain activity often requires expert knowledge and judgment. Small organizations might lack the incentive or resources to invest in research to find out what the optimal care level would be. Also, there would be little incentive to carry out intensive research if the results were automatically available to competitors in the market—the well-known free rider problem. Legal instruments granting an intellectual property right to the results of the research can partially counter this problem. However, the problem remains that it may not be possible for small companies to undertake studies on the optimal technology for preventing environmental damage. Therefore, it is often more efficient to allow the government to do the research on the optimal technology (e.g., in a governmental research institute). The results of this research can then be passed on to the parties in the market through the regulation.

Hence, the setting of standards in regulation can be seen as a means of passing on information on the minimal technology required. Obviously it is more efficient for the government to acquire information on the optimal standard than it would be for, say, an individual firm to find out what additional reduction in risk would produce an optimal reduction of the expected damages from the particular CCS-related risk. There are undeniable economies-of-scale advantages in regulation.

6.1.3 Information-Revealing Mechanisms

Regulations could provide truth-revealing mechanisms that could reveal information not otherwise available to tortfeasors, victims, or the marketplace. Glachant has argued that traditional public regulatory information collection methods might not rectify the market failure in information;[35] insufficient information would be collected to make a meaningful decision on which method of governance to employ.[36]

Truth-revealing mechanisms leverage incentives for various parties to bring forward privately held information that otherwise would have been withheld, particularly between tortfeasors and public authorities.[37] Glachant was concerned that information searches are transaction costs that should be included in the optimization function; he introduced game-theoretic methods of information exchanges to the communications between regulators and tortfeasors.[38]

Glachant demonstrated that both regulations and rules of civil liability have substantial transaction costs and that those transaction costs may determine which rule set is more efficient: "we are especially suspicious towards the zero administrative costs assumption."[39] The regulatory authority must balance individualized (decentralized) policy plans for each tortfeasor or a common centralized policy plan for all of the tortfeasors together to attain an efficient policy strategy. This can be implemented as a game by the regulatory authority: "Communication between agents is subject to strategic manipulation if (i) the objectives sought by the emitter and the receptor differ and (ii) the receptor's decision influence emitter's gains."[40]

The regulatory authority can play the strategic game to gain the revelation of information from tortfeasors and other parties that would not otherwise be available from the market or from the operation of civil liabilities,[41] especially before the incident of accidental harms. This potential gain in information enables efficiencies not obtainable from the operation of civil liability rules.[42]

Within the context of CCS, there are two problems of concern: the depth and breadth of information to be obtained and the strategic importance of much of that information to be obtained. For example, in determining the risks of certain CCS activities, one might need to gather a variety of information on technology, experimental experiences, and best practices. However, the revelation of that information could transform the perceived standards for duty of care or for regulatory standards and thus raise the costs of the operators. Without a regulator, such CCS information might well remain guarded or secret. But a public regulator can induce data sharing with appropriate incentives. That revealed information could then be put to use to either better govern and monitor a specific CCS installation or be provided to the broader CCS community to raise industry standards. Thus, regulations can enable critical CCS information to be revealed when that information would otherwise be prevented from disclosure.

6.1.4 Operating beyond Monetary Incentives

Regulation can operate without the need for financial incentives, so it can operate when tortfeasors are insolvent or otherwise beyond financial incentives.[43] Insolvency frustrates rules of civil liability because damages have limited or marginal impact on those without substantial assets. And as discussed in chapter 5,[44] there are reasons to suspect that many operators might be challenged to simply remain in existence for the long time periods that the CO_2 would be in storage; thus, there are limits to the incentives that can be provided from rules of civil liability.

If the potential damages can be so high that they will exceed the wealth of the individual injurer, liability rules will not provide optimal incentives. The reason is that the costs of care are directly related to the magnitude of the expected damages. If the expected damages are much greater than the individual wealth of the injurer, the injurer will only consider the accident as having a magnitude equal to his wealth. He will therefore take only the care necessary to avoid an accident equal to his wealth, which can be lower than the care required to avoid the total accident risk.[45] This is a simple application of the principle that the deterrent effect of tort liability works only if the injurer has assets to pay for the damages he causes. If an injurer is protected against such liability, a problem of underdeterrence arises.[46] Environmental liability rules may therefore not generate efficient precaution and effective compensation when enterprises run insolvency risks.[47]

Safety regulation can overcome this problem of underdeterrence caused by insolvency.[48] In that case, the efficient care will be determined by preexisting ex ante regulation and will be affected by enforcement instruments that induce the potential injurer to comply with the regulatory standard, irrespective of his or her wealth.

In that case, a problem might still arise if the regulation were also enforced by means of monetary sanctions. Again, if these were to exceed the injurer's wealth, the insolvency problem would remain.

Regulation can act beyond raw financial incentives, and is thought to remain effective against insolvent and avoidance factors by providing other forms of punishment and deterrence.[49] Hence, if safety regulation is introduced because of a potential insolvency problem, the regulation itself should be enforced by nonmonetary sanctions.[50] Shavell found that regulations were superior to rules of civil liability for the conditions of insolvency.[51] He also found that regulations were superior to rules of civil liability for the conditions of strategic avoidance, such as incorporation and limited liability partnerships.[52]

In the context of privately managed CCS operations, it would be expected that CCS operators would generally be commercially minded and sensitive to monetary incentives. To boot, the delivery of performance bonds and other similar financial assurances are all predicated on the assumption that the operator would act responsibly to ensure a return of the capital deposited to secure the performance bonds. However, some events might be better addressed with additional nonmonetary incentives. Concerns of reckless operator behavior, for example, might be more efficiently prevented with threats of jail time or other nonmonetary punishments.

CCS operations also need to address circumstances that persist long beyond traditional tort time frames; many of the victims and operators will not persist as long as the stored CO_2 and its potential for injuries and harms will remain. Thus, other instruments will be required to ensure performance across the long-term time frames. Regulations can better provide those tools than the rules of civil liabilities for CCS operations.

Also, the insolvency argument points in the direction of regulation. CCS-related risks can be caused by individuals or by firms with assets that are generally lower than the damages they can cause as a result of the risk.[53] In this respect it should not be forgotten that even a small firm could cause harm to a large number of individuals. Consider the example of terrorist attacks. The insolvency problem is very clear there (and liability rules could hardly deter such a behavior). But the same is true for firms. They can also cause damage resulting from systemic risks that may largely exceed their individual assets. Moreover, most firms have been incorporated as a legal entity and therefore benefit from limited liability. Hence, the individual shareholders are not liable to the extent of their personal assets, but a creditor of the firm can lay claim only to part of the total assets purchased in the firm by the shareholders.

6.1.5 Ensuring Resolution of Injuries and Damages

Regulation as administered by a public regulator could ensure that torts are detected and responded to, particularly when victims might be hesitant or unable to litigate. Civil liability rules might fail to be operative if there is in fact no lawsuit pending; thus, insufficient litigation could result in a poverty of standard settings or in inaccurate standard settings.[54] Regulation can avoid those problems and both provide standards and ensure tortfeasor compliance and thus provide the necessary incentives.

Shavell demonstrated three reasons for a lack of lawsuits: disparate plaintiffs, a lack of evidence, and potentially missing parties.[55] Disparate plaintiffs with "rational apathy" could result in fewer lawsuits being brought,[56] as when many plaintiffs divide the damages so that each plaintiff's unique damage would not suffice to balance the costs of joining or starting litigation. Many environmental torts can be scientifically complicated and cause difficulties to evidence causation.[57] Landes and Posner, and separately Kunreuther and Freeman, have identified both the complexity of general science and the difficulties of specific causation within environmental cases as reducing the likelihood of lawsuit.[58] Finally, the evidence might not be readily discoverable until some passage of time after the initial incident. The injuries might lag far behind and plaintiffs might no longer be traceable or alive.[59]

Even so, Schäfer and Schönenberger reported that not all parties desired to bring litigation even when they lacked Shavell's problems to litigation; a variety of transaction costs might discourage them.[60] Thus, incomplete litigation of tort complaints could result in less than full damages being assessed against the tortfeasors, resulting in insufficient incentives for the tortfeasor. Schäfer and Schönenberger argued that the public regulatory authority could bring other measures to bear on tortfeasors to ensure both costs of compliance and the derivative provision of *ex ante–provided* incentives.[61]

The injuries from CCS leakages or venting accidents could be widely dispersed, could persist at low levels for long periods, and could be difficult to trace or detect for a variety of scientific reasons. Injuries might occur at light levels across agricultural fields or ranch animals. Human health might be harmed by contaminated drinking waters or by noxious air volumes, but again at levels difficult to detect. A central regulatory authority would be better positioned to amalgamate the broadly dispersed injuries and determine when the sum of the impact on public welfare merited review, detection, prosecution, and punishment. CCS operations are the perfect example of when the risky activity itself is complicated and sophisticated, with the potential to cause widely dispersed and potentially difficult-to-identify injuries over both short and long time frames. Thus, CCS is an ideal model for when regulations can address such injuries better than the rules of civil liabilities can.

Some activities can cause considerable damage, but even so, a lawsuit to recover these damages may be never brought. If this were the case, there would of course be no deterrent effect of liability rules. Therefore, the absence of a liability suit would again be an argument to enforce the duty of efficient care by means of safety regulations rather than through liability rules.[62] There can be a number of reasons that a lawsuit is not brought, even though considerable damages have been caused.[63]

Sometimes an injurer can escape liability because the harm is thinly spread among a number of victims. As a consequence, the damage incurred by every individual victim is so small that the victim has no incentive to bring suit. In particular, this problem will arise if the damage is not caused to an individual but to a common property, such as the surface waters in which each member of the population has a minor interest. In addition, a long time might have elapsed before the damage becomes apparent; in this case, much of the necessary evidence may be lost or never obtained. Another problem is that if the damage manifests itself only years after the activity, the injurer might have gone out of business.

A related problem is that it is often hard to prove that a causal link exists between an activity and a type of damage.[64] The burden of proof of a causal relationship becomes more difficult with the increasing passage of time since the damaging incident took place. Often a victim will not recognize that the harm has been caused by a tort; she might think that her particular ailment, perhaps cancer, has a "natural cause," associated with general ill health. For all these reasons, a liability suit might not be brought. Hence safety regulation is necessary to ensure that the potential injurer takes efficient care.[65]

An additional problem is that the liability system cannot always incorporate all social costs of harm as damages that have to be compensated for by an injurer. This is typically the case for so-called nonpecuniary losses, such as pain and suffering. It is well known that these losses may be difficult to estimate. Think in this respect of the deterioration of image and reputation, but also of the losses one suffers because of the death of a relative. The administrative costs for courts to valuate these losses may be huge. Economics generally makes the argument that tort law should, as far as possible, incorporate these types of nonpecuniary losses as well, especially in unilateral accident types.[66] However, the amount to which tort law succeeds in incorporating these nonpecuniary losses in damage awards may be suboptimal. In some of these cases, society might simply want to prevent an action it considered inappropriate. Recall the secondary economic consequences of bovine spongiform encephalopathy (a transmissible neuro-generative and fatal brain disease of cattle) or the foot-and-mouth outbreaks in the United Kingdom. In these cases, the role of the liability system is modest, and inadequate compensation can lead to underdeterrence. It is clear that in those cases, other types of intervention are necessary, more particularly the ex ante prohibition of certain activities. This requires a regulatory intervention with the threat of sanctions on those who would violate such a prohibition.[67]

Also, the chances of a liability suit being brought for damage caused as a result of CCS-related activities is naturally very low. Whether there will be particular victims who have an incentive to sue may very much depend on the potential damage that could be caused by a CCS project. This project can be of quite a differing nature.[68] Especially when there would be human health impact, the danger is that the damage may be spread over a large number of people who will have difficulties organizing themselves to bring a lawsuit. In addition, the damage could become apparent only many years after the wrongful act took place. This is, as we argued above,[69] the essence of CCS-related risks: they are long-tail risks. This will bring proof of causation and latency problems, which will make it difficult for a lawsuit to be brought against the one who caused the risks.[70]

6.1.6 Powerful When Court Institutions Are Weak

Where courts and institutions affiliated with litigation-based justice are institutionally challenged, regulations could use public budgets and resources to overcome these problems.[71] Court systems can be expensive to operate; not only are court matters themselves costly, but the litigants bear substantial costs in trial preparation and litigation. Public governments are generally assumed, via taxes and other revenues, to be able to afford what often its citizens might not; as such, the ministerial bodies of a government might be better capitalized to pursue justice through regulatory authority than its citizens through the courts.[72]

Even in advanced economies such as in Germany and the United States, the operators of CCS projects might very well dwarf the economic and institutional capacities of both courts and victims. A central regulator would be better placed to match the economic and institutional capacity of the operators and be more efficient in addressing the needs of detection, prosecution, and punishment than a court addressing them through the rules of civil liabilities.

6.1.7 Public Regulations Can Be Inefficient Too

While these six concerns have addressed the circumstances in which regulations are expected to be efficient for policymakers, it worth noting when theoretical literature expects regulations, especially public regulations, to perform poorly. This review is needed because policymakers should be aware of when regulatory efforts are less likely to succeed and to heighten awareness of when other means of governance might be attractive. There are four main concerns: (1) poor focus of regulatory goals, (2) large and burdensome transactional costs, (3) inefficient and sticky regulatory drafting methods, and a (4) lack of incentives for innovation.

A primary critique is more historical than empirical. Historically, public regulations focused on preventing bad acts when the goal should have been the optimal level of bad acts.[73] Regulators of earlier eras tended to focus on the complaints raised, not on the potential optimal results.[74] To the extent that regulations are drafted to avoid certain results instead of to attain certain results, they are less likely to result in efficient policy implementations.

While the processes of democratic deliberation and of regulatory transparency are correctly lauded, they can increase the transaction costs of regulatory drafting. Transaction costs of regulations are often large;[75] they must be balanced against the benefits of the regulatory goals or suffer from diminishing returns to policy efforts.[76] The tighter the policy goals are, and thus their benefits, the higher the transaction costs of implementation are likely to become.[77] Yet were the regulatory goals set too loose to reduce costs of implementation, then society would likely suffer from the fundamental lack of policy goal attainment.[78]

Regulatory drafting often fails to set efficient regulations because of the effects of agency capture, cost balancing needs, and strategic information games. And once the standard regulations are agreed to, the potential for each regulated actor to achieve personal efficiencies is reduced. Standardization of regulations defeats the efficiency of decentralization.[79]

Combining these problems often results in regulations that set low standards, enabling innovation for safety and environmental concerns to go static; combined with Tietenberg and Lewis's concerns on balancing transaction costs only increases the likelihood of policy efficiency.[80] Furthermore, overcompliance is costly for private actors, and they do not rationally do it, so a lack of incentives for innovation leads to minimal compliance.[81] Thus, it is possible for a society to retain an excess of risk even when each actor is at full regulatory compliance due to the problems with drafting correctly designed regulation that should attain efficient levels of risk.

Transaction costs of drafting regulations require costly political alignment for passage, so regulations are infrequently or even rarely updated so they become sticky,[82] creating a risk of path dependence and a loss of policy optimality.[83] Thus, unless otherwise overcome, regulatory measures can become outdated and inefficient and result in a lack of innovation. However, what might be called a regulatory malaise is in fact a result of certain strategies employed to address the transaction costs of redrafting; it is not endemic to the process of public regulation itself.

6.2 Evaluating Private Regulations for CCS

Traditionally in the economic analysis on which this study relies, regulation was almost implicitly equated with public regulation, that is, public standards set by government. More recently, economics scholars have also pointed to the fact that under particular circumstances, there may be strong arguments in favor of private standard setting by industry or to have some form of private standard setting supervised by government.

The most important argument in favor of self-regulation[84] relates again to the often mentioned problem of information asymmetry.[85] In this case, it refers to the information asymmetry between industry and government. In many cases, it is the operators who have much better information than government does (e.g., on optimal preventive technologies in the offshore sector). For that reason, self-regulation is advanced in the law and economics literature given the larger expertise and technical knowledge of industry. As a result, information costs to fix the standards are supposed to be lower than when the government would have to set standards.[86]

We next review arguments for and against the potential implementation of private regulation for CCS storage activities. We found that a balanced, complementary integrate approach that employs both public and private forms of regulation would be an efficient approach.

6.2.1 Theory of Private Regulation

In a perfect world full of zero-cost information, no regulations would be needed because the Coasian issues of rights reallocation would be readily accomplished.[87] The purpose of regulations is in part to redress issues caused by positive transaction costs to fix incomplete and imperfect markets.[88] While sometimes the least-cost regulatory measure would be a public regulation, that is not always the case;[89] sometimes private regulation might provide similar or superior results at lower costs to society.[90]

Private regulation is predicated on the idea that the party, or parties, with the best collection of relevant information might not be a public authority but a group of private actors already closely engaged with the risky activity to be regulated.[91] As a legal phenomenon, self-regulation enables a deliberate delegation of a state's lawmaking powers to a private agency.[92]

Private regulation can be defined as a private agency within a collective group of actors who are engaged in a specific risky activity, are well informed to determine the best available practices, and are capable of responding to recent innovations.[93] The agency need not be a collection of tortfeasors;[94] it could also include other parties from the community,[95] such as public interest or environmental groups,[96] green consumers,[97] financial investors,[98] insurance institutions,[99] or environmental consultants or academics.[100] This second type of private regulation has been labeled surrogate regulators.[101] The standards set by these private regulatory groups might be developed as industrial standards,[102] standards from professional associations,[103] or standards recommended as best practices from a coalition of informed parties.[104] There are reasons to suspect that CCS storage activities might benefit from both public and private forms of regulation.[105]

CCS storage activities likely would present circumstances wherein the operator would have the "best collection of relevant information."[106] There are potential benefits to be derived from delegating certain elements of the regulatory framework to the CCS operators in self-regulation. Delegations to the private agency, in this instance the CCS operators, could reduce the governing authority's overall cost of regulation.[107]

As needs arise and technology improves, it would be cheaper to amend rules due to the flexibility of the private agency.[108] "Self regulation may, therefore, be an appropriate solution where bargaining, at low cost, can occur between risk-creators and those affected."[109] In the case of CCS projects, it is foreseeable that while the permitting process is underway, procedures could be put into place to facilitate such ongoing feedback between the operators as self-regulators and the public authorities as representatives of the affected populace.

6.2.2   Advantages of Self-Regulation vis-à-vis Public Regulation

Under certain circumstances, private regulation would be expected to be more robust than public regulation. First, when a novel industry presents a high degree of uncertainty about the risks and hazards of its activities, private regulation would be able to react faster to changes in information and ensure superior standard setting.[110] This is certainly the case for CCS operations; the field is rapidly advancing, and it would benefit from standards that reflect the most complete and most up-to-date scientific and engineering advancements.

Second, when there is a substantial need for regulatory nimbleness (e.g., for an industry experiencing rapid changes in technology), private regulation would be expected to be more robust. Private enterprise could be more flexible than bureaucratic organs at adapting to change.[111] Regulatory authorities can be delayed by the requirements of due process in public deliberation, whereas private actors may be able to update regulations more quickly to ensure more efficient safety results. Also, if technological advance requires years of study and experience with the technology, then private enterprise may be more efficient than regulators at setting necessary standards. Similarly, the CCS industry, in accelerated development, would much benefit from standards that maintain and match that developmental speed. To have standards significantly lag behind technological know-how is an unnecessary concession to lower standards of safety and higher effective costs for those results. Private regulators, especially insurance providers and industry groups, particularly co-venturers or co-investors with audit rights, could better ensure the rapid uptake of cutting-edge improvements in safety and cost-reducing standards.

Third, if public law and regulation are well exposed to corruption and coercion, private regulation might be more robust.[112] These concerns are especially relevant for safety regulations.[113] Thus, for industries facing such problems, private regulation could be more relevant than in other scenarios. Ultimately the operators of CCS projects would be expected to be profit maximizing and thus desirous of continuing their revenues from CCS sequestration, meaning more greenhouse gas prevented from harming the atmosphere, and reducing their overall costs of damages. Corrupted officials or even corrupted procedures could result in regulatory goals that deviate from either goal. CCS in certain places might not be under such corruptive influences, but if "safe" countries foster the development of effective private regulatory measures, then when CCS activities spread to industrial areas where corruption or confusion might otherwise mislead regulatory efforts, the in-place private regulations could retain efficient governance of the risks of CCS operations.

Fourth, when the divergence of interests between producers and consumers is small, a pro-industry bias is likely well correlated to the biases of the consumers, and thus no welfare loss should occur.[114] In the case of CCS operations, this is expected to be the case. Both producers and consumers align on the goal of protecting the world from climate risks; moreover, neither seeks to see injuries from the project, especially given a backdrop of public regulations and civil liabilities. In the case of CCS activities, the suggestion is not that CCS operators should solely self-govern with no public regulations, but that higher efficiencies and improved risk governance might be gained from the complementary enablement of private regulatory efforts.

Miller presents two more arguments in favor of self-regulation.[115] First, industry would be able to organize in a less bureaucratic way than government and would therefore show more flexibility. This would more particularly be important in an area where regulatory standards (as a result of evolutions in research and development) are very volatile and hence change quickly. Government or administrative agencies are often not able to react quickly due to bureaucratic problems and slow decision-making processes. The costs of adapting norms to rapidly changing societal circumstances and technical evolutions can be relatively low compared to the situation where government intervention would be necessary.[116]

Second, industry would be better able to minimize the costs of regulation since in drafting the standards, they would also take into account costs of compliance and enforcement. Moreover, industry organizations bear their own costs and would have incentives for cost minimization. For government, self-regulation would have the advantage that it could save on enforcement costs. Moreover, self-regulation would also fit into "ecological modernization" and a reflexive approach to environmental law.[117]

In the context of CCS storage operations, the operator should already be aware of the costs and technologies involved in achieving the efficient level of CCS-related accidents and could therefore develop the new necessary standards more cheaply than bureaucrats could due to that informational advantage.[118] In addition, the CCS operator would need to bear its own costs of operations and maintain a profit-seeking optimand from its investors and shareholders, so the operator will be bound to achieve both an efficient level of accidents and safety and an efficient use of its capital resources.[119]

It is likely that the CCS operators will have better information than governmental actors on the technologies and best practices for operating CCS storage projects.[120] In that case, it would be simpler, cheaper, and more efficient for operators to develop the necessary guidance to achieve the efficient level of accidents once that level is determined.

In addition, to the extent that a grouping of CCS operators could develop the procedures, that is, some form of industry organization, there would be potential for private CCS standards to evolve and become privately enforced., Many oil and gas projects, for example, are joint venture projects with joint investment by several operators but managed and operated by a single specific operator. In an example where the operators agreed to certain standards and norms of operational procedures for a CCS storage project, then the nonoperating investors would want the rights and permissions to audit the operator to ensure that the operator was indeed enforcing the agreed-to standards and norms and that the co-venturers' investments were soundly within planning guidelines.

Another example of where a private party could bring such private enforcement to privately set standards is the situation found between an operator and its insurance provider[121] as a form of surrogacy regulation.[122] Contractually, they could agree to certain behavior requirements against the operator in exchange for the provision of insurance by the insurer. The insurance company could then obtain, under that contract, the affirmative rights to audit and inspect the behaviors and circumstances of the insured operators.[123] Operators could be provided financial incentives by the insurers to encourage performance at best-practice levels, even above public regulatory standards if set inefficiently low, and to encourage operators to demonstrate operational safety results in pursuit of future premium discounts.[124] Failure to comply with those agreed-to standards could cause the insurance coverage to become more costly or otherwise unavailable to operators.[125] This is an instance of how private regulation might work: "Mandatory insurance aligns the incentives of both insured and insurers in favor of learning about safety and trying to improve safety in the insured's operations."[126]

Substantial arguments have been raised that insurance-driven private regulation could be more robust than public regulation in certain circumstances. While public regulators might be susceptible to capture by industry or environmental interest groups, insurance providers operate in a different context and would be better placed to resist such influences.[127] While discussing the potential to privately regulate shale oil activities, Dana and Wiseman summarized the potential role of private insurance:[128] "Insurance provides a mechanism for reducing risk to the extent insurance premiums are set to reward behavior that creates less risk and penalize behavior that creates more risk."[129] Supporting the argument that private regulations could operate with similar efficiency to public regulations, they found that mandatory requirements for private insurance could address many of the theoretical advantages of public regulation—insolvency, strategic avoidance, and clouded causation issues—by providing a transfer of capital ex ante to third parties to address ex post funding concerns.[130] They also found that the greater the likelihood for funds to be available with such capital tools such as long-term insurance, the greater is the likelihood that future courts would find the operators liable. Thus, the very invocation of insurance would increase the likelihood of ex post liability and thus increase the impact of those insurance costs on the ex ante decision making of the operators.[131] There is no reason to suspect that the arguments drawn here would be any less functional in application to the potential private governance of CCS storage activities.

In conclusion, so long as the private regulatory standards and their means of enforcement were acceptable to both governmental and other agencies, the costs of enforcement and policing could be borne by those profiting from the ongoing operations of the CCS storage project, whether that would be co-venturers or insurers. This method of private regulation could generate a significant welfare effect for the operator/investors, the government-cum-regulator, and the public at large.

6.2.3 Disadvantages of Self-Regulation over Public Regulation

Against these advantages of safety regulation are many disadvantages mentioned in the literature.[132] One problem, according to Shover,[133] is that industry does not often show that it really deserves the trust that society puts in it, for example, by reacting effectively against violations of the standards. Also the private interest theory has warned that interest groups often have an incentive to abuse regulation to create barriers to entry and hence limit market entry for newcomers. This would increase the income for the existing industry.

Moreover, it is not always certain that the industry group will use the information advantage via self-regulation effectively to promote the public cause.[134] With self-regulation, in some cases industry could serve its own interests and not the public interest.[135] Enforcement of self-regulation is often weak, and deterrent sanctions may be lacking.[136]

More recently, questions have been asked with respect to self-regulation and private standards as far as the legitimacy and accountability of self-regulation is concerned.[137] Increasingly in some situations, government will make use of self-regulation but delineate the circumstances under which it will rely on self-regulation. Hence, in those cases it is not a pure self-regulation by industry but rather a "conditional self-regulation" whereby government still monitors self-regulation by industry.[138]

6.3 Regulations Benefit from Availability of Civil Liability Rules

The literature from law and economics supports the complementary implementation of both regulations and rules of civil liability. While each has its own areas of efficiency, it appears from the literature that regulation and rules of civil liabilities can be coordinated to mutually reinforce each other's efficiency and performance.[139] While this conclusion might not appear surprising, the literature can provide particular insight into which areas these different forms of governance can best interact. This discussion attempts to highlight those areas for policymakers.

It is perhaps surprising for some readers, after reading the previous discussions reviewing how regulations succeed when civil liability rules might fail, that one of the best ways to ensure the efficacy and viability of regulatory systems is the complementary functioning of a civil liability system. But this section makes exactly *that* argument: that another key reason to implement rules of civil liability is to ensure the very function of the regulatory mechanisms that most scholars are already supporting.

Thus, the role of civil liability rules is not limited to merely functioning in their own right but also as a buttress and support to the function of regulations. In that sense, rules of civil liability are central to the effective application of regulations, both public and private[140]

6.3.1 Civil Liability Rules Improve Regulatory Efficacy

Although there are strong arguments to make safety regulation the primary instrument aiming at the prevention of harm related to CCS, there may still be an important, although supplementary, role to be played by liability rules. There can be many reasons for this.[141]

The first point that is often stressed is that the fact that there are many arguments in favor of regulation of CCS-related risks does not mean that the tort system should not be used any longer for its deterring and compensating functions. One reason to continue to rely on the tort system is that the effectiveness of regulation is dependent on enforcement, which may be weak. In addition, the influence of lobby groups on regulation, to which public choice theory has rightly pointed, can be overcome to some extent by combining safety regulation and liability rules. Moreover, safety regulation (e.g., in licenses) can be outdated fast and often lacks flexibility, which equally merits a combination with tort rules.[142]

Also the literature with respect to CCS stresses the importance of liability rules in order to back up for regulatory failure. Especially where liability is targeted not at the long-term stewardship but rather at the near-term risks, it is as a useful complement in addition to regulatory requirements.[143] Precisely because regulation can also be vulnerable to information gaps (in the situation where operators possess better information on preventive measures than regulators), liability rules could have an important additional deterrent effect.[144] The most important reason for this supplementary role of liability is related to the informational advantage of operators and the rather static nature of regulation. Because CCS is a new technology, involving many uncertainties,[145] it is possible that, for example, after particular pilot projects have been executed and more experience on risk preventive measures is acquired,[146] information on optimal technology evolves rapidly. Ideally this would lead to a rapid adaptation of safety standards, which would dynamically follow the newest insights in technological developments. In practice, however, adaptation of safety standards (e.g., in permits set by administrative agencies) may not be that easy (depending on the legal system). The advantage of exposing an operator to strict liability is that this provides additional incentives for investments in research and development and hence for innovation of preventive technologies.

One consequence of the joint use of regulation and liability is that regulation can, in legal systems that would still employ a negligence standard, inform the judiciary about the optimal level of care. Indeed, in many legal systems, a breach of a regulatory duty is automatically considered a fault or unlawfulness. This is sometimes referred to as negligence per se. The advantage of this model is that in those cases, the regulator in fact solves the information deficiency on the side of the judge. The judge verifies only whether the regulatory standard has been breached. If that breach stands in a causal relationship with the damage, liability of the defendant will be established. Moreover, it provides incentives to victims to prove that the regulatory standard was breached. This facilitates the burden of proof for the victims (again, in a negligence setting) but also makes the victim de facto an enforcer of safety regulation. This is an example of a "smart"[147] combination of liability rules and regulatory standards. Of course, the question can also arise whether compliance with regulation will automatically free an injurer from liability; we will argue that this should not necessarily be the case.[148]

6.3.2   Civil Liability Rules Defend the Function of Regulatory Designs

It has been stated that "'single instrument' or 'single strategy' approaches are misguided," but that "in the large majority of circumstances (though certainly not all), a mix of instruments is required, tailored to specific policy goals."[149] Particularly for environmental hazards, such as those present within CCS storage activities, there is a standing recommendation for the complementary implementation of both.[150]

Several arguments for complementary implementation exist within the literature. Despite the benefits of public regulation, their overall effectiveness remains challenged, and thus could be buttressed by the complementary forces of civil liability rules, particularly when enforcement itself is institutionally challenged.[151] The existence of rules and the opportunity to engage rules of civil liability can act as a preemptory defense for authorities challenged by threats of agency capture and other sources of regulatory frustration.[152]

Arguably, if a tortfeasor is aware that its efforts to lobby and agency capture are likely to be upended in civil litigation, it would be less likely to expend the capital to lobby in the first place. Thus, rules of civil liability could relieve Nyborg and Telle's problem of "regulatory loss of control."[153]

The previous discussion covered how rules of civil liability might aid in the enforcement and protection of public regulation. But it is also the case that public regulations can aid in the efficacy and transaction cost efficiency of civil liability litigation. To begin, the existence of public regulations and the failure to comply with them could lead to a finding of negligence; the regulation serves in effect as a threshold event for failing a duty of care even when the courts might lack a previously determined standard and thus reduce the efforts required to demonstrate both an existing duty and the failure to meet it.[154]

In another sense of why both systems should be in place is an issue of social justice. The literature refers to the potentially adverse effect that public regulations can have on public morale when no alternatives are present.[155] In large measure, this is due to the relocation of the decision-making process away from either the tortfeasor or the victim, reducing their ability to control their private trajectories.[156] The decision process is handled elsewhere, by the bureaucratic and remote public authority that gets to set the standards.[157] Thus, one of the most powerful arguments on why rules of civil liabilities should be retained in the presence of regulations to address similar injuries is the need within a democracy for all citizens to remain informed, engaged, and part of the standard-setting process.

7 Compensation via Market-Based Measures

The study so far has mainly addressed how liability rules and regulations could provide incentives for prevention of accidents from CCS storage activities. However, from a policy perspective, the compensation issue is important as well, especially if the liability rules and regulations were designed with prevention as their primary goal. In such a case, the damages awarded by the liability and regulatory systems might not address certain externalities or other political or justice considerations. The remainder of this study hence focuses on the question, "Through what types of instruments could this compensation be provided?"

This chapter evaluates a variety of market-based measures. First, it considers the range of goals that compensation could provide. Next, it evaluates the potential for private forms of insurance to provide compensation. Concerns of feasibility, including impacts of risk aversion, barriers to entry, and ambiguity are reviewed. The potential for insurance to be provided from both capital capacity and diverse time frames is discussed next. Would the provision of insurance create moral hazards leading to inefficient levels of risk? That issue is addressed in three sections. We conclude that insurance products will likely be available for CCS operations, including storage.

Market-based forms of compensation planning beyond private insurance are reviewed in turn. The strategy of self-insurance and of captive insurance companies is evaluated. The potential to apply lessons learned from risk-sharing agreements from other industries is reviewed. The provision of capital through guarantees and deposits, as well as from letters of credit and performance bonds, is reviewed. Trust funds and escrow accounts are similarly explored, and the potential for a market in catastrophe bonds is investigated. The chapter concludes with a note on the potential to apply market-based measures in combination for enhanced efficacy.

7.1   Goals of Compensation for CCS Policy

Haan-Kamminga has suggested that one of the key issues clouding the development of clarity in CCS regulations is the question of "who should pay for the risks of CCS."[1] She raises a concern that the question has been intertwined with concerns of whether policymakers should expect a high-risk conjecture or a low-risk conjecture and, most important, that much of the policy discussions appear to have assumed that risk levels would be high.[2]

A preliminary difficulty that could arise is why one should even consider the compensation of damage caused by CO_2 storage. Using an economic perspective, one could argue that compensation of victims is not a singular pure goal of the legal system, but rather the broader idea of welfare maximization through the minimization of all accident costs. The question of how victims of CCS could be provided with appropriate compensation is, in that strict sense, a secondary consideration from an economic perspective.[3] However, some have argued that there may be strong normative beliefs that providing relief, especially for victims of disasters, is one of the principal functions of government.[4] Extreme damage resulting from a CO_2 storage site could lead to serious political and economic instability, which justifies thinking about how to structure efficient compensation for victims, including the role government should play in that respect.

Economics can still be a useful instrument in indicating how a form of compensation can be provided at the lowest possible administrative costs and with a minimum of unnecessary side effects (as far as affecting incentives for prevention is concerned) or negative distributional consequences.[5] Providing specific compensation for victims of CO_2 sites is indeed problematic from a distributional perspective for the simple reason that some victims (those that are suffering damage as a result of a CO_2 site) may receive preferential treatment compared to other victims (those hit by an individual accident). This differentiated treatment would be difficult to explain to the two sets of otherwise equally injured victims from the perspective of the equality principle.[6]

From this it can be concluded that providing specific attention to victims of CO_2 sites (differently than to other victims) might indeed constitute discrimination, but it may not only be unavoidable, given the pressure on politicians to act, but also desirable, given the potentially devastating effects of damage resulting from a CO_2 site. From a policy perspective, there are good reasons to analyze how to provide adequate compensation to victims.

Thus, the literature has not only called for clarity of liability rules and regulations; it has also called on governments to develop plans to enable compensatory measures in advance of accidents, and it has suggested a variety of measures. This chapter addresses the strategy option of insurance as means of providing compensation. Then it examines other private, market-based means to provide compensation, such as performance bonds, risk pools, and trust funds. Chapter 8 evaluates public means of providing compensation.

7.2 Insurance for CCS

Insurance as a means of compensation is a cost-reimbursement method,[7] and thus indirect, in that victims first need to establish their injuries and have damages rendered in court, or receive in settlement discussions, prior to the operator's claims to the insurer for coverage for the payment of damages. Also, insurers are likely to resist claims based on the technical details of coverage and conflicting perceptions of accidental events, their causations, and potential assignments of liability; thus, insurance is seen as potentially providing less-than-complete coverage for CCS-related damages.[8]

Dana and Wiseman reject that mandatory insurance requirements would create negative impacts on adverse selection or moral hazards.[9] Mandatory insurance would require all operators to carry such policies, so the opportunity for adverse selection would be limited or unavailable.[10] Moral hazard could be efficiently addressed by the needs of the insurers to audit and investigate the activities of their clients; also, moral hazard is addressed by informed pricing of premiums to accurately reflect the underlying risk profiles of the insurance customers-cum-operators.[11] Dana and Wiseman do not directly address the field of foreseeable operators of CCS storage sites, which is likely to remain a fairly small number of parties, and that limited population group could be efficiently observed by the major insurance providers.

Dana and Wiseman are also concerned with the arguments that the sheer novelty of certain industries could frustrate the policy implementation of mandatory insurance requirements, but they find little evidence to merit such concerns in historical cases.[12] They give three means that private insurers could use to overcome such novel industries: (1) pool resources to provide additional depth of coverage, (2) engage in reinsurance, or (3) use catastrophic bonds.[13] Attanasio provides an argument that financial instruments that diversify risk over large groups of operators, such as insurance or funding pools, can be efficiently responsive to those risks with "skinny tails."[14]

It is of note that Wyoming, a state that opted not to provide indemnification or a transfer of liabilities, does require CCS operators to carry both performance bonds and private insurance.[15]

Given these discussions in the literature, there are several key concerns with relying on insurance to provide for the risks of CCS storage activities. We look at each of these in turn, followed by a series of discussions on market-based alternatives to more routine forms of insurance.

7.2.1 Insurance and Risk Aversion

Economists have used the concept of risk aversion to explain that many persons will be averse to risks with a relatively low probability of occurring but with a possible large magnitude when they do occur. In principle, individuals can be risk neutral, in which case they do not care whether they are exposed to a risk with a low probability and a high damage, or the other way around: high probability and lower damage. They can also be risk lovers and thus seek risk. And the other possibility is to be risk averse. These individuals are particularly averse to the prospect of being exposed to a risk with a low probability of occurrence but very high damage when it does occurs. The attitude of individuals toward risk is to a large extent related to their financial situation. Risk aversion may hence be the usual attitude for most people in society given the limited spread of wealth.

The utilitarian approach with respect to insurance has demonstrated that risk creates a disutility for people with risk aversion. Their utility can be increased in case of loss spreading or if the small probability of a large loss is taken away from the injurer in exchange for the certainty of a small loss.[16] The latter is, of course, exactly the concept of insurance. The risk-averse injurer has a demand for insurance; he prefers the certainty of a small loss (the payment of the insurance premium) whereby the probability of a larger loss is shifted to the insurance company, thereby increasing the utility of the injurer.[17] It is remarkable that in this utilitarian approach to insurance, liability insurance is in the first place regarded as a means to increase the utility of a risk-averse injurer, not so much as a means to protect victims, as is sometimes argued by lawyers.

The reason an insurance company can take over the risk of the injurer is well known: because of the large number of participants, the risk can be spread over a larger group of people.[18] The insurer only has to pay attention to building relatively small risk groups in which the premium is as much as possible aligned to the risk of the members of that group. Gathering a sufficiently large number of insured to share a similar risk is therefore one of the crucial conditions of insurability.[19]

In addition to this utility-based theory of insurance, which sees insurance as an instrument to increase the expected utility of risk-averse persons through a system of risk spreading, Skogh has powerfully argued that insurance may also be used as a device to reduce transaction costs.[20]

The basic principle of insurance is therefore relatively simple: because individuals have an aversion toward risk, they will seek insurance coverage. The insurer is able to aggregate many risks of risk-averse individuals facing the same risk profile. Using the law of large numbers, the insurer is able to spread the risk over a larger risk group.[21] However, this supposes that the insurer is able to calculate an adequate premium based on the probability that a certain accident will occur and the potential magnitude of the damage. The latter is obviously important because the insurer has to make sure that there is enough capacity at the moment a certain risk with a potentially large magnitude emerges.

These principles can be used to explain the demand for insurance in case of CCS-related risks. Since the attitude to risk is strongly related to the wealth of an individual, the degree of risk aversion will also depend on the available assets of the CCS operator. For relatively small expected losses, a wealthy operator could be risk neutral and hence not have a demand for coverage. In that case, the demand for insurance would emerge only if insurers could manage, for example, claims handling more effectively. It would then mostly be for the reduction of transaction costs and not for risk aversion that insurance would be taken out.

The degree to which CCS operators have a demand for insurance may strongly depend on their own risk attitude, which in turn is related to their assets. Normally it is fair to state that especially for smaller operators, risk aversion is higher and, hence, a demand for some type of coverage (insurance or alternatives[22]) may emerge.

Generally a distinction is made between two types of insurances. One type is insurance that individuals take to protect themselves against the future losses that they may suffer themselves, in their income or their property. Fire insurance is a typical example. These types of insurances are referred to as first-party insurance. Insurance can also be taken for the risk that one has to compensate damage suffered by a third party. These types are therefore referred to as third-party insurance.[23] Liability insurance is a typical example. In that case, the potential injurer takes insurance against the risk he runs of having to compensate the potential victim.

In the case of CCS, both first-party and third-party insurance can be relevant. The operator will therefore obtain first-party insurance for the property damage he could suffer to his installations. However, more likely it will be a third-party liability insurance that plays the most important role in case of CCS. In that case, coverage is demanded for the risk that damage is suffered by a third party.

First-party insurance may obviously play a role as well in order to protect particular victims of CCS-related risks. In that case, victims do not take first-party insurance just for the CCS-related risks, but general accident insurance or property insurance, which may also serve to cover CCS-related risks as long as they are not excluded from the insurance policy.[24]

7.2.2 Barriers to Market Entry

An important condition for insurability (both first party and third party) is that insurers have information on the likelihood that the particular event will occur and the potential damage that may result from manifestation of the risk. Insurers generally acquire this information on the basis of a past loss history record—in other words, with statistics. For example, with respect to fire insurance, statistics are available that can inform insurers what the likelihood is that in a particular area, a building will be struck by fire. On the basis of this statistical information, the probability of the manifestation of the risk is calculated, multiplied by the loss in case of the manifestation of the risk. The result provides the so-called actuarially fair premium. Increased by the loading costs (the administrative costs for the insurance) and, depending on the market structure, a profit margin, this will constitute the premium to be paid by the insured. Since insurers will collect a large amount of insured within an insurance pool, the risks can be spread according to the law of the large numbers.[25]

However, in order to make the risk insurable, it is crucial that a sufficient number of insured parties are available since risk spreading otherwise may fail. Ideally, the number of insured is large enough to pay the premiums that allow compensation in case the risk materializes. This requirement also explains why there may be large barriers to enter into the insurance market. If a new enterprise enters the market and has, for example, only two hundred insured all paying a relatively small premium, the number of insured may not be sufficient to cover the risk.

An example can illustrate this. Suppose that the insured runs a risk of 1 percent of suffering a loss of 1 million during a particular time period. The actuarially fair premium in that case would be 1,000. Suppose that in that period, the newcomer to the market acquired only 500 insured. Together they would pay a premium of $500 \times 1,000 = 500,000$. The result is that the insurer does not have sufficient capacity to compensate the loss when it materializes. In this simple example, insurance would be possible only if at least 1,000 insured participated in the insurance pool in order to make sufficient risk spreading possible. Still, insurance in this example would be a risky business since in fact, a pool of 1,000 insured just barely covers the risk and is obviously very risky. If the risk occurred with, say, two insured during one year (instead of with one insured during two consecutive years), the insurer would already not be able to pay out.

This example shows why a large number of insured is necessary to make a risk insurable and hence to enter the market. This example also illustrates that huge factual barriers to market entry may exist in case of insurance, especially when it concerns new and unknown risks. This explains why, especially for those risks, only a few insurers are willing to provide cover, which may lead to limited supply. That problem could obviously also arise in case of new risks like CCS.

The example not only makes clear that entry into a new insurance market may be difficult; it also shows that the larger the number of insured, the more accurate the predictability of the statistics becomes. Therefore, insurance is especially attractive for insurers in case of a reasonable probability that harm will occur and a reasonable (not excessively high) damage when the risk manifests. Low-probability, high-damage events (often referred to as catastrophes) are therefore by definition difficult to insure since predictability may be lacking.

Another problem from the insurer's perspective is that the insurer may not always have statistics that provide accurate information on the likelihood that a particular risk will occur. With newly emerging risks like CCS, statistics on accident history may be lacking for the simple reason that it concerns a new technology. The literature has often indicated that uncertainty may endanger insurability, which could concern both factual and legal uncertainty.[26]

When these general insights of insurance economics are applied to CCS, both problems may threaten the insurability of CCS-related risks. Given the novel character of the technology, there may be uncertainties concerning the risks and, hence, the predictability of the risk. Moreover, the uncertainties concerning the scope of liability would create legal uncertainty that might endanger insurability, more particularly of third-party liability. However, the latter type of uncertainty can be removed when legislators provide a clear legislative framework, thus reducing at least legal uncertainty and increasing the predictability of CCS-related risks.

7.2.3 Dealing with Insurer Ambiguity

Before addressing the predictability of risk in the case of CCS, it should be stressed that insurance is very well possible even if statistics on accident records are lacking. After all, there are many examples of risks related to new activities that are insured although statistics with historical accident data are lacking. Indeed, while the industrial need is recent and actuarial data might be yet formative,[27] CCS insurance is available to some extent in the private markets.[28]

There are two explanations for this. One is that an insurer can use an alternative for statistics by relying on modeling of risk. Risk assessment tools can be used to acquire information on the likely risk posed by new technologies and the possible damages that would follow from them. These models, if sufficiently embedded in scientific studies, can also provide insurers information on the likelihood of the risk even in the absence of historical accident data.

Second, the literature has indicated that uncertainty concerning the probability or the damage is of course an element that the insurer can—in principle—take into account ex ante. Uncertainty because of a lack of reliable statistics should not necessarily lead to the conclusion that a particular risk is uninsurable. We are then dealing with the concept referred to as insurer ambiguity, addressed by Kunreuther, Hogarth, and Meszaros.[29] They argue that the insurer can react to this uncertainty concerning either the probability of the event or the magnitude of the damage by charging a so-called risk premium to account for this unpredictability. Hence, an insurer can in principle also deal with a hard-to-predict event by charging an additional premium. Although theoretically the additional risk premium is the answer to insurer ambiguity, in practice the insurer will need some information to make more than an educated guess concerning the risk premium to charge.

Moreover, given the fact that an insurer finds himself in a competitive environment, market forces may well drive him to engage in liability insurance even when an appropriate risk premium cannot be charged. Whether this will in practice actually be done is hard to predict. We have to remind that the insurance business in many countries is regulated to a large extent. This might limit the possibilities for insurers to engage in extremely risky activities by charging high premiums. The direct link between tarification and provisioning is indeed the subject of insurance supervision. Moreover, since many insurers are traditionally conservative and prudent, they might be inclined to be quite cautious with providing cover if uncertainty should be considered too large.

The stance of this argument is that predictability is a crucial requirement to keep a risk insurable, but that a lack of predictability should not immediately lead to the conclusion that risks are uninsurable. The crucial question is whether insurers are able to cope with uncertainty by charging an additional risk premium. However, in practice it may not always be possible to charge this additional premium. In some cases—think more particularly of risks such as genetically modified organisms or terrorism—there may be so little information (and hence the range of uncertainty can be quite high) that it may hardly be possible to calculate the additional risk premium to deal with insurer ambiguity. Moreover, the reaction of insurers to calculate an additional risk premium obviously works only if the insured are willing to pay. The willingness of potential policy holders to pay would depend on the extent to which those who are insured recognize that there are additional risks, albeit uncertain, for which additional cover needs to be extended. If, as a result of information deficiencies, the potential insured does not recognize these risks, they will not be willing to pay the additional risk premium and insurance cover will not take place.

The same problem could also arise in the reverse case: the insured hold that the new risk (say, CCS) is in fact manageable and reasonably calculable, but they are unable to convey this message to the insurer; as a result the insurer charges an excessively high risk premium that the insured is not willing to pay. More particularly, this lack of information (and the resulting unwillingness to pay) may explain why these additional risk premiums are seldom charged. The result is more often that a market for the particular new risk will simply not emerge or alternatives (e.g., risk-sharing agreements) will be developed.[30]

7.2.4 Predictability of CCS-Related Risks

Applying these insights to the risks created in case of CCS, many point to problems with uncertainty and lack of data. Klass and Wilson, for example, indicate that some of the conditions of insurability, particularly related to the predictability of the risk and a well-established time period, may not be met in the case of CCS;[31] moreover, there may be legal uncertainty in the situation as well.[32] Legal uncertainty may exist in legal systems that have not sufficiently clarified the exposure to risk of CCS operators in the different phases of the CCS project life cycle. It is a fear for liability in the phase of long-term stewardship that may create legal uncertainties that could endanger insurability. Trabucchi and Patton therefore point again to the CCS project life cycle and argue that in the phases before sequestration (hence the capture and transport), transfer of risk to a third party (more particularly, an insurer) may be possible,[33] but that especially as far as the risks related to the sequestration are concerned, the lack of readily accessible data may make predicting the risk difficult.[34]

Doubts are especially formulated with respect to the ability of insurance to cover long-term liability since this liability may endanger insurability.[35] However, others indicate that although uncertainties about risks remain even under the best of circumstances, this does not mean that an ex ante valuation of the risk would be impossible.[36] Site-specific risk analysis may be possible and could provide information on the specific risks related to one site.[37] It is even argued that it is possible to make probabilistic estimates of the expected loss values for every specific site based on a site-specific risk assessment.[38] This is confirmed in other research where it is even held that the risks related to CCS are "known, predictable and manageable."[39] These admittedly optimistic statements are very general and do not distinguish among the phases in the CCS project life cycle. One has the impression that those statements about the possibility of risk assessments, and hence the predictability of the risk, apply especially to the risks that may arise during the operation and, probably, the postclosure monitoring period but less to the uncertain period of long-term stewardship.

Therefore, in policy documents, it is also held that insurance can basically play a role only during the operation of the plant.[40] Also the EU CCS Directive still involves large uncertainties for operators[41] concerning the moment when the legal liability will be transferred from the operator to the state. Those factual and legal uncertainties may also endanger insurability.

7.2.5 Cover in Time

Traditionally liability insurance relates to sudden and accidental events that cause damage on a clearly established moment in time. Again, fire and traffic accidents are examples of these sudden events. Because of this requirement of a "sudden and accidental" event, gradual pollution has traditionally been excluded from environmental impairment liability insurance. Gradual pollution leads to difficulties at the level of insurability because it may not be possible for insurers to determine when the damage has actually occurred and therefore whether it was still covered within the time frame of a particular insurance policy. Precisely because of the potentially long tail of gradual pollution within the Environmental Liability Directive,[42] insurers argue that the environmental liability risks in this European directive are largely uninsurable.[43] Similar problems may also play a role with the insurance of CCS, particularly as far as this long-tail exposure (damage that occurs at an indefinite distant moment in time) is concerned

In such cases, it is of utmost importance to examine the precise period of cover under the particular insurance policy.[44] In that respect, a distinction is usually made among three systems:[45]

1. The act-committed system. The wrongful act must have taken place during the period of insurance cover.
3. The loss-occurrence system. The damage must have taken place within the period of insurance cover.
4. The claims-made system. The claim for damages must have been received by the insured or the insurer during the period of insurance cover.

A definite increase in claims-made policies throughout Europe is clearly apparent.[46] This is due to the obvious disadvantages, at least for insurers, connected with the two other systems in cases of long-tail exposure.

In the act-committed system, which was especially popular in the 1950s and 1960s, there is cover only if damage can be attributed to a wrongful act of the insured that took place during the period of insurance cover. Obviously this system is advantageous for the insured party since the potential claim remains in existence until a statute of limitation has passed. In addition, cover is in accord once, with the policy conditions valid at the moment that the damage was caused, meaning the moment that the wrongful act occurred. As a consequence, limitations of cover that might be introduced at a later stage cannot be invoked against the insured.

For insurers, this act-committed system can be highly disadvantageous. Indeed, if an insurer provided cover for product liability, for example, and the policy expired shortly after a certain product was brought to market, under an act-committed system of cover, the insurer can still be held liable if, several years later, it appears that this particular product caused damage to a victim. This may be problematic, especially when the insurer has not foreseen the possibility of this long tail and has therefore not charged an adequate premium.

These problems caused the international reinsurance market to introduce a loss-occurrence system of cover. Under loss occurrence, the damage must have taken place within the period of cover.[47] Therefore, cover is also provided if the wrongful act that caused the damage took place before the insurance contract entered into force. However, a loss-occurrence system does not completely eliminate the problem of the long tail; indeed, if an injury manifests itself during the period of cover, the insurance policy will apply for future consequences over, say, twenty years in the case of long-lasting diseases.[48] Therefore, insurers changed to claims-made cover. Although claims-made policies may exist in different forms,[49] it is essential to a claims-made system that the claim for damages should have been received by the insured party or the insurer during the period of insurance cover.

Although claims-made policies are probably the only way insurers can protect themselves adequately against long-tail risk, Abraham argued that these policies could dilute the deterrent function.[50] The argument is that policyholders under claims made would internalize only the costs that would arise during the year of coverage and not the future costs.[51] This argument is not totally convincing since, irrespective of insurance, the policyholder will be liable for those future costs as well. Nevertheless, in the future, an insolvency problem may arise if, as a result of the claims-made coverage, the cover would be limited in time.

In practice, many varieties of so-called claims-made policies do exist (e.g., as far as environmental impairment liability cover is concerned), but the bottom line is that the claim must have been received by the insured or the insurer during the period of insurance cover. This enables the insurer to exclude long-tail coverage. Precisely for risks that are hard to predict, insurers will therefore prefer claims-made coverage.

A consequence for the potential coverage of CCS-related risks is that coverage may be provided during the operation and perhaps (but even that is doubtful) during a period of postclosure monitoring. However, long-term liability (i.e., liability for damage that would occur after the site has been transferred to the government insurance) would, as a result of the claims-made coverage system, not be available. Also the US EPA Financial Responsibility Guidance mentions that insurance companies may not be willing to underwrite policies for the entire life of a project, which could last for a minimum of fifty years.[52] That is why most of the literature related to CCS holds that as far as the long-term stewardship of storage sites is concerned, the site should (under specific strict conditions[53]) be transferred to the state.[54]

As far as insurability is concerned, a similar conclusion can therefore be reached as with the exposure of liability in time. We noted that exposing operators to the long-tail risk, more particularly to the damage that might occur during the period of long-term stewardship, does not seem useful as it may not provide the incentives for prevention.[55] When discussing the same issue here from an insurability perspective, a similar conclusion is reached: liability for damage that could occur during the period of long-term stewardship should also be excluded because such liability exposure may be uninsurable.

7.2.6 Capacity

An important condition for the insurability of any risk is not only that the risk should be predictable; the insurer must also have sufficient capacity to have money available once a risk materializes.[56] That may be a problem when only a relatively small number of insured are interested in the product that the insurer wishes to sell. Not only the small number of insured may be problematic (especially in the initial phase of developing a CCS project); the same is the case for the high barriers to market entry created by the need for expert knowledge in CCS-related risks and creating sufficient capacity.[57] For CCS, that may limit the insurability of the risks. It could therefore lead to the situation that the total pool of insured is relatively too small and, hence, not enough insurance capacity can be generated to cover any loss that does occur.

The economic principle is simple: the total amount of premiums generated by the pool of insured should be large enough to cover any potential risk that materializes. This is obviously more difficult when the number of insured is small and, as in the case of CCS, the potential damage could be substantial. Obviously these problems are larger in the initial phase of the development of CCS and may disappear when more experience with CCS projects is obtained and more operators participate in the market. However, one should not conclude too quickly that the large magnitude of the potential damages makes the risk uninsurable. Insurers can turn to a number of techniques to make risks with a large potential magnitude insurable.

A first possibility is to use co-insurance. This amounts to the possibility of many insurers jointly covering one particular project. For example, if there were one CCS site to be covered, four insurers could decide to cover the risk together, each covering 25 percent, thus being able to generate higher capacity.

A second potential solution to deal with the capacity problem is reinsurance. Through reinsurance, an insurer effectively shifts part of the risk to a reinsurance company in exchange for a reinsurance premium. Large risks like CCS-related risks could be reinsured through the international reinsurance market, thus creating higher capacity.

A third possibility is pooling by insurers. This has to be distinguished from pooling by operators, which is usually qualified as a risk-sharing agreement. In pooling by insurers, insurance companies decide to cover a particular risk on a noncompetitive basis for an entire sector. An example is nuclear risk, which is provided through so-called nuclear pools. Since those risks were considered to be very large, the major national insurance companies in every nuclear country decided in the 1950s to pool their resources on a noncompetitive basis in order to be able to provide coverage for them.[58] These pools provide cover for the third-party liability and, to some extent, first-party insurance for the damage caused to the nuclear power plant as well. Pooling is also used with other risks, for example, with environmental liability insurance. Environmental pools exists in many countries.

From an economic perspective, pooling between insurance companies is at first blush problematic since insurance companies pool their resources on a noncompetitive basis, thus excluding the normal competitive process. However, the economic argument to justify such a restriction on competition is that without pooling, there would be no supply of insurance cover whatsoever. In order to provide any supply at all for those difficult risks, pooling by insurers could increase capacity.[59]

A problem from the perspective of competition is that every country has a separate national pool and there is no competition among these pools. This has led to the dissatisfaction of nuclear operators with the nuclear pools, arguing that their premiums are too high. They went looking for alternatives that could provide, for example, first-party cover for the nuclear installation itself. Those alternatives are now widely available on the market.[60] There is, in addition, criticism on the nuclear pools from the European Commission. The competition authorities have increasingly paid attention to those pools, especially examining the compatibility of pooling by insurers with European competition policy.[61] A recent study executed on behalf of the European Commission also examined the amount of co-reinsurance pools currently available in Europe. The study identified fifty-one pools, many of which focus on catastrophic risks such as nuclear-, environmental-, and terrorism-related risks.[62]

In summary, pooling by insurance companies can be a useful device to create higher capacity, but may be problematic from the perspective of competition law. That should not be a reason to reject pooling by insurers as a device to increase capacity for CCS-related risks; it would be important only to pay attention to remaining within the boundary set by competition policy. The CCS-related literature has also pointed to the possibility of pooling by insurance operators and in that respect has argued that antitrust waivers for participating parties may be necessary.[63] Antitrust waivers may perhaps be too strong a suggestion. Even within the boundaries set by competition policy, it should be possible to enjoy the benefits of cooperation from pooling and still have the benefits of competition, particularly when not just one insurance pool covering CCS-related risks would be created, but rather different pools among which competition would still exist.

7.2.7 Curing Moral Hazard and Adverse Selection

7.2.7.1 Curing Moral Hazard There are always two important dangers that are threatening the insurability of any risk, known as moral hazard and adverse selection.[64] We referred to the moral hazard risk when discussing the importance of liability rules for CCS.[65] Moral hazard in insurance refers to the phenomenon that the behavior of the insured injurer will change as soon as the risk is removed from him. Moral hazard can especially be a problem in case of liability insurance. The reason is that the disutility the insurer suffers because of exposure to risk is needed to give him correct incentives for caretaking. If the risk were fully removed from the injurer and completely shifted to the injurer, the operator may lack the incentive to take care. Under full exposure to liability rules, that incentive was provided to operators through the deterrent effect of having to pay compensation in case of an accident. Many economists, especially the early work of Nobel Prize winner Kenneth Arrow, have described the risk of moral hazard.[66] The key question is therefore whether instruments can be designed to provide the insured incentives to behave in exactly the same way as if no insurance were available. If that were possible, there would be an optimal control of moral hazard.[67]

In the literature, two ways of controlling the moral hazard problem have been identified.[68] The first best is a control of the behavior of the operator and a corresponding adaptation of the premium; the second-best solution is exposing the insured partially to risk. In a scenario of optimal control, the premium conditions would be exactly adapted to the behavior of the insured, and the premium would reflect the care taken by the insured. This would give the operator incentives to behave exactly as if no insurance were available and the premium would reflect the true accident risk. This first-best solution is of course possible only in the ideal world where the control by the insurance company would be costless and information on the behavior of the insured readily available.

In practice this is not the case. There are, however, some means for control of the insured, and a differentiation of premium conditions is possible according to certain groups of risk. This can either be an ex ante screening with a higher premium for certain high-risk groups or an ex post premium increase or change of policy conditions based on previous loss experience. This is the so-called experience rating. Much of insurance legislation is also aimed at reducing moral hazard. Think in this respect of the prohibition, contained in many insurance laws, of insuring accidents caused with intent.[69]

A second-best solution is exposing the insured partially to risk. This is considered second best because insurance should ideally aim at removing risk from the injurer. Exposing the insured to risk will mean that some degree of risk aversion will remain. This has the advantage that the insured injurer will still have some incentives for caretaking although he is insured. This exposure to risk can be at the lower level of damage or a higher level. One could indeed think of a system with a deductible whereby a lower threshold applies or one could introduce an upper limit on coverage whereby the insured bears the loss in case the damage exceeds the insured amount. This is often used in medical insurance (in some cases, referred to as a copayment). It means that the insured individual will always pay a particular amount (e.g., $50 or $100) .

In practice a combination of both systems for control of moral hazard will manifest. Usually there is some degree of differentiation within the policy conditions—a deductible and an upper limit on coverage. Of course, the methods used depend on the information costs but also the value of the insurance policy.[70] Obviously an insurer will more readily tend to invest resources in making a well-tailored insurance policy for a large company that pays a substantial premium than for consumer risks.

7.2.7.2 Remedying Adverse Selection A further condition for insurability is that the adverse selection problem can be cured. How can this problem of adverse selection briefly be described?

Insurance is, of course, based on a system of loss spreading. Therefore, the insurer needs a minimum number of similar risks to insure. At the same time, risk pools have to be constructed as narrowly as possible, meaning that the average premium in the risk pool should correspond to the risk of most of the members in the particular pool. Otherwise the average premium would be relatively high for low-risk members, who would then leave the group. In that case, the well-known phenomenon of adverse selection could emerge, which Akerlof described in his seminal paper on the market for lemons.[71]

Interestingly, the remedy for the problems of both moral hazard and adverse selection is to be found in adequate risk differentiation by insurers.

It follows from the economic principles of liability insurance that an adaptation of the policy conditions to the individual risk is essential to control both moral hazard and adverse selection. George Priest claimed that the adverse selection problem has caused an insurance crisis in the United States that can be cured only by an appropriate differentiation of risk.[72] If the insurance policy requires preventive action from the insured party and provides for a corresponding reward in the premium, this should give optimal incentives to the insured for accident reduction. Thus, risk pools should be constructed as narrowly as possible so that the premium reflects the risk of the average member of that pool.[73]

A further differentiation of the risk is obviously efficient only as long as the marginal benefits of this further differentiation outweigh its marginal costs.[74] Risk differentiation certainly does not mean that insurers would have to use an individual tariff in each case.[75] The possibilities for individual differentiation will inevitably also depend on the value of the insurance policy. For mass insurance products with a low premium, risk differentiation can take place only in general categories. In the professional liability insurance of enterprises, however, the benefit of detailed differentiation, rewarding an enterprise for preventive action, may well outweigh the costs.

7.2.7.3 Differentiating CCS-Related Risks

These principles apply in a similar way to CCS-related risks as well. We already mentioned that the CCS literature refers to the danger of moral hazard as follows:

"Moral hazard refers to the specific situation where the risks of an unplanned event increase, because the responsible party is (partially) insulated from being held fully liable for resulting harm. If CCS facilities are not held completely responsible for the consequences of their actions, arguably they will be less careful in their siting and operating decision. Therefore, the incentive to capture, transport, site/characterize, and inject carbon dioxide in an environmentally sound and protective manner may be diminished. The potential for risk increases because the chances of an unpredictable event occurring due to poor siting/operating decisions increase."[76]

Trabucchi and Patton apply this moral hazard to the lack of exposure to liability, but the same moral hazard could obviously arise if full insurance cover were available.

Although the CCS-related literature indicates that moral hazard may be an issue in case of CCS (thus stressing the need for providing appropriate incentives for prevention), the literature equally indicates ample possibilities of risk differentiation, which could thus remedy the moral hazard risk. It is very well possible to identify different risk profiles on the basis of the type and upkeep of a storage site.[77] As we also made clear, the CCS-related literature clearly shows that the technological knowledge is developed to such an extent that it is possible to differentiate the risk prevention measures that operators can and should take[78] and that, hence, differentiated premiums could be charged based on the differing technical characteristics of the site and the type of operation.[79] In principle, a comprehensive system of mapping and ranking of potential sequestration sites would be possible.[80] Insurers could obviously use such a mapping and ranking to apply an adequate system of risk differentiation.[81] As with ordinary environmental insurance, insurers can rely on compliance with third-party environmental management systems, certification, or International Organisation for Standardization (ISO) standards to verify the adequacy of the preventive measures taken by operators.[82]

It has often been mentioned that as far as environmental insurance is concerned, regulation can play an important role.[83] Regulatory requirements with respect to the siting of CCS facilities, operation, and postclosure monitoring[84] are important elements that can all be laid down in regulation. In that respect, properly tailored regulation can assist the insurer's risk differentiation.[85] Given the informational advantage of the regulator (most insurers probably not being specialized in CCS-related risks), regulation can have the major advantage that (as most insurers do) the insurer can primarily require the CCS operator to comply with regulatory standards. Violation of regulatory standards by the insured will often be a cause for refusal of coverage or for a recourse action by the insurer. This shows again the importance of smart collaboration between the regulator and the insurer whereby the regulator informs the insurer about minimum safety standards (laid down in regulation) and the insurer becomes de facto the enforcer of regulatory standards. The promulgation of regulatory standards can thus facilitate risk differentiation by the insurer and in that way contribute to an important extent to the insurability of the CCS-related risk.

7.2.8 Insurability of CCS?

Taking into account the criteria for insurability discussed, should CCS-related risks be considered insurable? Depending on the phase of the CCS project, there may be particular problems that could endanger the insurability of CCS-related risks, especially the risk of liability toward third parties for CCS-related damage. Looking at the criteria we have discussed for the insurability of risks, there could be a serious problem of insurer ambiguity: CCS is a new technology and so actuarially reliable information on CCS-related risks may be lacking to a large extent. There can be high uncertainty concerning the specific probabilities of damage, as well as the potential magnitude of the damage if the risk materializes.[86] There may hence be considerable information deficiencies, which might reduce the appetite of insurers for CCS-related risks.[87] Insurer ambiguity may lead to reduced supply or, if the risk is covered at all, high premiums corresponding to insurer ambiguity. If, however, that high-risk premium does not correspond to the risk perception of operators, there may be no corresponding willingness to pay those high premiums, and a market for covering the particular risk may not emerge.[88]

In addition, capacity may be a serious problem since the potential number of players is (at least at this moment) limited and the potential damage could be quite large. Some of the traditional remedies to create larger capacity (co-insurance, reinsurance, and pooling by insurers) could be employed, but still it is uncertain whether those would be able to generate the substantial amounts that may be necessary to cover the damage that could result from CCS-related risks.

Lacking information on the side of insurers is not only a problem as far as fixing an adequate premium and predicting the risk is concerned, but also for the necessary risk differentiation in order to cure the dangers of moral hazard and adverse selection. However, it was held that the regulator can in that respect help insurers and thus promote the insurability of CCS-related risks by providing a tailored regulatory framework on which insurers could largely rely to judge CCS-related risks and differentiate premiums accordingly.

The specific phases of the CCS life cycle have considerable differences. To put it simply, traditional (environmental) insurance is good at insuring sudden and accidental events but less suited for long-tail risks. This can be understood: given the asbestos nightmare, insurers dislike risks that could expand very long in time and create potential risk exposures for many decades. That makes premium calculations obviously very difficult. That also explains why insurers, as we already explained,[89] have largely moved to so-called claims-made policies. The advantage for insurers is that their risk exposure ends (with a few nuances) when the policy ends, but that may be problematic from the perspective of operators. Looking at the CCS project life cycle, it can therefore be held that (if the other problems mentioned could be solved) insurance may be available for the injection and operation phase, but become more difficult to obtain during postclosure monitoring (when probably other alternatives will have to be examined) and will be excluded in case of long-term stewardship (for which liability should anyway be transferred to the state).

CCS-related literature is equally pessimistic concerning the current possibilities of insurance cover for CCS-related risks. For example, Wilson et al. hold that "CCS might violate many of the conventional rules of insurability."[90] Conventional insurance requires (1) a sufficient number of similar and uncorrelated events to allow for risk pooling, (2) clearly calculable losses, (3) a well-established time period for potential losses, (4) frequent enough losses to calculate premiums, and (5) that the insured party has no incentive to cause the loss.[91] CCS may violate many of those conditions,[92] as we illustrated. Also, in policy documents, it is therefore held that de facto specific policy terms and conditions have not yet been made available for a cover for CCS-related risks, so precise terms and limits of coverage cannot be determined.[93]

As a result of those observations, there may be only a limited role for insurance to cover CCS-related risks. Insurance could play a role only in the phase of injection and operation and most likely not in later phases. Moreover, insurance could play that role only if particular problems were adequately addressed. Given those limits of the traditional insurance markets in providing cover for CCS-related risks, there may be good arguments to examine whether alternative compensation mechanisms may be able to overcome some of the problems of traditional insurance markets.

7.3 Alternative Compensation Mechanisms

Because traditional insurance markets may not be able to cover all of the CCS-related risks in a satisfactory manner, the question arises whether alternative compensation mechanisms could be used that could better deal with some of the problems that traditional insurers are confronted with in covering CCS-related risks.

The search for alternative compensation mechanisms is not new. The CCS-related literature has largely addressed whether other instruments, like bonding or a special fund, could better deal with CCS-related risks than insurance could.[94] Also policy documents show an understanding that insurance may have limits in providing financial security for CCS-related risks; as a result, funds, escrow, bank guarantees, letters of credit, and many other alternatives are examined (e.g., in Guidance Document 4 of the European Commission as well).[95] Also the US EPA Financial Responsibility Guidance Document mentions a variety of alternatives to insurance that operators or owners can use to satisfy financial responsibility, such trust funds, surety bonds, letters of credit, escrow accounts, or self-insurance.[96]

Thus, consideration of alternative means of compensation is well established in the literature and in practice.

Within the scope of this study, we address a few interesting alternatives that have also been discussed in the CCS-related literature: self-insurance risk-sharing agreements will be explored, guarantees and deposits, and alternative risk transfer. Finally, we look at possibilities of combinations of different instruments.

7.3.1 Self-Insurance and Captives

Self-insurance is a mechanism whereby larger players in the market do not take insurance coverage at all but run the risk themselves. In fact, self-insurance can take two different forms.

One is pure self-insurance, which in fact is nothing other than major companies' constituting a reserve for future losses. In a technical sense, this cannot be considered "insurance" for the simple reason that there is no risk spreading, no risk distribution, and hence no loss spreading after an accident.[97] Self-insurance in that sense is merely a reserve for potential losses whereby operators use their balance sheet to guarantee payment in case a major accident happens. Reserves do allow for a risk spreading in time, but not between various parties exposed to a risk.

A second possibility is the creation of a so-called captive.[98] A captive is in fact an insurance company that is created and owned by industry.[99] Many large oil and gas operators have created their own insurance companies. The reason is that in this way, they can satisfy a statutory duty, in the countries where this exists, to show financial coverage but avoid a transfer of funds to a third party (i.e., an insurance company) that would otherwise be necessary.

Self-insurance (with reserves, captives, or using the capital market) obviously has several strengths and weaknesses. The advantage from the industry's perspective is that it is a relatively low-cost solution; operators themselves can provide guarantees for future losses and do not have to transfer risks to an insurance company, which may create additional transaction costs. Moreover, especially for major operators, forcing them to shift risks to an insurance company may make little sense, especially when the credit rating of the operator is in fact higher than that of an insurance company. Using self-insurance at least partially also has an advantage in curing the so-called moral hazard risk that will always emerge in case of insurance: by taking a substantial retention of the risks, the operator will still be exposed to liability, and thus moral hazard could be more effectively controlled.[100]

The disadvantages of self-insurance, no matter what form it takes, may also be obvious: self-insurance is not necessarily a foolproof guarantee against insolvency.[101] That would be the case if only regulation could guarantee that the money set aside for covering the CCS-related losses would be used only for that specific goal. Moreover, self-insurance could lead to an externalization of risk in case of insolvency. Especially smaller operators without strong balance sheets or credit rating could run the risk of liability, and if the risk materialized, they would pass on the costs to taxpayers. Self-insurance can hence be considered effective financial security only if guarantees can be provided so that the reserves set aside will actually be used for the potential losses for which they were earmarked. Otherwise the risk would also exist that in case of, say, insolvency, the trustee in bankruptcy could collect the assets, and money may not be available to compensate victims.

Self-insurance could be a valuable strategy for major players engaged in CCS, such as large energy companies with substantial balance sheets. For smaller and medium-sized players, self-insurance could play a role only as a deductible, in addition to other hedging strategies. If operators offered self-insurance as a form of financial guarantee, serious controls should be imposed to verify the viability of the self-insurance as a reliable guarantee.[102] This is why in the EU policy document, self-insurance is in fact considered the riskiest option because no protection is provided from claims of creditors.[103]

Self-insurance is equally mentioned in the US EPA Financial Responsibility Guidance Document.[104] According to the document, owners or operators can expect to provide documents, such as annual financial statements, that show profits and losses for the year and statements verifying total net worth and networking capital, to be confirmed by an independent auditor. The UIC program director decides whether the information submitted is sufficient to make a determination on the owners' or operators' financial stability.[105] The US EPA Guidance Document considers self-insurance beneficial for owners or operators because it is likely to have the lowest overhead costs. However, there is a high risk of failure, especially in the postinjection site care period, since the injection site is no longer in operation and profitable.[106] The guidance document moreover provides recommended specifications concerning self-insurance, holding that the owner or operator should have a tangible net worth of at least $100 million; specific ratings by credit rating agencies would be required as well.[107]

There is some experience with self-insurance in a related area: the financial security demanded from operators of offshore oil and gas installations.

Financial cover is provided through the Offshore Pollution Liability Association (OPOL).[108] This association imposes stringent criteria for operators to be allowed to qualify as self-insurers. For example, an operator can qualify as self-insurer under OPOL only if it has an A or higher rating from Standard & Poor's, A minus or higher from A. M. Best, or an A3 or higher from Moody's.[109]

These examples show that it may be necessary to apply flexibility as far as the use of self-insurance is concerned. For some (major) operators, self-insurance could perhaps be the only financial mechanism; for others, it could at least partially be used in addition to other hedging strategies. The experience with OPOL, but also the recommended specifications in the US guidance document, shows that stringent rules must apply to control whether a particular operator should qualify as self-insurer.

7.3.2 Risk-Sharing Agreements

A risk-sharing agreement (RSA) or a pool is a system whereby operators mutually agree to share each other's losses. It resembles insurance, but with some fundamental differences. The basic difference is that insurance involves a third party (the insurance company), whereas in a risk-sharing system, the operators are both insured and insurer; there is no third-party involvement .

Unlike in the case of commercial insurance, where ex ante information about the probability of a certain risk and its magnitude should be available to allow the calculation of the ex ante premium, each member's contribution in a risk-sharing agreement policy can be agreed on beforehand and paid only ex post. This characteristic makes it possible for a risk-sharing agreement to deal with uncertain risk, for which the statistical data about the occurrence are rare or the probability and size are less predictable. As long as a risk differentiation can be made among the members, a risk-sharing agreement can be feasible, since charging premiums ex ante is no longer necessary. Only the relative contribution of each member to the risk has to be known.

Another difference between a risk-sharing agreement and insurance concerns the costs. In an insurance policy, the risk is shifted to the insurer at the price of a premium. The premium is not recoverable by the insured no matter whether the insured risk materialized or not. In a risk-sharing agreement, a member contributes only if an accident happens; the duty to compensate can either be postponed or the contribution can be carried over to the following year if there is no accident.[110] A member can also recover his contribution by stopping creating the risk and leaving the pool.

Summarizing, a risk-sharing agreement has a few important theoretical advantages (and differences) compared to insurance:

1. It creates strong incentives for mutual monitoring since the members are dependent on each other; that is, a bad risk can create the likelihood that the pool will have to intervene.
2. For highly technical and complicated (often new) risks, operators themselves may have better information (compared to insurers) on optimal preventive technologies, which they can reflect in a differentiation of the contribution to the pool (or excluding membership for bad risks).
3. A risk-sharing agreement does not require actuarial information ex ante on the probability of an accident and the scope of the damage for the simple reason that no ex ante premium has to be fixed. Only information is needed on the relative contribution of each member to the risk, but this does not necessarily have to be translated into a premium. Costs to administer a risk pool can hence be lower, especially if actuarial information may not be available (e.g., because the risk is new and statistical information is lacking).
4. Since ex ante premiums do not have to be paid, risk sharing creates fewer liquidity problems. It can be based on an agreement of the members to share in case the risk emerges.
5. Unlike insurance when the risk would not emerge, no premiums are paid to an insurance company that are (at least in the view of the operator) "lost." If the risk for which the risk-sharing agreement is concluded does not emerge, the members of the risk pooling scheme do not have to contribute.
6. This also points to the relative flexibility of a risk-pooling mechanism. If many accidents happened during a particular period, the risk pool can ask for additional ex post contributions from the member on an ad hoc basis.
7. A risk-pooling mechanism may have all these advantages if the number of members in the pool is relatively restricted; the comparative benefit (compared to insurance) mostly applies to highly technical (new) risks. When the members of the pool would be very large (e.g., all car drivers in a particular area), the administrative costs of running the pool would become huge and the comparative benefits over insurance would disappear.

Given the importance of some of these benefits, the following sections provide examples of RSAs, establish the requirements of successful RSAs, and evaluate if RSAs could be employed in the context of CCS risk management.

7.3.2.1 Examples There are many examples of risk-sharing agreements. Interestingly, they often emerge with hazardous (catastrophic) types of risk and in an environment where financial security has been mandated by government (e.g., as a result of international conventions). The typical sectors where pooling between operators is often used is the nuclear and marine pollution domain.

In the nuclear sector, different pooling constructions are used in different jurisdictions. Traditional insurance companies do not provide coverage for damage caused by a nuclear accident; therefore, their insurance policies exclude coverage for such damage. Insurance for nuclear damage is generally provided for by nuclear insurance pools.[111] The insurance of nuclear risks by nuclear insurance pools should be regarded as a bundling of resources at a national level. Such bundling allows the creation of a supply to the demand of insurance coverage[112] for damage resulting from nuclear incidents.

7.3.2.2 Nuclear Mutual Pools In both the United States and Europe, risk-sharing agreements have been created by the nuclear operators as a reaction to the pooling by insurers. In the United States, property insurance for nuclear installations is provided by the Nuclear Electric Insurance Limited (NEIL), a captive incorporated under the laws of Bermuda. This mutual insurance (also referred to as a captive) was created to provide nuclear operators with an alternative for the insurance offered by the American nuclear insurance pool, American Nuclear Insurers (ANI).

A similar evolution can be observed in Europe. Mutual insurance schemes have been created as a reaction of nuclear operators to nuclear insurance pools. In 1978, the European Mutual Association for Nuclear Insurance (EMANI) was created and aimed at reducing the insurance premiums of its members. EMANI offers cover for certain insurance risks relating not only to nuclear power stations but also to other nuclear facilities in several European countries.[113]

7.3.2.3 Protection and Indemnity Clubs Examples of risk-sharing agreements can also be found in the compensation for marine oil pollution, more particularly via Protection and Indemnity (P&I) Clubs.[114] A P&I Club is a nonprofit-making mutual insurance association established by shipowners and charterers to cover their third-party liabilities related to the use and operation of ships. Now thirteen separate and independent clubs cooperate as the International Group of P&I Clubs, accounting for approximately 90 percent of the world's oceangoing tonnage.[115] The P&I Clubs are hence mutuals that cover maritime, especially oil pollution, risks through mutual risk sharing.[116]

7.3.2.4 Pooling in the Price-Anderson Act It would draw too far to discuss those pooling arrangements in detail, but it may be interesting to briefly discuss the risk pooling in the Price-Anderson Act[117] and the absence of a similar nuclear risk pool for third-party liability in Europe. The reason to pay a bit more attention to those is that they may provide interesting lessons for the likely emergence of a risk-sharing agreement for CCS-related risks.

In the United States, nuclear liability is governed by the Price-Anderson Act of 1957,[118] which has been revised approximately every decade.[119] According to this act, each license issued should have and maintain financial guarantee to cover public liability claims.[120]

The existing capacity of the insurance market (provided by a monopolistic national pool, ANI) to provide is $300 million. If an accident creates damage in excess of that amount, a retrospective premium needs to be called on all American nuclear operators licensed by the Nuclear Regulatory Commission (NRC). This premium is payable in annual installments up to a certain maximum amount per incident per power plant and is determined according to the size and number of reactors each plant has.[121] In October 2008, the NRC adapted the amounts in the second tier to inflation and set the amount at $111.9 million, with a maximum annual retrospective premium of $17.5 million per reactor per year.[122] For the first tier, American nuclear insurers decided to make available from January 1, 2010, a maximum limit of $375 million for domestic nuclear third-party liability. That means an increase of 25 percent compared to the previous limit of $300 million, which was established in 2003.[123]

In July 2013, NRC adapted the amounts to inflation. As of September 10, 2013, the following amounts apply: $375 million in the first tier, and in the second (collective) tier, $121.255 million per reactor per accident, plus 5 percent for legal expenses. Thus, since that date, the total amount of compensation available in the United States is $13,616,046,000 [$375 million + (104 × ($121,255 million + $6,062,750))], with a maximum contribution of $18.963 million per reactor per calendar year.

In the event of a catastrophic accident that needs the collection of retrospective premiums over years, the victims do not have to wait until the operators pay all the premiums. The NRC guarantees those retrospective premiums; in other words, it advances the compensation in the second tier and later collects this from the operators.[124]

The premium under the second layer of the Price-Anderson Act is financed through a so-called retrospective premium scheme. Hence, premiums are paid only after a nuclear accident has materialized. This retrospective premium scheme was introduced in 1975. Initially the financial requirement was satisfied with the coverage from private insurance and a government indemnity agreement. This was because it was thought that the nuclear industry was not capable of bearing all the burdens at its infancy and the Price-Anderson Act intended to encourage the development of the nuclear industry. However, after years of development, it was believed that the industry should take on its own responsibilities.[125] This was achieved by phasing out federal indemnity and establishing a system of retrospective premiums paid by nuclear operators.

In a simple scheme, the compensation regime under the Price-Anderson Act since 2013 could be sketched as follows:

Operators' liability	$375 million
Retrospective premiums	$121,255 million per operator
Total compensation	$13.6 billion[126]

Under the retrospective premiums scheme, if an accident leads to damage in excess of $375 million, all qualified nuclear operators are obliged to pay the retrospective premiums up to $121.255 million. This arrangement is a form of a risk-sharing agreement. However, different from the usual understanding of risk-sharing agreements, this arrangement is a mandatory system imposed by statute. Rather than the voluntary pooling of operators, the retrospective premiums scheme is established according to the legislative requirement of the amended Price-Anderson Act. This arrangement ensures that all nuclear power plant operators participate in the system and provide strong capacity. Besides, this system is used only when the primary instrument, the insurance market, fails to cover the full damage. In other words, the retrospective premiums scheme provides an upper layer of compensation for victims.

7.3.2.5 Conditions for a Successful Risk-Sharing Agreement

The examples of risk sharing in the nuclear sector showed that the theoretical advantage of a risk-sharing agreement (strong incentives for mutual monitoring) materializes only if there is a minimum level of harmonization of safety regulations. If the operators can at least rely on the fact that all members will have to comply with minimum safety regulations, their additional task, mutual monitoring, will be relatively limited. Since such safety regulation is enforced in the United States (via the NRC), risk sharing is easier in the United States than in Europe, where there was large reluctance given the absence of mandatory safety requirements.

That may be an important lesson at the policy level: if the policymaker (like the EU Commission) would like to stimulate risk sharing by operators, it can play an important role in providing a facilitative strategy, that is, providing minimum safety standards, thus reducing the need for intensive mutual monitoring. In the absence of minimum safety standards, there will always be a risk of negative redistribution and adverse selection, since the risk-sharing agreement will then be most attractive for the bad risks, and as a result, the good risks will not be willing to join the pool.

In that respect, yet another interesting lesson may be drawn from the nuclear example (although the policy implication may not be that straightforward). In the Price-Anderson Act, substantial amounts are generated in a second layer (differently than in the nuclear liability regimes under the international conventions, where large subsidies to the nuclear industry are generated). However, the risk sharing is in fact mandated by the Price-Anderson Act and hence is not voluntary. The way it is arranged, however, has the advantage that funds do not have to be available upfront as a result of which one avoids immobilizing important financial capacity. Moreover, the NRC again prefinances the compensation in the second layer and then asks for contributions from all operators through annual retrospective premiums. Hence, funds should be made available only ex post and, moreover, guarantees are provided that the operators will be able to meet their obligations.

This could hence perhaps work for providing a second layer of compensation that would go above the individual capacity of one CCS operator (or his insurer). One could also imagine that especially for the phases of the CCS project cycle where insurance is more difficult to obtain (particularly for postclosure monitoring), a pooling by operators (eventually by a mandated retrospective premium scheme) could be indicated. However, one has to keep in mind that the system in the Price-Anderson Act limits the liability of operators to the extent that the damage is larger than the amount provided in the second layer. What we propose as far as compensation for CCS is concerned is obviously not that there would be any limit on the liability of the CCS operator (in amount), but rather that the pooling by operators in a second layer would be used to generate compensation and protect risk-averse operators. However, to the extent that the amount generated through such a risk-pooling system would not suffice to compensate the loss, operators still should remain liable.

Yet another lesson from the nuclear and marine pollution models is probably that the risk sharing among operators in those cases emerged only in the framework of a regime of mandatory financial security and in a regime (like the United States) where safety standards have been largely harmonized. Moreover, the example of the Price-Anderson Act shows that for the regulation of a retrospective pooling scheme, some prepayment is necessary via a regulatory agency that can claim back compensation from the individual operators.

7.3.2.6 A Risk-Sharing Agreement for CCS? Just looking at the problems that emerge with the insurance of CCS-related risk and the theory and practice with risk-sharing agreements, it is not difficult to argue that a risk-sharing agreement may be an attractive solution to cover CCS-related risks, particularly where traditional insurance markets may fail. A risk-sharing agreement does not necessarily require ex ante actuarial information, which would be necessary if premiums are required (as in an insurance model). Moreover, premiums may not necessarily have to be paid, so risk sharing does not create liquidity problems. An optimal risk-pooling mechanism would also create strong incentives for mutual monitoring and could contribute to the sharing of information on this new and technically complicated risk, thus raising the safety levels of all operators in the sector. In that sense, a risk-sharing agreement could even increase general safety. However, the experience with the nuclear sector shows that an important condition for a risk-sharing agreement to work is that risks are relatively comparable. Moreover, to an important extent, mandatory safety regulation can facilitate risk sharing, thus reducing the need for mutual monitoring. The voluntary creation of a risk-sharing agreement is, moreover, more likely in a regime of mandatory financial security.[127]

The CCS-related literature is, not surprisingly, enthusiastic about the potential of risk-sharing agreements. Many refer to the Price-Anderson Act,[128] and at the policy level, there is some interest in risk-sharing agreements.[129] In sum, if conditions are met, a risk-sharing agreement among operators may be a valuable instrument to cover CCS-related risks.

7.3.3 Guarantees and Deposits
Another means of providing privately sourced capital would be for a private third party to commit to delivering such capital when needed on behalf of

the operator.[130] Obviously many parties could provide a financial guarantee for operators. Such a financial guarantee could be provided by a mother company or another third party that presumably would have a stronger financial capacity than the operator. It could also be provided by a financial institution such as a bank. In particular cases, the guarantee could take the form of a letter of credit.

All of those guarantees have a simple theoretical basis: a presumably stronger third party, either a corporate entity or a financial institution, directly puts its balance sheet at risk by guaranteeing that it will cover the liabilities of the operator in case a particular risk materialized.

For policymakers or regulators, the advantage may be that a stronger guarantee is provided than when only the operator's assets were at stake, both because there are two entities and because the additional party is arguably better capitalized. However, the costs of those guarantees can be quite high. For guarantees by a related corporate entity, that should not necessarily be the case, but it may be different if guarantees have to be provided by a bank or other financial entity.[131]

An alternative is to require the operator to provide an ex ante deposit into a guarantee account. The idea would be that a direct guarantee is provided by depositing a sum of money corresponding to the potential damage in order to cover future losses that could result from the CCS-related activity. Comparable instruments are trust funds, stand-by trust funds, or escrow accounts. In all those cases (although there are differences), money is set aside for a specific purpose in the future.

The enthusiasm in the sector for these will logically be relatively small. The obvious reason is that they lead to an immediate immobilization of capital for losses that may not ever materialize, which leads to a large liquidity problem. If the deposit is large enough to cover all potential losses, it would indeed be substantial and could create liquidity problems for the operators involved. For that reason, it may not be very popular. In some cases, corporate guarantees can be used. The problem is that a guarantee would provide security to victims only if a third party (guarantor) would directly accept obligations toward the authorities or potential victims to compensate in case of harm.[132] Also, bank guarantees may not be an attractive alternative. The reason is that they are often considered relatively expensive, compared to insurance.

Guarantees and deposits have been discussed in the CCS-related literature but not with great enthusiasm.[133] The deposit is apparently used only as a guarantee in Germany where a draft CCS Act of 2009 mentioned that an operator must deposit with the competent authority the equivalent of 3 percent of the emissions trading allowances that the storage saves each year.[134] However, this may provide security only for the climate-related risks (the risk of future emissions), not the other types of damage that could result from CCS. In the EU Guidance Document 4, the corporate guarantee and escrow accounts are not considered reliable financial securities since they may all be subject to claims of creditors. The document is more enthusiastic concerning a deposit, which has the advantage that if the money is deposited to the competent authority, it is no longer subject to claims of creditors of the operator; the disadvantage is obviously the high cost of a deposit.[135] Another disadvantage of a deposit is that substantial amounts of capital would have to be immobilized for a potentially longer period of time. That capital cannot be used for other valuable societal activities, whereas the likelihood that it should ever be used to satisfy claims of victims may be small.

The US EPA Guidance Document discusses guarantees and deposits under different headings. Corporate guarantees can be used within the framework of self-insurance. A corporate guarantee can be used if it is owned by a parent corporation to benefit a corporate subsidiary in which it owns at least 50 percent of the subsidiary's voting stock and has been in business for at least five years.[136] With corporate guarantees, the relative financial risk to the government is considered to be high.[137] Trust funds are considered useful for activities that are relatively certain in terms of occurrence and costs, such as postinjection site care and site closure demonstration, but not for activities of uncertain frequency and costs.[138] The relative financial risk to the government may be low, but the costs of a trust fund can be high.[139]

In conclusion, guarantees and deposits could be functional forms of private compensation, but those funds must somehow be made secure and isolated from other claims. As such and given the potential scale of CCS accidents, the provision of guarantees and deposits could be more capital intense for the operators than other private means of providing compensation. Nevertheless, in certain cases, such guarantees could be effective.

7.3.4 Financial Provisions: Letters of Credit and Bonding Instruments

There are alternatives to both full ex ante payments or deposits and to insurance. These alternatives can be clustered as letters of credit, performance bonds, surety bonds, and catastrophe (cat) bonds; all of these alternative forms provide alternative means of ex ante financial provisions.

Letters of credit are representations provided by banks or other financial intermediaries to provide capital for a specific party for a specific purpose, should that party be short of capital funds when those funds might be called for in a future time period.[140] The operator could thus pay a small annual carrying cost, perhaps 1 to 3 percent, against the provisioned funds, and avoid paying the capital sums required by either escrow accounts or trust funds.[141] Letters of credit have been brought up in policy discussions in liability planning for CCS projects in Texas[142] and by the EPA.[143]

Bonding is the instrument form that is most often mentioned for covering CCS-related risks based on the capital markets.[144] The principle is that bonds are issued whereby the interest rate on the bond would reflect the accident rate. For example, investors have the opportunity to buy a bond creating a warrant in favor of the operator of their choice. If, during the period of the bond (say, one year), no accident happened, the amount of the guarantee provided by the bond would be paid with interest. If the risk materialized, the bond posted would be used to cover the damage.[145]

Performance bonds, or assurance bonds, fall between letters of credit and insurance, in that the bank or financial intermediary promises to deliver capital funds if the operator is found to have failed to properly operate the carbon storage facility.[146] Dana and Wiseman defined it in this way:

Assurance bonds are one kind of market mechanism whereby the operator of a facility is required to post upfront funds or other proof of committed financial resources, which the bondholder can return to the operator once it provides assurance that it closed the facility in a safe way. The incentive to recover the bond motivates, at least in part, responsible conduct.[147]

Dana and Wiseman observed that the use of performance bonds could enable greater alignment of the polluter-pays principle with the need to ensure that funds are available at the time the injuries are detected and, prior to that, ensure that the operators undertake efficient operational decisions and then comply with requirements to address the injuries in a timely manner so that the performance bonds could be released back to them, preventing both the need for litigation and delay in remediation.[148] They also discussed the need to ensure that bonding levels were set sufficiently high so that operators do not find bond forfeiture financially attractive.[149] It is worth noting that setting the bonding level at such sufficiently high levels would be tantamount to affecting the results of a strict liability rule prior to the actual event of litigation. Another concern they raised, and perhaps a hint as to why they appear in their research to prefer private insurance, was that public regulators had appeared to release performance bonds based on less-than-rigorous inspections of the underlying details of performance.[150]

A concern for the use of performance bonds would be the appropriate time frame for operators to be bonded before the operator's promise to follow regulatory guidance could be deemed as fulfilled and thus enable the return of funds.[151] Additional concerns are that performance bonds, by their very denomination, are limited in the amount of capital available in the case of damages. This concern is exacerbated when the operator has gone insolvent.[152]

A possible effect of performance bonds is to shift the regulatory burden of proof from the regulator who otherwise needs to prove a regulatory breach to the operator who would need to provide evidence that no such breach has occurred—the costs of evidence are shifted to the party with the best access to it and simultaneously enables cost savings to the taxpayers supporting the regulatory infrastructure.[153] Additionally, performance bonds, and similar instruments, can facilitate public regulatory goals. Dana and Wiseman found that in the case of shale fracturing liability governance, implementation of performance bonds and mandatory insurance requirements could be complementary to the governance provided by public regulations.[154] They found that such requirements could be especially effective in reducing the challenges posed by operator insolvency and by clouded causation problems.[155] Thus, bonding can help ensure a means of financial provisions while simultaneously improving on the efficiency of regulatory oversight.

The surety bond has been discussed in the US EPA Financial Responsibility Document as a guarantee related to either a specific performance or a financial guarantee.[156] However, the document also recognizes that a bond may not be useful to cover long-term liabilities and would therefore not be available for postinjection site care and site closure.[157]

Other capital market solutions, often brought together under the heading of alternative risk transfer (ART), can be used to cover CCS-related risks. Forms of so-called ART or securitization have been used to transfer environmental risks.[158] The idea is to buy catastrophe bonds at the stock exchange.[159]

Initially these alternative mechanisms to hedge environmental risks were mostly developed in the United States and were not widely used in Europe.[160] Although it is still held that the capital market will never totally replace traditional insurance products,[161] capital markets, especially alternative risk transfer markets, are increasingly used for environmental risks.[162]

A problem with any use of capital markets, including bonding, is that most of the so-called catastrophe bonds are always used for catastrophes that are essentially sudden events. These bonds usually have a short period of cover within which either the catastrophe happens, and hence the bond is lost, or nothing happens, and the profit on the bond is made. These bonds are thus not traditionally used for long-tail latent risks. That is why the CCS-related literature mentions that bonding is problematic given the potentially long-tail character of the CCS-related liability risks. For that reason, catastrophe bonds may not be appropriate instruments to cover CCS-related risks.[163]

Various forms of bonding could at best be used during the exploration, development, and operational injection periods, but not for liability during monitoring after closure. A model for bonding requirements has been suggested from the requirements set out in the US Surface Mining Control and Reclamation Act (SMCRA).[164] The SMCRA requires that bonding cover the costs of cleanup and remediation after setup of the mining activity, and it requires that bonding to be in place before mining operations begin.[165] CCS regulations could require that an operator file a bond prior to obtaining an operational permit; this bond would then provide sufficient capital to cover both surface and subsurface concerns.[166] Additional new activities could be matched with additional provisions of performance bonds.[167]

Within the United States, calls for financial provisions have often either broadly included all of the options noted or provided for smaller clusters from the same group. Kansas's Carbon Dioxide Reduction Act requires that permit holders for carbon storage facilities need to be able to demonstrate to authorities that the permit holder has the financial ability to address the costs and potential damages associated with the long-term closure of the storage facility.[168] Montana requires operators to provide "furnishing of reasonable bond or other surety" prior to receipt of a permit.[169] It also requires similar proof of "bond or other surety furnished" prior to the transfer of liability at closure;[170] a similar requirement is for "adequate bond or other surety" prior to receipt of the certificate of completion.[171] At the time of transfer of liability to the state, "any bonds or other surety posted by the geologic storage operator must be released."[172] Wyoming's regulations require proof of bonding or financial assurance,[173] public liability insurance,[174] and payment of a "per ton injection fee or a closure fee" into a government-maintained fund.[175] A state working group clarified the details of the "bonding or financial assurance" to include traditional performance bonds.[176] Thus, states have generally found letters of credit and various forms of bonding to be effective and desirable in their planning for future CSS contingencies.

7.3.5 Capital Accounts: Trust Funds and Escrow Accounts

Trust funds have been brought up in the United States as a means to afford compensation.[177] A trust fund transfers funds from the operator to a third-party trustee, and those funds are generally set to meet the expected future needs.[178] Thus, the trust fund is a secure source of funding, as it is established up front prior to the incident of harm and the capital funds are held by a third party. However, this method does displace a substantial amount of capital and is generally seen as inefficient.[179] Due to the loss of capital use and the loss of growth on those funds,[180] this method is generally reserved for short- or near-term reserves.[181]

An escrow account is similar in concept to a trust fund, except that the funds are paid in over a period of time instead of as an upfront lump sum.[182] The escrow account is less burdensome on the operator of a carbon storage facility as it could pay into the escrow account as it earned operational revenues; should damages occur earlier in life than forecast, the escrow account might well be short of funding.[183] Therefore, escrow accounts are risky regarding problems of operator insolvency and strategic liability avoidance.

7.3.6 Conclusion

This overview of alternative compensation mechanisms shows that there are several financial instruments, and they all have their own strengths and weaknesses.[184] Rather than expressing a preference for the use of one exclusive instrument, it may be a more robust policy to look for the optimal combination of instruments that could be employed as means of financial security by particular operators for specific phases within the project life cycle of their CCS projects.[185]

Aldrich, Koerner, and Sloan advocated the implementation of a three-tiered scheme that combines private insurance, industry pooling, and governmental indemnification.[186] They say that the EPA has considered such a plan, based on the Price-Anderson Nuclear Industries Act of 1957.[187] They find the approach attractive because of the way that liability is shared by the individual operator, the CCS industry, and the public.[188]

Similarly, as far as the phase of operation and injection is concerned, self-insurance (or eventually even captives) may be effective for larger operators. For smaller operators, self-insurance could be used for a first layer (as a kind of retention), depending on their balance sheet. In both cases, in a regulatory environment where financial security would be demanded, regulatory authorities need to carefully verify the adequacy of the self-insurance that has been offered; moreover, it should be equally guaranteed that the self-insurance remains available as long as the risks can materialize.

The amount of self-insurance (or retention) in this phase of operation and injection insurance may play a role, provided some of the difficulties we have identified could be remedied.[189] Eventually guarantees could be used as well, although important conditions would have to be met to verify the viability of the guarantee. The same would be true for bonding. There may not be much enthusiasm on the capital market for these types of bonds.

As far as the second phase of postclosure monitoring is concerned, the alternatives become more limited. Given the potential long-tail character of postclosure monitoring, insurance may not be an option. Self-insurance may play a role, but the difficulty is that the monitoring requirement may extend over many years. This requires a regulatory agency (under the assumption there would be a mandatory financial security) to monitor the adequacy of the provided self-insurance on a regular basis. However, the option of self-insurance should not be ruled out since it remains a low-cost option for operators. Provided specific conditions (as far as the reliability of the offered balance sheet protection is concerned) are met, this may be an adequate instrument.

Bonding may not be viable for postclosure monitoring given the long-tail character (whereas bonds are typically suited for short-term risks). The most appropriate instrument to deal with the financial security for postclosure monitoring is probably a risk-sharing agreement among operators. Provided specific regulatory conditions are met (minimum safety standards) and a reasonable comparability of the risk, the advantage would be that this could be a low-cost alternative for operators. It would also provide incentives for mutual monitoring and thus contribute to improved safety and investments in technological innovation.

As far as the phase of long-term stewardship is concerned, we have ruled out liability for this long-tail risk as a result of which for that phase, alternative compensation mechanisms would not be needed from the operator.[190] That may, however, require a role for the government in facilitating compensation. That discussion is covered in the next chapter.

8   Compensation Using Public Resources

The previous two chapters explored the potential for private market forces to provide compensation to victims from CCS-related accidents. The potential for insurance-based compensation and alternative modes of financial provisions were reviewed. However, historical experience readily reveals that governments have, in certain occasions, provided compensation for large-scale accidents, such as earthquakes and certain nuclear accidents. This chapter expands on the previous two chapters by exploring the efficiency and expediency of government compensation measures.

We have stressed in previous chapters that government might have an important role in facilitating compensation for CCS-related risks. However, one has to be careful; such "facilitation" does not necessarily mean that government would need to be the source of the compensation funds. That could, as the literature has rightly mentioned, amount to an undue subsidization of CCS operators and create a moral hazard risk that could enable them to assume higher levels of risk.[1] That is why we argued in previous chapters that a liability regime is important, particularly as an instrument to prevent moral hazard.[2]

Government nevertheless can play an important role in facilitating financial security and, depending on the various phases in the CCS project life cycle, this facilitative role could take different forms. An important role of government is obviously to organize facilitative strategies to stimulate insurability.[3] In this respect, we argued in previous chapters[4] that an important task of government would be to create an appropriate regulatory framework providing minimum safety standards for CCS. Such a framework could stimulate risk differentiation and control of moral hazard by insurers[5] and create risk-sharing agreements among operators.[6]

As a further instance, depending on the various phases of the CCS project life cycle and especially as far as long-term stewardship is concerned, there may be an argument for government intervention not only in a facilitative role but in taking over liability from operators. This specifically concerns the phase of long-term stewardship. In that respect, the intervention of government to provide relief would certainly not be limited to CCS. Governments intervene, sometimes on the basis of international conventions[7] and in other cases on the basis of national law,[8] in a more or less generous manner. Also, in the literature concerning CCS-related risks, government compensation programs, such as the National Flood Insurance Program in the United States, are called on as one of the potential solutions to financing CCS-related risks.[9] Hence, in addressing the role of government in facilitating compensation, some comparison with a similar role of government in providing relief for victims of catastrophes can be instructive.

We first discuss a simple but effective facilitative strategy: the provision of compulsory financial security via regulation. Next, we ask to what extent direct compensation by government should be provided for particular phases in the CCS project life cycle.

Then the question is whether the creation of a compensation fund, financed by operators, would be a viable alternative to some of the compensation mechanisms discussed previously in the chapter; this is to explore the potential of privately sourced capital administered by public authorities. Could the public authorities provide some advantage over market-based means of capital administration? In examining this question, we review the potential for a public authority to oversee a compensation fund, explore the potential for public authorities to support private insurance markets by serving as a reinsurer of last resort, and examine the option to establish a public utility to manage private operators or private capital.

8.1 Compulsory Financial Guarantees?

In the law and economics research literature, several criteria have been advanced to analyze where mandatory financial security may be indicated. The most important reason for introducing compulsory insurance is insolvency. Insolvency may pose a problem of underdeterrence without additional measures.

If the expected damage largely exceeds the injurer's assets, the injurer will have incentives to purchase insurance only up to the amount of his own assets. He is therefore exposed only to the risk of losing his own assets in a liability suit. The judgment-proof problem may therefore lead to underinsurance, and thus to underdeterrence. Jost has rightly pointed out that in these circumstances of insolvency, compulsory insurance might provide an optimal outcome.[10] By introducing a duty to purchase insurance coverage for the amount of the expected loss, better results will be obtained than with insolvency, whereby the magnitude of the loss exceeds the injurer's assets.[11] In the latter case, the injurer will consider the risk only as one where he could at most lose his own assets and will set his standard of care accordingly. When he is under a duty to insure and exposed to full liability, the insurer will obviously have incentives to control his behavior. Through the traditional instruments for the control of moral hazard, the insurer can make sure that the injurer will take the necessary care to avoid an accident with the real magnitude of the loss. Thus, Jost and Skogh argue that compulsory insurance can, provided that the moral hazard problem can be remedied adequately, yield better results than under the judgment-proof scenario. If that the moral hazard problem can be adequately controlled, compulsory insurance could also provide adequate incentives to operators to prevent environmental harm.[12]

Indeed, this economic argument shows that insolvency may cause injurers to externalize harm: they may be engaged in activities that might cause harm that largely exceeds their assets. Without financial provisions, these costs would be transferred to society and would hence be externalized instead of internalized. Such internalization can be achieved if the insurer is able to control the behavior of the insured. As we have shown when discussing how risk differentiation can be applied, the insurer could set appropriate policy conditions and an adequate premium. This shows that if the moral hazard problem can be dealt with adequately, insurance leads to an even higher deterrence than a situation without liability insurance and with insolvency.

The literature has formulated a few conditions and warnings when introducing compulsory financial security.[13] First, the moral hazard problem should be controlled. If it cannot be controlled, the regulator should consider a prohibition of liability insurance.[14]

Second, if one were to introduce only compulsory insurance (as compared to mandatory financial security, which is obviously broader), this should be done only when no restrictions on the insurance market exist. In the case of high concentration, premiums would be too high, and this could equally reduce the incentives of insurers to control the moral hazard risk.

Third, from a policy perspective, it may not be wise to limit the duty to provide financial security to insurance. If the policymaker were to introduce compulsory insurance, it would become totally dependent on insurance to fulfill the duty to insure. This could create an undesirable situation whereby insurers would become de facto licensors of the industry, which could be problematic from a policy perspective.[15] Insurers under compulsory insurance de facto become surrogate regulators who at least would have the advantage that it provides them strong arguments for effective risk management.[16] That may be a strong argument to a flexible approach: not to limit the provision of mandatory security necessarily to insurance, but to allow the market itself to suggest a wide variety of financial and insurance instruments as long as they can guarantee compensation when the accident happens.

To a large extent, some of the arguments in favor of mandatory financial security may apply to CCS-related risks as well. Especially where smaller and medium-sized operators may also be involved in CCS, they could create a risk of major damage with a magnitude that could exceed their personal assets. In that case, an insolvency risk would emerge and, hence, a danger of externalization of the damage. Also in the CCS-related literature, some arguments have been made in favor of compulsory financial guarantees, especially by Trabucchi and Patton.[17] However, the warnings we have formulated may apply to CCS as well. As mentioned, it may be dangerous to mandate financial security when it is not certain that the market can deliver the required financial security. Especially when it concerns new risks such as CCS, this calls for caution. One way of dealing with this cautious approach is to allow sufficient flexibility as far as the form of financial security is concerned and not to limit this necessarily to insurance. An important condition would be that only the regulator would accurately verify whether the form and amount of the financial security offered by the operator would be adequate to cover potential damage emerging from the CCS operation.

When referring to the possibility of introducing mandatory financial security for CCS, we once more differentiate among the phases of the CCS life cycle. Recall that insurance may be available during the first phase of CO_2 injection but may already become more problematic during the second phase of postclosure monitoring and may not be available at all during the final phase of long-term stewardship. However, the mere fact that insurance is difficult to obtain (e.g., during the phase of postclosure monitoring) should not be an argument against imposing compulsory financial security. The key issue is that compulsory financial security should not automatically be equated with insurance. As we have discussed in detail, operators should be provided freedom to make use of alternative compensation mechanisms. It does indeed not make sense to force, say, large and well-capitalized operators to take out insurance that would not substantially improve their financial situation and lead only to the costs of paying premiums. In those situations, alternatives like controlled self-regulation[18] could be preferred.

One way of approaching this issue is shown in the European Union, where a guidance document has been issued describing the possible financial security and financial mechanisms that could be used to cover CCS-related risks.[19] The advantage of such a guidance note is that it provides information to licensing authorities on the type of financial security that could be accepted when offered by CCS operators. Such an approach has the advantage of allowing sufficient flexibility and avoiding unnecessary costs (e.g., forcing major energy companies that would be engaged in CCS to transfer risks to an insurance company). The model followed in the EU seems to allow for sufficient flexibility by mandating financial security while leaving flexibility to local regulators to determine the amount and form of financial security, also taking into account the specific risks posed by the site and the specific features of the operator. That may be a model for other jurisdictions as well.[20]

A similar model is followed in the United States as well: owners or operators of geological sequestration wells must demonstrate and maintain financial responsibility for performing corrective actions on wells in the area of review, injection well plugging, postinjection site care and site closure, and emergency and remedial response. The way in which operators and owners can demonstrate financial responsibility has been worked out in the US EPA Guidance Document on Underground Injection Control (UIC) financial responsibility guidance.[21] The approach followed in the United States is hence similar to that followed in Europe by providing guidance on the types of instruments that could be considered a sufficient demonstration of financial responsibility to be reviewed by the UIC director.[22] Also the US EPA Guidance Document has a similar type of flexibility as far as meeting the specified financial test criteria is concerned. The document does list qualifying instruments, but this list is not considered exhaustive or absolute, so owners or operators may also use other financial instruments if the UIC program director finds them satisfactory.[23]

8.2   Direct Compensation by Government?

A second way in which government could play a role in facilitating compensation for CCS-related damage is to provide outright compensation to the victims. In that case, the role of government would not (as when it mandates financial security) be merely facilitative; government would intervene directly to compensate based on public revenues. These types of government-based compensation models are well known from the experiences with the compensation by government of victims of natural disasters.[24] Usually this involves ex post relief after the disaster through lump-sum payments to victims financed by the general taxpayers.[25]

There are sound arguments in favor of direct government supports. Although economic criticisms can be formulated and have been in the literature on this type of government charity,[26] there are some positive aspects as well. A positive aspect of government intervention is that the prospects of large-scale payments in the aftermath of a disaster, like a flooding event, might encourage the government to take ex ante cost-benefit-justified precautions in benefit of the general public, such as building dikes and dams.[27] However, whereas that argument may be a valid justification for ex post relief after natural disasters, it is unclear how this could justify government relief in case of a man-made risk like CCS.

A related argument in favor of government intervention would be that government has the capacity to diversify the risks over the entire population and spread past losses to future generations, thus creating a form of cross-time diversification that the market could not achieve. The argument can be made that the government is in a better position to adapt to risks than individuals.[28] Hence, in some cases, the government may be in the best position to prevent disasters. Its intervention from this perspective would provide incentives to politicians to invest in preventive measures. In the broader domain of adaptation to climate change, one could imagine again the case of a government building dikes against the risk of flooding or rising sea levels resulting from climate change or effective zoning and planning decisions to avoid locating residences in flood-prone areas.[29] But again, while this argument may justify government expenditure in adapting to climate change, it would not justify compensating victims of CCS-related risks, which should remain the primary obligation of the operator.

Hence, there are substantial arguments against direct governmental intervention. We already indicated that the theoretical arguments presented in the literature in favor of government compensation in cases of natural disasters do not necessarily apply to CCS-related risks. Moreover, the literature points to the high costs and disadvantages of this type of government intervention as well.

A first major disadvantage of the lump-sum payment under government relief is that no incentives are provided to potential victims to take effective preventive measures. Since the payments under government relief usually do not relate to risk, they offer no incentives for taking preventive measures. Whether this moral hazard on the side of victims is realistic depends strongly on the nature of the damage. In case of CCS-related damage, the moral hazard on the side of the victim may not be that serious a problem since there may be little that the potential victims could do in terms of preventing or mitigating the damage.

A second problem is that victims may be counting on government compensation, which may create an incentive not to purchase insurance.[30] The problem of government-provided compensation is indeed that it may dilute incentives to purchase insurance since victims could simply free-ride on the state.[31] In the words of Gollier, "Solidarity kills market insurance."[32] Coate has identified the lack of insurance resulting from the generosity of the government.[33] This problem has been referred to as the charity hazard.[34] Ex post government compensation is therefore generally seen as problematic by law and economics scholars, which is nicely expressed by the title of a contribution by Epstein: he qualifies this relief as "catastrophic responses to catastrophic risks."[35]

A third problem would especially play a role in case the government compensation does not concern a natural disaster (e.g., flooding, on which most of the literature is focused) but a man-made disaster. Damage resulting from CCS clearly has to be qualified as a technological or man-made disaster. In that case, the primary instrument to use should be liability law. An exposure of the potential operator to liability is, as we have repeatedly argued,[36] necessary in order to expose the CCS operator to the social costs of its activity. From that perspective, compensation by government would lead to an externalization of the risk to society (particularly taxpayers) and would thus create a moral hazard on the side of operators.[37]

Thus, the question looms whether CCS accident victims should be provided direct public funds. From that perspective, one can understand that much of the CCS-related literature is strongly against any type of financial intervention by government, arguing that this would amount to an undesirable subsidy of CCS. A transfer of risk to the public (i.e., the taxpayers) would proffer a competitive disadvantage to environmentally superior operations.[38] Public financing is therefore considered to distort or eliminate the impact of market forces.[39] A transfer of liability to the state could amount to an undesirable subsidy,[40] and therefore most of the CCS-related literature has argued that limiting the liability of CCS operators (and thus transferring the risk to the state) would be undesirable.[41] In that respect, it could also be argued that government-provided compensation would violate the principles of fair and efficient compensation already discussed,[42] arguing that the duty to contribute financially should in principle be related to the amount in which a specific activity or entrepreneur contributed to the risk.

Adelman and Duncan, among others, take a different approach: they argue in favor of a transfer of the long-term liability to the government.[43] This follows logically from their position that liability rules have only a limited role to play as far as CCS is concerned and can certainly not play a role as far as the long-term liability is concerned since this liability would not deter CCS operators.[44]

The question of where government should financially intervene is unavoidably linked to the question of whether the liability of CCS operators should be limited. That again depends on the question of what precise goals have to be fulfilled with a liability regime. We have argued that there is no reason to put a financial limit (a so-called cap) on operators' liability.[45] However, it has equally been argued that there are reasons to limit the liability of CCS operators in time. The simple economic reason is, as especially strongly argued by Adelman and Duncan,[46] that long-tail liability will be discounted to today's present value and can therefore have no deterrent effect whatsoever. From that perspective follows the argument that a transfer of liability for the long-term stewardship from the operator to the state can be defended on economic grounds.[47]

This makes also sense, even taking into account the critical economic literature concerning the intervention of government in cases of disaster relief. In this respect, a distinction has to be made (as repeatedly held) among the different phases in the CCS project life cycle. Government intervention would take place only for damage occurring during the phase of long-term stewardship. During the previous phases (operations/injection and postclosure monitoring) operators would still be fully liable and hence have sufficient incentives to internalize social costs. Therefore, some of the arguments that are mentioned in the literature on disaster relief do not apply to government intervention for long-term stewardship in the case of CCS. The argument that it would create disincentives to victims to take preventive measures or to purchase insurance does not hold for the simple reason that also those victims may belong to future generations and therefore cannot automatically count on government relief. Moreover, the negative incentive effect for operators also does not hold since liability for those long-tail risks does not provide any incentives at all. In that respect, government compensation for the last phase of long-term stewardship is also in line with the principles of fair and efficient compensation since in this case, liability cannot be considered a useful instrument in providing optimal incentives for prevention to stakeholders.

The positive arguments in favor of transferring liability for long-term stewardship to the state seem convincing. It can make optimal use of the capacity of government to distribute risks over time and even over future generations.[48] Moreover, operators would, even for the long-term stewardship, not completely be off the hook since an advance payment could still be required for the costs of monitoring during the long-term stewardship. However, that means that once these costs are paid, operator liability ends and shifts to the government. Monitoring costs are of course substantially lower than the total potential damage, which keeps the liability exposure of operators within reasonable time limits.

In addition to the argument that liability for long-tail risk would not generate positive incentive effects, one can also point once more to the positive externalities that can be generated by CCS technology. As many have stressed, the most important barrier to successfully developing CCS is precisely the potential liability for long-term stewardship.[49] Hence, as long as that problem would not be adequately regulated, CCS projects would not be developed and the positive externalities (more particularly the benefits for mitigating climate change) could not accrue to society either.

Finally, society has (at the international level but also in legislation of national legal systems) often provided subsidies for starting industries. For example, the goal of the nuclear liability conventions as they emerged in the 1960s was clearly to protect nuclear operators from potentially broad liability.[50] In that case of these conventions, there were, moreover, financial limits on the liability of power plant operators, which is not at all what has been proposed for CCS. Government intervention would apply only to the phase of long-term stewardship, which is much more limited than in the case of the nuclear liability conventions. Moreover, that transfer occurs for the long-tail liability risk, which could not create any positive incentives for prevention at all, whereas in the case of the nuclear liability conventions (still in force today), there are overall limits on liability of operators that are absent in the case of CCS. Also, in the case of nuclear energy, the argument that positive externalities justify a limit on liability is much less convincing than in the case of CCS given the much larger risks generated through the nuclear risk.

8.3   A Compensation Fund for CCS-Related Damage?

A third possible intervention by government would be setting up a compensation fund. The role government could play in this respect could differ depending on the financing.[51] A first option, our theoretical preference, would be for the fund to be financed by operators, with the government playing merely a facilitative role, as, say, the fund organizer or manager. Or taxpayers' money could be used to finance the fund; in that case, the fund would amount to a subsidy and would look very much like direct compensation by government.

Given the potential scope and size of the eventual necessary compensation funds and, separately, based on the estimate of the likely necessity of providing such a fund, a variety of compensation schemes have been discussed in the literature. We thus examine the theoretical discussion and review and survey that literature. The survey begins with those who attempted to scope the nature and purpose, and thus the scope, of such compensation plans. Then the survey turns to the literature's discussion of the pros and cons of such plans. Thereafter, support is provided for a limited form of public provision of such a fund, with a preference for a majority of the funding for such compensation schemes to be derived from private contributions.[52]

Havercroft and Macrory highlighted the concerns of CCS planning that the sites might pose hazards long after the operators have vanished over time. The storage sites would be revenue neutral or negative and thus be challenged to respond to their own future needs without specific anticipatory planning.[53]

Flatt raised a variety of concerns with the ability of policymakers to address De Figueiredo–type notions of risk assessment,[54] particularly with regard to providing assessments storage site by storage site in connection with industry pool contributions for postclosure activities, including potential compensation for resultant damages.[55] Flatt challenges whether the very long-term periods required for successful sequestration might elude administrative capabilities; even the science is potentially challenged by a lack of precedence of similar engineering challenges.[56] Given (hopefully) standard and consistent regulations for CCS storage sites within a common jurisdiction, Flatt advocated for a volumetric basis to impose whatever fees might be charged in conjunction with CO_2 storage.[57]

Haan-Kamminga wrote that industry compensation funds should be used when activities are highly specialized and when that same activity lacks market access to sufficient private insurance, primarily because such a fund provides a single viable financial entity for long-term liability needs.[58] She stated that three concerns must be addressed to ensure the successful operation of such funds: (1) ex ante clarification of which events are compensable, (2) elaboration of the means and methods to finance the fund, and (3) determination of the methods and metrics to provide victims with their awarded damages.[59]

Barton, Jordan, and Severinsen found three methods to provide capital to the funding pools.[60] First, funds could be created based on a volumetric basis or a risk assessment basis against the operator of the injection facility;[61] these methods could be complemented with insurance pooling or compensation trust fund models.[62] Second, a public actor could charge rent on the use of the underground pore space.[63] Third, in a purely public-funded option, the public actor could rely on taxpayers to cover any necessary costs.[64] In New Zealand's case, they argued for a strong limitation to the amount of postclosure capital requirements to be drawn from operators because they held that the basic provision of CCS storage was in essence an act of government seeking private contracting for the provision of a public good.[65]

Flatt stated that a theoretical question would be how best to determine the specific contributions from each operator. Should it be based, for example, on volumes injected, qualities of volumes injected, adjacency to population clusters, or the quality and character of the reservoir into which the CO_2 was injected?[66] While Flatt reviewed a variety of options, he ultimately distanced himself from De Figueiredo's preference of a nuanced risk model for a more simple volumetric approach.[67]

Jacobs has also called for a volumetric-based fee to be imposed on operators to provide for a fund that could cover the costs that arise after a postclosure transfer of assets or an operator's insolvency, or to cover other similar liquidity problems.[68] Her recommendations are in alignment with the literature on oil and gas fracturing technology.[69] Jacobs primarily called for a federal funds plan but remained open to similar funding arrangements at the state level.[70]

Many also mention the regime of the Price-Anderson Act[71] as an interesting model for compensation of CCS-related damage.[72]

Before discussing the possibilities of a CCS fund, also in light of the literature, we will look at the pros and cons of a compensation fund, especially when compared to insurance and at some conditions that should be fulfilled for the efficient functioning of a compensation fund.

Flatt advocated the use of a compensation fund because it would be beneficial to injured parties by reducing the complexity of receiving compensation.[73] Flatt stated as well that injured parties would not need to assert causation to a specific CCS site or operator but instead need only demonstrate that their injuries were caused from CCS activities in general.[74] Furthermore, the funds would be available for dispensation and would not need to be collected, as would need to be in a routine civil lawsuit, for example.[75]

Barton, Jordan, and Severinsen have argued that it must be recognized "that companies do not last forever." Legal measures that fail to consider this likelihood, given the long time frames involved, are likely to "leave an injured party without any recourse."[76]

However, applying the principles of law and economics, as we have suggested, it can be shown that there are substantial doubts, if both a compensation fund and insurance options are present, that a compensation fund could provide better protection against insolvency than the private insurance markets.[77] One can assume that an insurer is better able to differentiate risks since the insurer specializes in risk differentiation and risk spreading. Insurers therefore have techniques for determining in what way insured parties contribute to the risk. This assumes that insurance markets are competitive. In the absence of competition on insurance markets, either the supply of insurance coverage could be too limited or premiums could be excessively high, which could justify a preference for a compensation fund.[78] But if insurance markets are competitive, insurers can be assumed to be better able to deal with classic insurance problems such as moral hazard and adverse selection than the administrators of a compensation fund. One cannot see as a matter of principle why a government agency running a compensation fund would have better information on risks than an insurer.

This might, however, be different if highly technical risks are involved where operators of certain facilities are in a much better position than the insurance company to monitor each other. This point has been made, for instance, concerning compensation for nuclear damage. One could argue that a risk-sharing agreement between nuclear plant operators could lead to optimal monitoring among the operators since they would possess much better information on prevention and good and bad risks than an insurance company would.[79] In maritime insurance too, the Protection and Indemnity Clubs already discussed, which are based on mutual risk sharing by tanker owners, play a crucial role.[80]

With respect to these highly specialized matters, one could therefore argue that the operators themselves might in some cases be better equipped than an insurance company to control moral hazard since they are better able to process information on the risks. However, the examples given show that these risk-sharing agreements do not involve the use of a government-run compensation fund.

A structural fund solution for disasters has also been criticized in the economics literature. For example, Gron and Sykes have argued that a structural fund may provide the wrong signal to the market. If market participants are aware that when a catastrophe occurs, the financial consequences will be covered through government intervention, they will have little incentive to develop financial solutions themselves.[81] A structural fund solution may also lead to the charity hazard, which occurs with any compensation by government: it can dilute the incentives for victims to self-insure by or take preventive measures in the framework of an effective risk management.[82]

The possibility of a compensation fund for CCS would be financed by operators, and the contribution would be related to the tons of CO_2 injected.[83] The fund created under the American Oil Pollution Act (OPA), the Oil Spill Liability Trust Fund (OSLTF),[84] is a potential model for CCS.[85] However, others are much more critical toward a fund solution, arguing that if a fixed fee per ton were charged, as some of the literature advances,[86] the issue of moral hazard would again arise, since no differentiation would take place with respect to site selection and operation.[87]

That situation reveals an important problem with a funding mechanism: it is efficient only when it can apply appropriate risk differentiation. In the absence of such a differentiation, perverse incentives would be provided since good risks would subsidize bad ones, and hence the funding mechanism would not contribute to, for example, proper site selection and operation. It would also create negative redistribution since bad risks would benefit from good risks paying the same contribution.

In sum, what could be the potential role of a compensation fund in the different phases of the CCS project life cycle? It could theoretically be set up in the second phase of postclosure monitoring.[88] It should then (in order to respect economic principles of risk differentiation) not be financed through a flat fee on the amount of CO_2 emissions injected, but should take into account risk-related financing criteria, such as site selection and management. In that respect, the question arises whether government would be better suited than the operators themselves to apply such a risk differentiation. Experiences with government-run funds show that usually they are financed by means of a flat tax without sufficient risk differentiation. That would have negative incentives for prevention (creating moral hazard) and would lead to negative redistribution. There should be a role for government in facilitating (managing, but certainly not financing) such a fund in this phase only if it could be argued that the manager of a government-run fund would be better able than operators to adequately differentiate risks. Normally one would assume that the technical information on optimal preventive measures is better available with industry than with government. That would be a strong argument to favor a risk-sharing agreement[89] instead of a government-run compensation fund. Only when for particular reasons[90] appropriate risk differentiation through a risk-sharing agreement between operators would not be possible would there be a role for a government compensation fund, provided that contributions can be adequately differentiated according to risk.

Finally, one could also think of a role for government (in organizing a compensation fund) in the final phase of long-term stewardship. In that case, government intervention would not merely be facilitative but financing, since the risk has been transferred to government. However, in that case, the government fund would amount to the direct compensation by government.[91] The comparative benefits of creating a fund solution in that case are not very obvious. A fund would have an advantage if the fear were that the government may not have funds available in the future to finance damage that would result from the long-term stewardship. In that case, it would (e.g., in case of developing countries) make sense to reserve part of the public budget for damage that would occur during this long-term stewardship.[92]

A legal argument would also be made, and has been made by some scholars,[93] that direct compensation by government is often ad hoc and hence can lead to inequality: some victims may be compensated and others not, and victims might not have a guarantee of compensation. That equality argument could be a reason in favor of organizing the government compensation for damage during the long-term stewardship through a compensation fund rather than ad hoc. On the other hand, as we have argued, some economists have argued that a structural fund may provide the wrong signals to the market since victims would be able to count on government compensation.[94] That would be an argument against a structural compensation fund to the extent that the charity hazard (moral hazard on the side of the victim) would be a serious problem, which is not immediately obvious.

8.4   Compensation Fund: A Second Best Solution?

The position defended in this study is that liability for long-term stewardship should be transferred to the state. The logic behind this reasoning is
that liability for long-term stewardship has, given the discounting factor,
no positive incentive effects. However, at the policy level, decision making
concerning CCS may not only be influenced by considerations with respect
to incentives for prevention, deterrence, and economic efficiency. Some
may (and as we indicated, there are voices clearly heard in the literature)
also stress that excluding liability from operators would be unacceptable
from a fairness perspective, and the argument would then be that operators
reap the benefits from CCS and should therefore also bear the costs of
future liability. Shifting liability to the state would amount to shifting the
costs to future taxpayers, and hence subsidizing CCS, which may be difficult to accept at a policy level.

Within that perspective, the question could arise whether other instruments could still be employed as a second-best solution to deal with those
future losses for the long-term risk. One obvious solution that comes to
mind is a compensation fund.[95] As indicated, the problem with that is that
if one wishes to construct a compensation fund in an efficient manner, the
risks should be sufficiently differentiated, with higher-risk operators paying
higher contributions to the fund than low-risk operators. This supposes
that the government (in the assumption that government would run the
fund) has better information than operators in order to apply a correct risk
differentiation. This may be particularly difficult given the fact that one
would have to pay a contribution for losses that may (or may not) occur in
a very distant future. If risk differentiation is absent, a compensation fund
would simply be financed by a tax on current operators without any beneficial effect on incentives for prevention. Moreover, large amounts of funds
may (via the contributions to the fund) be immobilized, whereas it is uncertain that the money will actually be necessary.

An attractive solution in that respect is presented by the Price-Anderson
Act,[96] which regulates nuclear liability in the United States. The major
advantage of the model in the act is that, in addition to the first layer of
compensation provided by nuclear operators, a substantial amount of
money is generated ex post. The money is prefinanced by the Nuclear Regulatory Commission (NRC) and collected from all operators through a system of so-called retrospective premiums. The advantage of this model is
that money is not immobilized ex ante (if an accident never happens) but
only ex post. A similar model could be employed for CCS as well. The

disadvantage is that the retrospective premiums would be paid only by operators still in business when the damage occurs. Operators that went insolvent or out of business do not contribute. However, since the retrospective premium scheme makes compensation a collective responsibility of industry, this creates excellent incentives for mutual monitoring. After all, the likelihood that one operator will have to compensate the scheme is also dependent on the risk created by the others. Operators will therefore monitor the solvency of the others and also lobby the government to impose stringent safety regulations in order to reduce risk.

In other words, if, as a second-best solution, it would still be desirable to have compensation for the long-term stewardship paid by operators, a model of retrospective premiums like the one followed in the Price-Anderson Act may be worthwhile to examine.

8.5 Reinsurer of Last Resort

In various legal systems, a fourth model has been developed whereby the government acts as reinsurer of last resort. Under this approach of government involvement, the state takes at least some of the risk for losses from catastrophes. Even though government intervention is required,[97] the underlying philosophy of this approach is that private insurance should keep on playing a significant role in allocating compensation for victims of catastrophes. This option then usually takes the form of a multilayered insurance program. Such a program is normally administered by private insurance companies, meaning that they sell insurance, collect premiums, and pay claims.

In the developed world, there are many examples of government acting as a reinsurer of last resort with the goal of promoting capacity for disaster risks, both natural and man-made.[98] Terrorism insurance coverage is an example of government reinsurance backing a man-made risk. The current schemes for terrorism insurance in France (GAREAT),[99] Germany (Extremus), the United Kingdom (Pool Re), the Netherlands (the Dutch Terrorism Risk Reinsurance Company, NHT[100]), and the United States (the Terrorism Risk Insurance Act, TRIA)[101] all have features of a government acting as a reinsurer of last resort, at least in general terms.[102] In the British example, the government provided a form of reinsurance for terrorism-related events.[103] The Pool Reinsurance Company Limited (Pool Re) was created in 1993 to provide reinsurance for when the capacity of private insurers would be exceeded.[104] However, the reinsurance is to be repaid from future insurance earnings. According to the Pool Re website:

If losses ever became so large as to exhaust its reserves, Pool Re would draw funds from the UK government to meet its obligations. Pool Re, in turn, pays a premium to government for this cover and would be required to repay any funds drawn down in this way from its future income.[105]

For natural hazards a well-known example is the Caisse Centrale de Réassurance (CCR) in France. Insurers can reinsure the risk of natural hazards with this CCR, which benefits from a state guarantee in the event that the CCR exhausts its resources.[106] Similarly, Japan has provided reinsurance for earthquake-related events.[107] The Japan Earthquake Reinsurance Company (JERC) can provide reinsurance beyond a 5 trillion yen cap; the goal was to enable burden sharing between private insurers and the national government, with the government bearing approximately 85 percent of catastrophic claims.[108]

Also in many developing countries, insurance solutions against natural disasters have been developed.[109] A variety of insurance schemes in developing countries often rely on (more or less large) government support to make the risk insurable.

There can indeed be arguments to favor of such reinsurance by government, assuming that capacity on the private insurance market is severely falling behind. It can be held that without state intervention, insurance coverage for disasters would not have developed.[110] Reinsurance by the state can then be considered an adequate method to resolve the uninsurability problem.[111] A condition is, of course, that the government charges an actuarially fair premium for its intervention.[112] This type of government intervention also has the advantage that ex post relief sponsored through the public purse can be avoided. Where the government acts as reinsurer, this at least has the advantage that a premium can be paid by those who actually cause or run the risk. It can thus facilitate market solutions, still provide incentives for prevention to potential victims, and avoid the negative redistribution discussed.[113] A state intervention as reinsurer may avoid the "catastrophic responses to catastrophic risks."[114]

Thus, not only has the policy option of publicly provided reinsurance been historically tested, it has also received attention in the CCS literature as a potential means of enabling private insurance markets to function. Haan-Kamminga advocated the role of private liability insurance for CCS; her arguments include its role as a risk-transferring function, a risk-spreading function, and a risk-allocating function.[115] She also documented the availability of such policies for CCS operators,[116] addressing in part the concerns that such might not be available given both the risks of CCS itself and the yet-undeveloped liability rules. To the extent that private insurers are unable or unwilling to provide complete coverage as the market expands, Haan-Kamminga advocated for governments to provide back-end supplemental risk bearing[117] in what appears to be a form of reinsurance.

However, criticisms have been formulated on the facilitative role of a government stimulating insurance markets. For example, Gron and Sykes argue in several papers that it would be unjust for the government to provide (re)insurance at a lower price than the market price.[118] This would give a wrong signal to the market as far as stimulating insurability is concerned. It is a criticism that Levmore and Logue share; they argue that such a regime (of acting as reinsurer of last resort) has its desired effect of encouraging the purchase of commercially provided terrorism coverage only when it involves a substantial subsidy.[119] They are skeptical of these types of interventions in the market (for terrorism insurance), holding that without government intervention, "the market would likely have been able to provide the necessary coverage."[120]

There is not a lot of discussion on the role of government as reinsurer of last resort in the CCS-related literature. Some authors refer to the National Flood Insurance Plan in the United States as an example for CCS,[121] and Makuch et al. refer to public and private liability funds for CCS.[122] To some extent, the intervention of government as reinsurer of last resort is also considered a public/private partnership.[123] There is not yet sufficient information available to evoke that this type of intervention would, in this stage of the development of CCS technology and the knowledge about the risks and potential damage, be necessary. Indeed, such an intervention of government as reinsurer of last resort typically happens for catastrophic risks where it is argued that the traditional commercial (re)insurance market cannot provide sufficient capacity to cover the risks. Government then steps in to provide additional supply. Although we have indicated that there may indeed be potential capacity problems in the provision of insurance,[124] we have equally argued that a variety of alternative compensation mechanisms could be developed, as well as such risk-sharing agreements between operators,[125] or eventually a government-facilitated compensation fund financed by risk-dependent contributions from operators.[126] Hence, it seems that at this stage, calling for a role for government as reinsurer of last resort simply comes too early, since it may be that market solutions (such as risk-sharing agreements) can still be developed.

Only when it would appear that those capacity problems would be real and that government should step in to supply coverage could this be considered. In that case, however, the conditions for an efficient intervention as mentioned in the literature should be respected:[127]

1. Government intervention would be necessary only when a market solution did not and could not develop without it. Government intervention should hence be nondistortive.
2. Government should charge risk-based premiums for its supply of reinsurance.
3. Reinsurance by government should be organized in such a way that market solutions are still stimulated.
4. Insurers should be left free to choose the state reinsurance, and in principle, government intervention should also have a temporary character (and thus have a so-called sunset provision).

The literature has indicated that such an intervention by government should be nondistortive in the sense that the government should intervene only when market solutions are not available.[128] Moreover, government intervention should always have a temporary character in order to stimulate the development of market solutions.[129] In this stage of the development of the CCS technology, it is not yet known whether there would be a problem with the supply of sufficient capacity. Hence, it seems that it is too early to plead in favor of a role for government as reinsurer of last resort for CCS-related risks without other precautions in place.

8.6   Creation of a Public CCS Utility

A fifth approach that could reduce the concerns of long-term liability given the potential absence of a revenue-managing operator would be to operate CCS storage facilities within a public utility.[130] A geologic sequestration utility (GSU) could be employed to facilitate the earlier adoption and long-term governance of subterranean injection facilities.[131] This institutional solution was originally proposed by the Midwestern Governors Association (MGA) in the United States;[132] thus, it is likely a politically viable option in some jurisdictions.

The MGA proposal was sweeping, in that the GSU would both regulate and administer the facilities that would gather, inject, and store the CO_2 fluids.[133] The GSU would be capable of surviving as long as it continued to receive the support of the local government and would thus be able to support its liabilities over a longer time period than many private operators might be able to support liabilities.[134]

Another option would be a GSU that regulates but does not operate the facilities.[135] It is suggested, but not explained, that this limited form of a GSU would remain capable of addressing a variety of liability concerns.[136]

Such a utility could possibly be the recipient of transferred long-term liabilities that it had managed and governed since the beginning of any CCS project.

A third proposal, mentioned but not developed, is that the GSU could be an amalgam of the two ideas, with limited operational capacity but some sort of operational capacity.[137] A possible way of implementing this suggestion might be to include the regulatory body or GSU in a mode similar to joint venturers who invest in a nonoperational role.[138] In the oil and gas industry, this is referred to as an operated-by-other (OBO) relationship and retains many of the aspects of control and audit and could provide for future period transfers of operational control, as might be needed at CCS sites. Such an organization could also include private regulators, such as private insurers or investors in carbon trading permits, if appropriate.

8.7 Conclusion

There could be a governmental role in facilitating compensation; when certain circumstances make it likely that private markets could fail, public authorities might consider undertaking certain efforts to ensure that compensation could be provided efficiently.

This could occur in various ways. First, there seem to be strong arguments in favor of demanding mandatory financial security from operators, while at the same time providing flexibility to local regulators to determine the optimality of the amount and type of the security provided by the operator. Not only does this retain the proper alignment of incentives to efficiently minimize risks, but it also aligns with the polluter-pays principle.

Second, in the phase of long-term stewardship, compensation can be directly provided by government through either ad hoc payments or a structural compensation fund, financed by taxpayers. This is in alignment with the public goods aspect of the prevented anthropogenic climate change.

Third, during the phase of postclosure monitoring, there could be a role for a government-run compensation fund on the condition that it would be financed through risk-dependent contributions by operators. There would, however, be an argument in favor of such a compensation fund only when the government, as manager of the fund, would be better able than industry to differentiate risks. If that is not the case, a risk-sharing agreement between operators seems preferable.

At this initial stage of the development of CCS technology, there does not seem to be a role for government as reinsurer of last resort. If capacity problems appear to be serious and insurance and its alternatives could not supply sufficient financial security, such a role could be envisaged on the condition that market principles are respected by charging risk-dependent premiums and stimulating market solutions.

However, it remains important that the goals of compensation be coordinated so as not to disrupt the prior goals of efficiently minimizing the social risks of CCS operations. To avoid is more efficient than to cure.

9 Policy Recommendations

Since conclusions or summaries have been provided at the end of most chapters, we can be relatively efficient at this point since the policy recommendations follow logically from the points already discussed. The crucial question at this stage is what policymakers could do to stimulate solutions, especially for the long-term CCS liability.

Without other technological innovations to provide carbon-free energy supplies, the provision of CCS-coordinated energy supplies may be critical to enabling reduced carbon emissions to prevent or defer the risks of anthropogenic climate change. Yet despite decades of research and policy discussions, commercially operating CCS storage sites remain few.

The motive for this book was the recognition that one of the key problems facing the developers of CCS storage operations is the uncertainty facing them from potential future liability exposures. This book has endeavored to identify leading liability questions, review optimal liability and governance options, and provide policymakers with clear explanations of why those policy options would be the most robust options to enable the future development of CCS storage facilities.

9.1 A Review of the Policy Recommendations

This section provides an executive summary of the findings of this book. Fuller explanations of the arguments summarized here can be found within each sub-section's correlated chapter.

9.1.1 Foreseeable Risks and Potential to Mitigate or Remedy

A starting point for designing an optimal liability and compensation mechanism is obviously identifying and distinguishing the specific risks involved in CCS.[1] Chapter 2 found a variety of potential injuries and characterized them as accidents that occur at or above ground and those that occur primarily underground. While injuries to human health and ecological settings were found to be possible, the key foreseeable injury was found to be risks of contamination to subterranean water reserves.

It was found that most of the events leading to injury were predicated in either poor storage site selection decisions or poor stewardship of injection activities. Yet careful site selection is readily feasible given existing geological knowledge and modern technology; similarly, a variety of monitoring technologies exist that could enable real-time feedback on emerging problems with the injection process. Thus, there are decision processes that could change the resultant injury levels, and those processes would likely be influenced by the type of incentives provided by rules of civil liability and by public and private regulations.

Furthermore, while those incentives might efficiently govern the level of damages resulting from CCS storage activities, some injuries and damages would remain likely to occur. It was shown, based on historical experiences, that most of the likely injuries could be mitigated in early stages or could be remediated. While it is nice to know that damages can be physically addressed and potentially cured, another important aspect is that the potential to provide such remedies would enable more accurate damages to be set by the courts, reinforcing the potential accuracy and efficiency of civil liability rules.

The legal policy recommendations developed in this book can be broadly summarized: a liability and compensation regime should first provide incentives to stakeholders to optimize the prevention of harm from CCS operations, but to recognize that such incentive mechanisms may have limits of effectiveness. A rule of strict liability in joint implementation with both public and private means of regulation could provide such incentives to various CCS stakeholders. Second, compensation planning should efficiently address damages that could not be prevented by those measures.

9.1.2 Choice of Strict Liability over Negligence

Rules of civil liability are well known in the law and economics literature to provide a variety of incentives that can affect the decision processes of actors engaged in risky activities to attain efficient levels of risk. After reviewing the theoretical advantages and weaknesses of both the rule of strict liability and of the rule of negligence, it was found that a rule of strict liability would be expected to be more robust in the circumstances of CCS storage activities.

While some might have expected our argument to be primarily, or exclusively, predicated on notions of "ultrahazardous" or "abnormally dangerous" characterizations of CCS activities, our analyses rested primarily on the informational mechanisms within the rule of strict liability. The unilateral character of the CCS storage activities, the need to enable decentralized safety planning in contrast to a jurisdictionally set uniform due care standard, the need to respond to technological novelty in the underlying risks, and the potential to reduce jurisprudential transactions costs all contributed to the robust nature of the strict liability rule for CCS storage activities.

To the extent that a strict liability rule might lack an advantage found under a negligence rule (e.g., informational disclosures), it was found, in chapter 6, that a regulatory mechanism could often provide similar or better benefits. Thus, we concluded that a rule of strict liability should be the default rule of civil liability for CCS storage, to be complemented by public and private forms of regulation.

9.1.3 Determination of Liable Parties

The implementation of civil liability requires more than just a choice of rule from a small menu of options; there are many subsidiary questions of implementation. Chapter 4 provided a review of those options and suggested optimal settings for the circumstances of CCS storage activities.

It was first found that there is much uncertainty in the "who" of liability, an issue we referred to as stakeholding. In the operations, ownership, and oversight of CCS projects, there are potentially multiple parties who might bear liability risk to some extent. Questions of joint and several liability, on the one hand, and of liability channeling, on the other, complicate the potential set of liable stakeholders. Similarly, property law presented several unique challenges to identifying the ownership of the injected volumes at different temporal stages of the storage project.

Our recommendation can be boiled down to this: ex ante standards are needed to clarify this situation before CCS developers and affected communities can fully understand their roles and potential liabilities for CCS storage activities. Our recommendations in these areas of concern focus first on the need to provide ex ante clarification of the future liability and property rules and second on the specific form of clarification.

We found the assignment of stakeholding to be critical, but that liability channeling should be avoided. The imposition of joint and several liability has been traditionally applied within the energy sector without substantial controversy and might similarly serve the needs of clarity for

CCS stakeholders. However, policymakers should be cautious with the introduction of joint and several liability regimes because that could increase the potential scope of liability for CCS operators and could endanger the insurability of the liability.

The risk of causal uncertainty should not be shifted to the operator; rather a proportional liability rule should be applied to deal with causal uncertainty.

Property law is much more dependent on local jurisdictional precedence, so our recommendation remains focused on the need for ex ante clarifications to enable proper exploration and development of CCS storage sites.

9.1.4 Disarming Regulatory Defense and Force Majeure

Some jurisdictions allow compliance with public regulatory measures to suffice as evidence of meeting a minimal duty-of-care requirement. This is known as a regulatory defense. Our recommendation is for implementing a rule of strict liability, within which such a defense would be contextless and useless. But even if a rule of negligence were to be implemented within a certain jurisdiction, as it is well recognized that certain jurisdictions display an aversion to implementing rules of strict liability, then this means of defending against a negligence claim should be prevented by the courts.

Primary in our arguments is the notion that regulations tend to set necessary but perhaps insufficient standards and that exposure to rules of civil liability retains the economic capacity to incentivize for the correct, optimal levels in standards on preventive behavior. Furthermore, the exercise of civil liability provides a functional defense to certain weaknesses known to plague public regulatory efforts, such as lobby capture. Thus, our primary recommendation would obviate the notion of a regulatory defense, but in cases where negligence avails, this defense should be prohibited.

Similarly, operators should not be held liable when the damage is caused by force majeure. However, natural disasters such as flooding or earthquakes do not automatically qualify as a force majeure that would free the operator from liability. To an important extent, operators have to foresee the likelihood of those natural hazards and take that into account in their siting decision and operation management. Force majeure hence excludes liability only when a disaster would have an exceptional and unforeseeable character.

Given the role of site selection and awareness of potential natural events that might affect safe sequestration, as with Japan's nuclear rules, force majeure ought to be minimized as a defense in preference of incentivizing careful exploration and development planning.

9.1.5   No Caps to Liability Exposure

The operations of CCS storage sites will involve large capital expenditures and potentially place large areas or substantial populations at some level of risk. Thus, the development of CCS operations could place their operators at substantial levels of capital risk.

Arguments could be made that by reducing some of the capital risk, operators could be incentivized to invest in more CCS storage capacity. Given the global benefit to reduced carbon emissions, it might be reasonable to see this as a form of public support of CCS storage activities. However, we recommend against implementing liability caps for operators in that those caps would create inefficient conditions for managing the risks from CCS operations.

This recommendation is limited to the periods when the operators maintain oversight and operations at the storage sites.

9.1.6   Planning for Operator Extinction

In order for CCS to be a successful means of preventing the harms from climate change, its storage systems need to endure for centuries, maybe millennia. From historical evidence, it is highly foreseeable that very few privately organized operators could manage to survive for the duration of the storage activity, especially over the very long periods of postclosure storage. An operator that is not in existence cannot be governed by rules of civil liability or regulatory measures; the risks would simply be uncontrolled without additional actions to replace the operator. Thus, planning is required to enable the eventual transition of responsibility from private operators to some form of public oversight.

Our recommendation, in alignment with the EU's CCS Directive, is for some form of public authority to take over whatever operational activities are required to secure and maintain the storage site after some successful period of injection cessation, well plugging, and fluid stabilization within the geological storage system. Furthermore, the public authority should make provision to address liabilities from CCS-related damages that arise after this transition of oversight.

After exploring several potential options, our key recommendation is that policymakers clarify, from before the start of development and permitting, their preferred plan of transition to best enable operators to make sound development plans for their CCS storage sites.

9.1.7 The Need for Both Public and Private Forms of Regulation

In chapter 6, we reviewed the need for public regulations to govern the risks and hazards of CCS storage activities. The legal literature on CCS concerns is practically unanimous in its call for public regulation of CCS storage activities. We concur that public regulations are needed, but we emphasize where public regulations might be efficient tools and where they might not be efficient.

Most important, regulations enable the governance of behaviors prior to the onset of injuries, whereas rules of civil liability remain dependent on causes of action (i.e., injuries). Also, the development of public regulations enables a public deliberative process of standard setting. The ability to prevent injury by correcting behaviors before accidents arise and the ability to publicly set standards are two key advantages of public regulations that will be critical for the future development and operation of CCS storage facilities.

In addition, tools of public regulation are robust where concerns are to collate and interpret information that might be difficult via private interests, rules of civil liability fail to result in its predicate litigation, and court institutions might not be effective in ensuring resolution of damages. Furthermore, in alignment with our concerns on causation, public regulatory oversight could do much to ensure the collection and public availability of the necessary scientific data that would facilitate courts in processing civil liability cases. Thus, public regulations could directly achieve their own policy goals while indirectly facilitating the efficiency of civil liability systems.

Also governments could draft public regulations, in addition to designing an optimal liability regime, in order to stimulate an optimal compensation for CCS-related risks:

1. Governments should introduce tailored and detailed safety regulations for the prevention of CCS-related risks.
2. Regulations should be proactive and dynamic. This implies that regulations should precede the potential risks and dynamically change according to new insights and technological developments.
3. Providing a clear, consistent, and dynamic regulatory framework is important in order to stimulate the insurability of risks by facilitating risk differentiation by insurers.
4. Providing such a regulatory framework is an equally important condition for stimulating successful risk-sharing agreements among operators.

We add to the discussion that due to the pace of technological innovation in CSS, the knowledge intensities required of the professionals involved with CCS, and the potential informational efficiencies obtainable from those closest to the action being engaged, among other arguments in favor, we recommend the active development of private regulatory efforts to coordinate and facilitate the efficacy of public regulations.

9.1.8 Private Market Means of Compensation

Chapter 7 explored the idea of leveraging the marketplace to provide compensation for damages resulting from CCS storage activities. The tools of insurance, self-insurance, guarantees and deposits, letters of credit and performance bonds, capital accounts, and the emerging tool of catastrophe bonds were reviewed.

Our baseline observation is that all of these tools would appear to be foreseeably available to CCS operators and could be used to provide compensation in the event of a CCS-related accident. Our recommendation was straightforward: market-based compensation schemes are preferred to public resource–based schemes due to their informational advantages that would retain robust incentives for the operators to efficiently manage their CCS-related risks. We recommend that each jurisdiction find its optimal blend of private market means of compensation, highlighting that we encourage a blended product approach and not reliance on a single means of compensation.

We also found reason to not fear the availability of insurance for CCS projects. Key among them were the existing market for such policies and the existence of policies for similarly functioning risks in the energy industries. Over the long run, we found that risk-sharing agreements should be highly considered for providing the capital for postclosure needs.

For large operators, we recommend that self-insurance should be allowed if the regulator routinely verifies the adequacy of the self-insurance provided. For smaller operators, self-insurance could be used during the operational years as a first layer (retention) in addition to other instruments.

9.1.9 Public Resource Means of Compensation

As we stated in chapter 8, many readers will reasonably hold that a key role of public authorities, in general without specific address to CCS projects, is to provide a sense of social cohesion by supporting the public in times of crisis. Is it reasonable to include the needs of CCS projects within such a scope? Yes. In certain circumstances, CCS projects might be supported by public resources. However, we provide guidance on the limits of that assistance.

In alignment with our earlier recommendation for operators to bear as much liability as possible while actively engaged in CCS storage activities, we recommend public support of such pre-/postclosure damages in only the most limited cases where there is simply no alternative.

We recommend that the public authority attempt, where possible, to prefer the role of steward, rather than that of provider, of the capital funds for postclosure needs. However, we recognize that a reasonable argument can be brought that secured long-term storage provides a sense of a public good and thus might well be entitled to taxpayer support at that time. Our recommendation is for public authorities to secure as much of the anticipated capital from the operators prior to the transition of the storage site, or otherwise coordinate the creation of the relevant risk-sharing agreements, and provide public resources only when those collected resources fall short.

9.1.10   CCS Storage Is Welfare Enhancing

Our set of recommendations aims to minimize the overall costs of CCS risks and harms, both the costs of avoiding injuries and the costs of restoring the victims of injuries, while enabling the welfare goal of preventing catastrophic climate change.

The ability to govern the risks from CCS storage operations has been presented in several steps: (1) the technological capacity to respond, (2) use of civil liability to provide incentives to prevent injuries and thus minimize damages, (3) the use of public regulations to manage risky behaviors prior to injury events, (4) the use of public regulations to set standards prior to commercial development, and (5) the use of private regulatory efforts to draft up-to-date regulatory standards as quickly and efficiently as possible. With this assembly of governance mechanisms, we hold that the risky activities of CCS storage could be efficiently managed and thus be welfare enhancing.

We recommend that operators be generally required to provide for the compensation of damages resulting from CCS activities. From the portfolio of private market solutions that can be employed, we recommend that a mixed set of tools be engaged in providing for compensatory needs. The role of public authorities should remain primarily focused on facilitating the operators' needs in achieving those private tools. However, we do recognize that in certain circumstances, it could be efficient and responsive for public resources to facilitate compensation needs.

9.2   Climate Change Liability

An important risk for operators consists of so-called climate change liability.[2] It could indeed happen after the injection-phase CO_2 escapes from the storage site. That will (e.g., in the EU) lead to an obligation for the operator to compensate and purchase the allowances for the CO_2 that was emitted. There may be uncertainty related to the question of whether such a release may occur.[3] Moreover, prices of CO_2 allowances may change. This price volatility could lead to a situation that operators may have to purchase allowances in the future at a much higher price, which may be difficult to foresee.

However, climate change liability should not be a serious objection to CCS projects, since the principles mentioned in this study can readily be applied to climate change liability, especially if liability for the long-term stewardship of postclosure risks is excluded.[4] The uncertainties of risks and price volatility an operator would be confronted with would then be limited to those of the short and middle to long term.

The mere fact that the damage consists of climate change liability rather than more traditional losses, such as property damage or personal injury, does not matter. A risk-averse operator could use any of the hedging strategies discussed in chapter 7, varying from commercial insurance to alternative compensation mechanisms such as self-insurance, risk-sharing agreements, or guarantees.[5] Hence, risk-averse operators can follow the same strategies: take (partial) coverage for those future climate change liability risks, depending on their attitude to risk and corresponding demand for protection.

9.3   Position of Developing Countries

In these policy recommendations, we should also mention the position of developing countries. They may be attractive targets for CCS projects since the geological conditions in those countries may make them attractive for international parties to develop CCS projects there. The principles and recommendations of this study apply equally to developing countries, albeit with a couple of particular caveats.

First, some of the hedging strategies for industry may not always be available in developing countries. For example, insurance markets may not be sufficiently developed, so commercial insurance may not always be available. That may be an argument for industry to look for alternative solutions such as captives or risk-sharing agreements. After all, operators in developing countries might not necessarily come from those developing countries, but might be financially sophisticated multinational corporations that could use hedging strategies other than insurance.

Second, as far as developing countries are concerned, the expectations concerning the role of government should be more modest in some cases. For example, direct compensation by government may, as was already mentioned,[6] not be an option in the case of developing countries where governments may lack the funds to step in in case of disastrous damage. The same caveat applies when it concerns a role for government as a reinsurer of last resort.[7]

Third, implementation of our recommendations might need to take into account North–South relationship planning. Governments could play an active role in promoting CCS projects, for example, through a transfer of information on CCS technologies or on the type and amount of security required from operators. Governments can also determine safety standards. To facilitate the governments of developing economies, international organizations could take up this role in promoting CCS to the assistance of local governments.

9.4   Concluding Observations

This study started from the assumption that CCS is an interesting technology worthy of investigation—one that might enable a bridging technology to defer or mitigate the onset of anthropogenic climate change until carbon-free energy resources can be brought to market to replace fossil fuels. That assumption is based on the work of the IPCC—that there are potentially great benefits in CCS as a mitigating strategy for climate change and that the long-term risks with CCS are relatively restricted.[8] Yet not everyone is convinced of the blessings of CCS. and some have even qualified it as "false hope."[9]

The goal of this study was not to take any position in that debate. Instead, the goal has been to resolve how policymakers might best address the necessary question of risk management for whatever CCS policies they ultimately decide to implement and, in particular, focus on how legal policies might best be brought to bear on these concerns.

This study has used the law and economics literature and literature dealing with CCS-related risks to provide an indication of how policymakers could provide certainty concerning the potential liability risks by both limiting liability in time and providing adequate financial security for future contingencies. In this way, the potential positive externalities, in terms of reducing climate change, could be appropriately balanced with the potential negative externalities from potential CCS-related accidents and injuries.

We recommend that several well-tested mechanisms be employed to efficiently balance the risks and benefits of CCS projects. A combination of civil liability, public regulations, and private regulations can be an effective combination that could balance the strengths and weaknesses of each approach. We also recommend that CCS operators be required to bear as much of the responsibility as possible for the provision of compensation. If CCS projects are to be privately operated, and one assumes profitably, then the externalities of future consequences should be integral to operators' decision processes to ensure the optimal level of efficiency and welfare. We found that by leveraging the techniques of mechanism design, along with efficient compensation schemes, reasonable and prudent CCS operators should be able to steward CCS projects in a manner that provides an optimal level of carbon sequestration and minimal burden to taxpayers.

Notes

Chapter 1

1. See chapter 2, sections 2.1 and 2.2. See also Steven Chu, "Carbon Capture and Sequestration," *Science* 325, no. 5948 (2009): 1599. Carbon capture and storage is also frequently restated as carbon capture and sequestration; both are abbreviated to CCS.

2. See chapter 2, section 2.4, and chapter 3, section 3.1.

3. See chapter 2, section 2.4.

4. See chapter 1, section 2. See also chapters 3, 4, and 5.

5. See chapter 6.

6. See chapters 7 and 8.

7. See chapter 9, section 9.1.

8. See Roy Andrew Partain, "Is a Green Paradox Spectre Haunting International Climate Change Laws and Conventions?" *UCLA Journal of Environmental Law and Policy*, 33, no. 1 (2015): 64.

9. See ibid., 65.

10. See ibid.

11. See Alexandra B. Klass and Elizabeth J. Wilson, "Climate Change and Carbon Sequestration: Assessing a Liability Regime for Long-Term Storage of Carbon Dioxide," *Emory Law Journal* 58 (2008). They qualify carbon capture and sequestration (CCS) as a promising technology. Some use the abbreviation *CCS* for "carbon capture and sequestration," whereas others use it for "carbon capture and storage." In this study the terms are used as synonyms.

12. See David Hawkins, George Peridas, and John Steelman, "Twelve Years after Sleipner: Moving CCS from Hype to Pipe," *Energy Procedia* 1, no. 1 (2009): 4403.

13. But see Emily Rochon et al., *False Hope: Why Carbon Capture and Storage Won't Save the Climate* (Amsterdam: Greenpeace International, 2008).

14. While concerns on technological development might once have been a concern in the legal literature, CCS injection technologies are based on preexisting natural gas and CO_2 injection systems already in place in a variety of oil and gas industry settings. See chapter 2, section 2.1. Furthermore, large-scale CCS projects have been in commercial operation for over a decade in a few locations, such as at Sleipner offshore of Norway. See chapter 2, section 2.1, n. 125. See also Berend Smit et al., *Introduction to Carbon Capture and Sequestration* (London: Imperial College Press, 2014), 371–372.

15. See chapter 3, section 3.1.

16. The means to set efficient or incentivizing prices for carbon or carbon injection services is beyond the scope of this book.

17. See chapter 3, section 3.3.

18. See chapter 2, section 2.2.

19. See chapter 2, sections 2.4 and 2.5.

20. See chapter 3, section 3.1.

21. See chapter 4, section 4.6.

22. See chapter 5, section 5.2.

23. See chapter 5, section 5.2. See also chapters 7 and 8.

24. For an introduction to this school of literature, see the research of founders Calabresi and Shavell: Guido Calabresi, *The Costs of Accidents: A Legal and Economic Analysis* (New Haven: Yale University Press, (1970), and Steven Shavell, *Economic Analysis of Accident Law* (Cambridge: Harvard University Press, 1987). For recent surveys of the literature, see Hans-Bernd Schäfer and Frank Müller-Langer, "Strict Liability versus Negligence," in *Tort Law and Economics*, ed. Michael G. Faure (Cheltenham: Edward Elgar, 2009). See also Michael G. Faure, "Regulatory Strategies in Environmental Liability," in *The Regulatory Function of European Private Law*, ed. F. Cafaggi, H. Watt, and H. Muir (Cheltenham: Edward Elgar, 2011).

25. See, for example, David E. Adelman and Ian J. Duncan, "The Limits of Liability in Promoting Safe Geologic Sequestration of CO_2," *Duke Environmental Law and Policy Forum* 22 (2011). See also Alexandra B. Klass and Elizabeth J. Wilson, "Climate Change and Carbon Sequestration: Assessing a Liability Regime for Long-Term Storage of Carbon Dioxide," *Emory Law Journal* 58 (2008).

26. For a summary of the literature on the economic analysis of liability rules, see Faure, "Regulatory Strategies in Environmental Liability."

27. Chapter 3, section 3.3.

28. Nigel Bankes, "The Legal and Regulatory Issues Associated with Carbon Capture and Storage in Arctic States," *Carbon and Climate Law Review* 6 (2012): 24.

29. Ibid.

30. Ibid.

31. Ibid.

32. For a study of liability for CCS under US law, see Adelman and Duncan. See also Klass and Wilson, "Climate Change and Carbon Sequestration," 123–158.

33. Council Directive 2004/35/EC of April 21, 2004, on Environmental Liability with regard to the prevention and remedying of environmental damage. Official Journal 2004 L143/56 and Directive 2003/87, Establishing a Scheme for Greenhouse Gas Emission Allowance Trading within the Community and Directive 2009/29/EC as to improve and extend the Greenhouse Gas Emission Allowance Trading Scheme of the Community.

34. For a discussion of the CCS under the United Nations Framework Convention on Climate Change, see Sven Bode and Martina Jung, "Carbon Dioxide Capture and Storage—Liability for Non-Permanence under the UNFCCC," *International Environmental Agreements: Politics, Law and Economics* 6, no. 2 (2006). For a discussion of state liability in case of CCS under international law, see Justine Garrett and Brendan Beck, *Carbon Capture and Storage: Legal Regulatory Review Edition*, 2nd ed. (Paris: OECD/International Energy Agency, 2011), 92–93.

Chapter 2

1. Sally Benson and Peter Cook, "Underground Geological Storage," in *IPCC Special Report on Carbon Dioxide Capture and Storage*, ed. B. Metz, O. Davidson, H. De Coninck, M. Loos, and L. Meyer (Cambridge: Intergovernmental Panel on Climate Change, 2005), 199. See also Nicola Swayne and Angela Phillips, "Legal Liability for Carbon Capture and Storage in Australia: Where Should the Losses Fall?" *Environmental and Planning Law Journal* 29, no. 3 (2012): 189–216.

2. Benson and Cook, 199; Mike Bickle, Niko Kampman, and Max Wigley, "Natural Analogues," *Reviews in Mineralogy and Geochemistry* 77, no. 1 (2013): 36, with reference to an older CO_2 field that dates back approximately 40 million years.

3. Benson and Cook, 199; see also Bickle et al., 21.

4. Benson and Cook, 199; Bickle et al., 26.

5. Benson and Cook, 210; Bickle et al., 15.

6. Benson and Cook, 210; Bickle et al., 16–17.

7. Benson and Cook, 210. The McElmo Dome (CO), the St. John's field (AZ), and the Jackson Dome (MS) each contain approximately 1,600 million tons of CO_2.

8. Ibid., 211. There are also lakes in Africa (e.g., Lake Nyos) that contain large volumes of dissolved CO_2, but the functional differences of those lake systems versus the preferred means of geological storage for CO_2 obviate a need to evaluate them here. In addition, those lakes tend to erupt and kill both humans and livestock, preventing their model from becoming a practical model of CO_2 sequestration.

9. Ibid., 212. In fact, the geochemistry of CO_2 appears to make it more likely to remain within underground traps than hydrocarbons.

10. Benson and Cook, 200; Helge Hellevang, "Carbon Capture and Storage (CCS)," in *Petroleum Geoscience: From Sedimetary Environments to Rock Physics,* ed. K. Bjørlykke (Berlin: Springer-Verlag, 2015), 600. Similarly listed as depleted oil and gas reservoirs, EOR projects, deep saline aquifers, and deep unminable coal seams. Berend Smit et al., *Introduction to Carbon Capture and Sequestration* (London: Imperial College Press, 2014), 365–368.

11. Benson and Cook, 200; Smit et al., 384.

12. Benson and Cook, 205. For example, Sleipner's CCS project injects into high-porosity sediments. Hellevang, 599.

13. Benson and Cook, 205; Smit et al., 392.

14. Crude oil and natural gas are often found in similar pore space conditions, and their reserves are an expected target of CCS storage locations.

15. Benson and Cook, 205; Smit et al., 361. Nevertheless, sufficient capacity must be in place. Hellevang, 593.

16. Benson and Cook, 205, Smit et al., 383. Volume, porosity, and permeability have a complex physical interrelationship. See Hellevang, 599, equation 24.1.

17. Benson and Cook, 205.

18. Ibid. Hellevang, 593–94. Storage capacity is proportional to CO_2 density. Hellevang, 593. Smit et al., 361.

19. Benson and Cook, 205; Hellevang, 595; Smit et al., 361.

20. Benson and Cook, 208. Hellevang, 595, listed four means of subsurface trapping.

21. Benson and Cook, 208. Examples are granite or salt. See Hellevang, 593.

22. Benson and Cook, 208; Smit et al., 389, adding the detail that often such structures repeatedly overlap, providing multiple lines of containment.

23. Benson and Cook, 208; Kurt Zenz House et al., "Permanent Carbon Dioxide Storage in Deep-Sea Sediments," *Proceedings of the National Academy of Sciences* 103, no. 33 (2006): 12292.

24. Benson and Cook, 208; House et al., 12292.

25. Benson and Cook, 208; Qi Li, Zhishen Wu, and Xiaochun Li, "Prediction of CO_2 Leakage during Sequestration into Marine Sedimentary Strata," *Energy Conversion and Management* 50, no. 3 (2009): 505.

26. Benson and Cook, 208.

27. Ibid.; Smit et al., 437.

28. Benson and Cook, 208; Smit et al., 437; see eq. I.

29. Benson and Cook, 208; Smit et al., 437; see eq. II.

30. Benson and Cook, 208, present the molecular equations for those transformations. See Smit et al., 437.

31. Benson and Cook, 209. Intriguingly, it appears that cap rock is more susceptible to carbonate mineralization than the deeper injection zones, so should CO_2 escape toward the cap rocks, those same CO_2 fluids would be even more likely to become mineralized at the edge of the cap rock and decrease the overall likelihood of their escape. See also Smit et al., 437–38.

32. Benson and Cook, 209.

33. Ibid., 210.

34. Ibid. For an exploration of the potential to coordinate CCS storage with hydrate formation in an offshore setting, see H. Koide et al., "Deep Sub-Seabed Disposal of CO_2—The Most Protective Storage," *Energy Conversion and Management* 38 (1997): S253–258. See House et al., 2006. See also Jordan K. Eccles and Lincoln Pratson, "Global CO_2 Storage Potential of Self-Sealing Marine Sedimentary Strata," *Geophysical Research Letters* 39, no. 19, (2012).

35. Benson and Cook, 200. See also Daniel P. Schrag, "Storage of Carbon Dioxide in Offshore Sediments," *Science* 325, no. 5948 (2009): 1658, for a discussion on the suitability of offshore sediments for CCS storage.

36. Benson and Cook, 200.

37. Ibid.

38. Benson and Cook, 217; M. Bentham and M. Kirby, "CO_2 Storage in Saline Aquifers," *Oil and Gas Science and Technology* 60, no. 3 (2005): 560; Mark Anthony De Figueiredo, "The Liability of Carbon Dioxide Storage" (PhD diss., MIT, 2007), 33–34.

39. Benson and Cook, 217; Bentham and Kirby, 560; Smit et al., 367; De Figueiredo, 33–34.

40. Benson and Cook, 217; Smit et al., 368.

41. Benson and Cook, 217; Hellevang, 599.

42. Benson and Cook, 217; Hellevang, 599.

43. Benson and Cook, 217.

44. Ibid. Integral to the offshore CCS technologies explored (see n. 68) is the phenomenon of negative buoyancy. See House et al. See also Eccles and Pratson.

45. Benson and Cook, 217. Currently there is one abandoned coal mine in Colorado that serves as a storage facility for natural gas. Laura Chiaramonte et al., "Geomechanics and Seal Integrity for the Geologic Sequestration of CO_2," in *Carbon Capture and Storage: R&D Technologies for Sustainable Energy Future*, ed. Malti Goel, Baleshwar Kumar, and S. Nirmal Charan (Oxford: Alpha Science, 2008), 112. See also Ajay K. Singh, "R&D Challenges for CO_2 Storage in Coal Seams," in *Carbon Capture and Storage: R&D Technologies for a Sustainable Energy Future* (Hyderabad: Alpha Science International, 2008), 143, with application to such assets within India.

46. Benson and Cook, 217: "Gaseous CO_2 injected through wells will flow through the cleat system of the coal, diffuse into the coal matrix and be adsorbed onto the coal micro-pore surfaces, freeing up gases with lower affinity to coal (i.e., methane)."

47. Ibid.

48. There is also an arguable economic advantage to storing CO_2 in coal seams in that the placement of the CO_2 would likely enhance recovery of methane from the coal and potentially provide either energy support for the injection operations or enable revenue offsets, both of which could support the economics of the CCS operations. Christopher Bidlack, "Regulating the Inevitable: Understanding the Legal Consequences of and Providing for the Regulation of the Geologic Sequestration of Carbon Dioxide," *Journal Land Resources and Environmental Law* 30 (2010): 205. Smit et al., 368, with reference to the nomenclature of "enhanced coalbed methane (ECBM) recovery." See also Singh, 144.

49. Benson and Cook, 217.

50. Ibid., 218; Smit et al., 368; Chiaramonte et al., 112–113.

51. Benson and Cook, 219; Yousif K. Kharaka et al., "Geochemical Monitoring for Potential Environmental Impacts of Geologic Sequestration of CO_2," in *GeoChemistry of Geologic CO_2 Sequestration*, ed. Donald J. DePaolo, David R. Cole, Alexandra Navrotsky, and Ian C. Bourg (Washington DC: Geochemical Society, 2013), 404.

52. Benson and Cook, 220.

53. Ibid. Hans Plaat, "Underground Gas Storage: Why and How," in *Underground Gas Storage: Worldwide Experiences and Future Development in the UK and Europe*, ed. D. J. Evans and R. A. Chadwick (London: British Geological Society, Keyworth, 2009), 29.

54. Benson and Cook, 220; K.-H. Lux, "Design of Salt Caverns for the Storage of Natural Gas, Crude Oil, and Compressed Air: Geomechanical Aspects of Construction, Operation, and Abandonment," in *Underground Gas Storage: Worldwide Experiences and Future Development in the UK and Europe*, ed. D. J. Evans and R. A. Chadwick (London: British Geological Society, Keyworth, 2009), 93; Plaat, 30.

55. Benson and Cook, 220. With regard to additional safety concerns related to long-term operation of salt mines as gas storage facilities, see Plaat, 30–31.

56. Benson and Cook, 220. With a particular view toward overall safety and integrity, see Lux, 127.

57. Ibid., 197. See Smit et al., 368, stating a global minimum storage capacity of fifty years of active injection.

58. Smit et al., 480–481. See also Benson and Cook, 200: "To turn technical geological storage capacity into economical storage capacity, the storage project must be economically viable, technically feasible, safe, environmentally and socially sustainable and acceptable to the community. Given these constraints, it is inevitable that *the storage capacity that will actually be used will be significantly less than the technical potential* [italics added]."

59. Benson and Cook, 220–221; Smit et al., 363.

60. Smit et al., 365; Benson and Cook, 213. See also Hellevang, 601.

61. Benson and Cook, 213. Hellevang, 592, commented that these conditions are similar to those needed for more ordinary hydrocarbon systems.

62. Benson and Cook, 214.

63. Because the terms *economic* and *reserves* are defined by then-current levels of technology and market prices, it could be difficult to ascertain when a field is sufficiently beyond its useful application as an energy source. See Smit et al., 480–481, for a discussion on the complications of applying the term *reserves* to CCS projects.

64. Smit et al., 366. Benson and Cook, 215, highlight that CCS storage might enhance reservoir conditions for currently active fields. There are limiting factors, they note; for example, the injection zone would need to be deeper than 600 meters in order to remain safe. The available volume for storage is closely related to the volume areas left vacant by the removal of hydrocarbons from the pore space (221). However, the storage potential would also be limited by leftover hydrocarbons, injected waters, and other fluids injected in tertiary recovery efforts.

65. Ibid., 215. While such fields would be expected to contain many plugged and abandoned exploratory or production wells within the injection field, and thus pose a potential risk, experience has shown that a similar experiment at the Weyburn Project did not result in any detectable leaks. Ibid. See also Smit et al., 366.

66. Benson and Cook, 216; Stefan Bachu, "Sequestration of CO_2 in Geological Media: Criteria and Approach for Site Selection in Response to Climate Change," *Energy Conversion and Management* 41, no. 9 (2000): 962–963.

67. Benson and Cook, 213–214; Bachu, 961.

68. Benson and Cook, 214. See also Bachu, 966–968, for the seminal article on point.

69. Benson and Cook, 222. Based on 126 to 400 $GtCO_2$ for crude oil reservoirs and 800 $GtCO_2$ for natural gas reservoirs. They also report global hydrocarbon storage estimates at 900 $GtCO_2$ and 800 $GtCO_2$. See also De Figueiredo, 31.

70. Benson and Cook, 223. Over fourteen studies were recorded by their research team. See also De Figueiredo, 31.

71. Benson and Cook, 224.

72. David P. Flynn and Susan M. Marriott, "Carbon Sequestration: A Liability Pathway to Commercial Viability," *Natural Resources and Environment,* 24 (2009): 37. In the United States, onshore estimates suggest that depleted hydrocarbon fields might support 98 $GtCO_2$,1 that saline reservoirs might support up to 500 $GtCO_2$, and that coal seams might enable an additional 60 to 90 $GtCO_2$. Benson and Cook, 224. It would thus appear that the remainder of the EPA's estimate might include offshore CCS storage in addition to onshore assets.

73. In Europe, estimates suggest that depleted hydrocarbon fields might support 47 $GtCO_2$ and that saline reservoirs might support 30 to 577 $GtCO_2$. Ibid., 222–223. In Canada, estimates suggest that depleted hydrocarbon fields might support 10 $GtCO_2$ and that saline reservoirs might support in excess of 4,000 $GtCO_2$. Australian estimates suggest that depleted hydrocarbon fields might support 15 $GtCO_2$ and that saline reservoirs might support up to 740 $GtCO_2$.

74. Howard Herzog, "Carbon Dioxide Capture and Storage," in *The Economics of Climate Change*, ed. D. Helms and C. Hepburn (New York: Oxford University Press, 2010), 264.

75. Ibid., 265. For example, CCS injection activities at Sleipner began twenty years ago, in 1996. Hellevang, 598. Smit et al., 375.

76. Benson and Cook, 199. It would appear that the earliest suggestions for CCS were made in the 1970s.

77. Ibid., 200.

78. Ibid., 204. They list the historical Frio Brine Project of Texas and then-current plans for projects in Ketzin of Germany, in the Otway Basin of Australia, at the Teapot Dome of Wyoming, and at Nagaoka in Japan, among others. A more recent review reported a dozen active CCS projects in operation. Hellevang, 598.

79. Benson and Cook, 244. They note that 200 $MtCO_2$ has been stored in the Pisgah Anticline in Mississippi for 65 million years. Other petroleum systems suggest natural gas storage periods in the dozens of millions of years. Ibid., 245. See also Smit et al., 361.

80. Philip M. Marston and Patricia A. Moore, "From EOR to CCS: The Evolving legal and Regulatory Framework for Carbon Capture and Storage," *Energy Law Journal* 29 (2008): 423.

81. Ibid., 424.

82. Ibid.

83. Ibid. Nine EOR-CCS projects are reported in operation, targeting the permanent sequestration of most of the injected CO_2. Hellevang, 600.

84. Marston and Moore, 424.

85. Ibid. The term *acquis* here is a direct quote from Marston and Moore. Similarly, they refer to climate change policymakers drafting CCS rules as Molière's *bourgeois gentilhomme* who has just realized that he has been speaking prose, vis-à-vis the forty years of EOR-related CO_2 injections and their legal frameworks.

86. Herzog, 265; Benson and Cook, 212.

87. Benson and Cook, 212.

88. Benson and Cook, 212; Herzog, 265. The percentage of CO_2 within those reinjection streams tends to be 90 percent, thus very similar to the potential injection of "unpure" CO_2 streams within CCS activities. Due to the return of the gases into formations either identical or similar to their source within a short time of extraction, their replacement in subterranean storage is not usually described in terms of "storage" but as "returned" volumes. Ibid.

89. The role of AGI is different from the injection of hazardous wastes, in that AGI injects supercritical gases and the other injects fluids primarily based on contaminated waters. Benson and Cook, 212. In the United States, extensive use of hazardous waste injection facilities enables the disposal of approximately 34.1 million cubic meters of wastes annually.

90. Herzog, 265. In Canada's Alberta and British Columbia provinces, for example, 200 million cubic meters of acid gas streams were reinjected into the fields from which they were produced. There are over forty such injection projects in those areas. Benson and Cook, 201, 212. Also, AGI has found application at Natuna in

Indonesia, at In Salah in Algeria, and at Gorgon in Australia. Benson and Cook, 199. See also Smit et al., 374.

91. Benson and Cook, 212; De Figueiredo, 231, 239.

92. Benson and Cook, 212; De Figueiredo, 232.

93. Benson and Cook, 212.

94. Herzog, 267; Benson and Cook, 200.

95. Ibid., 245.

96. Helium was once required to be stored within the United States by federal statute because it was seen as a high-priority resource for rocketry and other security-related purposes; the Federal Helium Program was begun in 1925. Helium was stored on a planning frame of decades, although currently, the US federal government is drawing down its helium reserves per the Helium Stewardship Act of 2013; H.R. 527 (113th Congress), enacted at 50 U.S.C. §167.

97. Herzog, 267; Benson and Cook, 211.

98. Herzog, 267.

99. Ibid.

100. Benson and Cook, 211. Of nine known documented cases, five were wellbore integrity problems, three cases were due to cap rock problems, and one involved a project failure based on poor site selection. Ibid., 245.

101. Ibid. The source states $10^{(-5)}$, or $\frac{1}{100,000}$, which is equivalent to 0.001 percent when stated as a percentage.

102. Ibid.

103. Ibid., 246, based on models developed for the Sleipner CCS project in Norway. Other models found similar results. Zhou et al. found no foreseeable release in 5,000 years and Walton et al. found a probably release of 00.1 percent if abandoned wells are included as a possible migration pathway.

104. Ibid.

105. Herzog, 267. The first known use of EOR-related CO_2 injections dates to 1972. Benson and Cook, 199, place those first 1970s-era EOR projects in Texas. Other sources (e.g., Marston and Moore, 428) pinpoint the field as the Scurry Area Canyon Reef Operators Committee (SACROC) unit in the Permian basin.

106. Herzog, 267.

107. Benson and Cook, 203. While most of these projects used CO_2 derived from natural reservoirs, a few employed anthropogenic sources. However, while Benson and Cook (211) reported that the injected gas at the Rangely Weber project used

anthropogenic CO_2 provided from the Schute Creek processing facility for the LaBarge gas field, in fact that gas is from a natural CO_2 reservoir that presents 65 percent CO_2 by content. In addition, "The LaBarge gas reserve is 'multi-component gas that is primarily CO_2 [carbon dioxide] but also contains other valuable components, including methane and helium. This reserve is also a sour gas because it contains H_2S [hydrogen sulfide], and this reserve is well suited for the development and sale of products of CO_2 for CO_2 flooding, methane for energy and helium as a helium source.'" State Board of Equalization, Wyoming, Docket No. 2000–142, Finding of Facts, B.3: http://taxappeals.state.wy.us/images/docket_no_2000142.htm. See also De Figueiredo, 262–263.

108. However, the initial EOR project at SACROC obtained its CO_2 volumes from coincidentally produced natural gas that would have otherwise been flared (Marston and Moore, 428), so to a certain extent, the original project did provide a reduction in greenhouse gas emissions, at least locally to the field. Also, as discussed above (see Marston and Moore, 424), it is likely that less than 50 percent of the injected CO_2 returns to the surface at most fields. Thus, EOR permanently sequesters a great deal of CO_2, but the bulk volume of that CO_2 is not industrial in source and results in additional crude oil production, so this form of CCS storage is not generally regarded as functional CCS, although the technology and environmental risks are quite similar.

109. Benson and Cook, 203. While data for storage are not a normal feature of EOR metrics, tracking the performance of EOR injections would provide data on pressure levels and potential migrations.

110. Ibid., 230. In particular, the applications of EOR technology and AGI technology are expected to be close mappings to the needs of CCS activities. Some adjustments might be necessary in some locations for higher pressure ratings or corrosion resistance. Furthermore, existing CCS projects currently sequester, on an annual basis, a sufficient volume of CO_2 to offset the emissions of either New Zealand or Singapore. Smit et al., 370.

111. Benson and Cook, 230.

112. Ibid., 233; Hellevang, 591.

113. Herzog, 268; Hellevang, 592.

114. Kole reviews literature that suggests there might be a three-to-one factor problem: if three coal-fired power plants are connected to a capture and sequestration system, an additional power plant would be needed to power the CCS system itself, including capture, transport, and injection. Allison Kole, "Carbon Capture and Storage: How Bad Policy Is By-Passing Environmental Safeguards," *Journal Environmental and Sustainability Law* 20 (2013): 109. That she separately discusses the costs of the whole CCS industry as enabling an 80 to 90 percent reduction in CO_2 emissions implies that CCS might work but would simultaneously increase the need for coal.

Similar analysis can be found at Arnold W. Reitze Jr. and Marie Bradshaw Durrant, "State and Regional Control of Geological Carbon Sequestration," *Environmental Law Reporter News and Analysis* 41 (2011): 10349.

115. Herzog, 264; Benson and Cook, 233.

116. Herzog, 265.

117. Ibid. These pipelines transport CO_2 from the giant CO_2 reservoirs that lay just east of the Rocky Mountains, such as ExxonMobil's Schute Creek Field, processed by the LaBarge Gas Processing Facility. See Marston and Moore, 429.

118. Susanna Much, "The Emerging International Regulation of Carbon Storage in Sub-Seabed Geological Formations," in *Shipping, Law and the Marine Environment in the 21st Century: Emerging Challenges for the Law of the Sea—Legal Implications and Liabilities,* ed. Richard Caddel and Rhidian Thomas (Witney: Lawtext Publishing, 2013), 230. See also Bidlack, 202. Shipping is expected in either international or transboundary contexts or in domestic settings where pipelines might be more complicated than boats, such as in Indonesia. Vessel shipments could be handled by fluids stored in tanks.

119. See Roy Andrew Partain, "Avoiding Epimetheus: Planning Ahead for the Commercial Development of Offshore Methane Hydrates," *Sustainable Development Law and Policy* 15 (2015): 19–20.

120. Benson and Cook, 233. Transport lines often also include a "slug catcher" to absorb the impact of uneven fluid flow accumulation to protect downstream pipes from kinetic shocks.

121. Ibid., 200, with reference to their glossary definition: "At a temperature and pressure above the critical temperature and pressure of the substance concerned. The critical point represents the highest temperature and pressure at which the substance can exist as a vapour and liquid in equilibrium." See also Hellevang, 593, for a description of the associated CO_2 phase stability diagrams.

122. Benson and Cook, 200; Hellevang, 593.

123. Marston and Moore, 426; Hellevang, 593.

124. Benson and Cook, 230. AGI wells routinely contain an additional backup valve below the land surface that is automatically triggered when the two surface-mounted valves fail.

125. Ibid.

126. Ibid.

127. Bidlack, 203, presents an argument that preexisting regulations on injection wells from before the advent of CCS considerations are likely sufficient to enable safe injection within the CCS context.

128. Benson and Cook, 231. Such measures could include periodic wellbore inspections, improved blowout preventer (BOP) maintenance, installation of redundant BOP on wells of concern, ongoing improvements to crew awareness, and provision of contingency plans and emergency response training.

129. Ibid., 232.

130. Ibid. For example, regulatory bodies have demonstrated historical success with specific oil and gas fields. They can rely on scientifically accepted standards for the range of tolerable injection pressures.

131. Ibid., 231.

132. Ibid. While some concerns exist on the corrosive nature of the injected materials, AGI has also faced this issue with impure CO_2, often mixed with other acids as well, and thus the technical nature of the problem is well understood. For example, CO_2-resistant cements and special metallic alloys exist for these purposes.

133. Ibid.

134. Jennifer L. Lewicki, Jens Birkholzer, and Chin-Fu Tsang, "Natural and Industrial Analogues for Leakage of CO_2 from Storage Reservoirs: Identification of Features, Events, and Processes and Lessons Learned," *Environmental Geology* 52, no. 3 (2007): 457, tables 1 and 2.

135. Ibid. See table 1, 459; and table 2, 460.

136. Roger D. Aines et al., "Quantifying the Potential Exposure Hazard due to Energetic Releases of CO_2 from a Failed Sequestration Well," *Energy Procedia* 1, no. 1 (2009): 2422. See also Lewicki et al., table 2 (460).

137. Ibid., 458.

138. Ibid. See table 3 (462). Table 3 presents a column that documents how the vents were mitigated or otherwise addressed. The natural venting events are addressed passively with monitoring; the anthropogenic events were actively mitigated and often resulted in plugging.

139. F. J. Gouveia and S. J. Friedmann, *Timing and Prediction of CO_2 Eruptions from Crystal Geyser, UT* (Washington, DC: US Department of Energy, 2006), 2. See also Elizabeth J. Wilson, S. Julio Friedmann, and Melisa F. Pollak, "Research for Deployment: Incorporating Risk, Regulation, and Liability for Carbon Capture and Sequestration," *Environmental Science and Technology* 41, no. 17 (2007): 5948. The well is currently known as the Crystal Geyser and is located near the Green River in Utah. See Lewicki et al., table 2 (460) at B2.

140. Gouveia and Friedmann, 2. See also Wilson, Friedmann, and Pollak, 5948.

141. Ibid. See also Wilson, Friedmann, and Pollak, 5948.

142. Ibid., 5, See figure 3. Youtube.com contains many tourist home videos of visitors standing next to the erupting Crystal Geyser. See, for example, http://www.youtube.com/watch?v=uJkgFxr68K8 or http://www.youtube.com/watch?v=VLTEiQQSwCc.

143. Ibid., 2. See also Wilson, Friedmann, and Pollak, 5948.

144. Wilson, Friedmann, and Pollak, 5948.

145. Gouveia and Friedmann, 2, 5.

146. Ibid., 5.

147. Wilson, Friedmann, and Pollak, 5948.

148. Gouveia and Friedmann, 6.

149. Wilson, Friedmann, and Pollak, 5948.

150. See Lewicki et al.; see table 2, 460, B1.

151. Wilson, Friedmann, and Pollak, 5948.

152. Ibid.

153. Ibid.

154. Ibid.

155. The CO_2 leakage at Sheep Hill was so intense that the CO_2 formed blocks of dry ice that were thrown through the field. Also, it bears noting that the comparative flow rates of the Sheep Hill versus those of the Crystal Geyser are roughly 100:1 or 1,000:1, demonstrating that both small and large CO_2 eruptions have been observed in the petroleum industry and safely addressed. See the discussion on Crystal Geyser in note 142.

156. Wilson, Friedmann, and Pollak, 5948.

157. Ibid.

158. G. Buonasorte et al., "Results of Geothermal Exploration in Central Italy (Latium-Campania)," in *Proceedings of the World Geothermal Congress* (Florence, Italy, 1995), 1296; A. Annunziatellis et al. "Gas Migration along Fault Systems and through the Vadose Zone in the Latera Caldera (Central Italy): Implications for CO_2 Geological Storage," *International Journal of Greenhouse Gas Control* 2, no. 3 (2008): 354–356.

159. Buonasorte et al., 1293. The saline water contains several gases at about 3 to 6 percent of mass, with CO_2 representing 98.35 percent with traces of methane and hydrogen sulfide. Annunziatellis et al., 356.

160. Lewicki et al., table 2 (460), at B4.

161. Ibid.

162. Aines et al., 2422. The underlying source for the CO_2 volumes is predominantly volcanic. Buonasorte et al., 1293, 1296.

163. Lewicki et al., table 2 (460), B4.

164. Ibid., table 3 (462), B4.

165. Ibid., See table 2 (460), B5.

166. Aines et al., 2422. Lewicki et al., table 2 (460), B5.

167. Lewicki et al., table 3 (462), at B5.

168. Ibid., table 2 (460), at B8. The well was identified as Edmund Trust #1–33 well. Aines et al., 2422.

169. Lewicki et al., table 3 (462), B8.

170. Based on 0.9 kg/sec. Aines et al., 2422.

171. Lewicki et al., table 2 (460), B7. Aines et al., 2422.

172. Ibid., table 2 (460), B7.

173. Ibid.; See table 3 (462), B7.

174. Ibid.

175. Aines et al., 2422.

176. Lewicki et al., table 3 (462), B6.

177. Ibid., table 2 (460), B6.

178. Ibid.

179. Ibid., table 3 (462), B6. Geophysical monitoring devices were also installed.

180. Aines et al., 2422.

181. Ibid.

182. Ibid.

183. Ibid.

184. Lewicki et al., 463–464.

185. Ibid., 463.

186. Ibid., 464.

187. Ibid.

188. Ibid.

189. Ibid.

190. Ibid.

191. Wilson, Friedmann, and Pollak, 5945. See Smit et al., 485–486, with particular concerns to the overall "scale at which geological carbon sequestration must be carried out to reduce CO_2 emissions."

192. Allan Ingelson, Anne Kleffner, and Norma Nielson, "Long-Term Liability for Carbon Capture and Storage in Depleted North American Oil and Gas Reservoirs: A Comparative Analysis," *Energy Law Journal* 31 (2010): 436. They cite Sally M. Benson, *Carbon Dioxide Capture and Storage in Underground Geologic Formations* (Berkeley, CA: Lawrence Berkeley National Laboratory, 2004).

193. Bidlack, 208, citing De Figueiredo without pin-cite, but presumably to De Figueiredo, "The Liability of Carbon Dioxide Storage," 154–191.

194. Ibid., 208–209. See the discussions in this chapter in sections. 2.4.2C and 2.4.2B. This is similar to the organization proposed by Wilson, Friedmann, and Pollak, 946: (1) surface leakage, (2) groundwater quality, (3) regional impact, and (4) permanence. Their listing included clarifications on liability rules, but that is handled as a separate type of issue item within this study. Ingelson et al., 436, cite from a study on the Norwegian Sleipner gas field when they describe the five types of harm: hazards to human health and safety, contamination hazards to groundwater, hazards to terrestrial and marine ecosystems, hazards of induced seismicity, and hazards of other hidden toxic gas streams within the CO_2 injection volumes. They cite Semere Solomon, "Carbon Dioxide Storage: Geological Security and Environmental Issues–Case Study on the Sleipner Gas field in Norway," *Bellona Report* (2007).

195. Swayne and Phillips.

196. Ibid.

197. Ibid.

198. Ibid.

199. Ibid.

200. Bidlack, 209. See also Smit et al., 487: "Nothing is intrinsically unhealthy about carbonated water."

201. David E. Adelman and Ian J. Duncan, "The Limits of Liability in Promoting Safe Geologic Sequestration of CO_2," *Duke Environmental Law and Policy Forum* 22 (2011): 4.

202. Chiara Trabucchi, Michael Donlan, and Sarah Wade, "A Multi-Disciplinary Framework to Monetize Financial Consequences Arising from CCS Projects and Motivate Effective Financial Responsibility," *International Journal of Greenhouse Gas Control* 4, no. 2 (2010): 390. Chiara Trabucchi and L. Patton, "Storing Carbon:

Options for Liability Risk Management, Financial Responsibility," World Climate Change Report 170 (Washington, DC: Bureau of National Affairs, 2008): 10.

203. Swayne and Phillips. However, the literature does indicate that although the probability is small, pressure changes as a result of CCS could potentially cause ground heave and even trigger seismic events. Elizabeth J. Wilson, Alexandra B. Klass, and Sara Bergan, "Assessing a Liability Regime for Carbon Capture and Storage," *Energy Procedia* 1, no. 1 (2009): 4576.

204. It is important to keep in mind that accident and tort law respond primarily to anthropogenic sources of causation or negligence to provide for foreseeable natural sources of causation. Given the very long time frame of CCS storage, it might be necessary to remember that geological events can often be unforeseeable but still provide the necessary causation of latter accidents and potential injuries.

205. Benson and Cook, 205. See also Mahmut Sengul, "Reservoir-Well Integrity Aspects of Carbon Capture and Storage," in *Carbon Capture and Storage: R&D Technologies for Sustainable Energy Future*, ed. Malti Goel, Baleshwar Kumar, and S. Nirmal Charan (Oxford: Alpha Science, 2008), 188.

206. See Jan Martin Nordbotten, Michael A. Celia, and Stefan Bachu. "Injection and Storage of CO_2 in Deep Saline Aquifers: Analytical Solution for CO_2 Plume Evolution during Injection." *Transport in Porous Media* 58, no. 3 (2005): 339–360.

207. Benson and Cook, 205. The conditions that enable mixing of the injected fluids and the in situ fluids are called miscibility.

208. Ibid. Given sufficient time, say, a century or so, most CO_2 fluids are expected to become miscible. Hellevang, 595–596.

209. Benson and Cook, 205. The buoyancy results from the difference in density of the supercritical CO_2 and the saline reserves, the saline being denser.

210. S. Ide et al., "CO_2 Leakage through Existing Wells: Current Technology and Regulations," in *Proceedings of the Eighth International Conference on Greenhouse Gas Control Technologies* (IEA Greenhouse Gas Programme, 2006), 1. Hellevang, 595–596.

211. Ide et al., 1.

212. Benson and Cook, 205. Hellevang, 593, contrasts the conditions at the Utsira CCS Unit against those of Snøhvit. Hellevang.

213. Benson and Cook, 206; Hellevang, 595.

214. Benson and Cook, 206; Hellevang, 594.

215. Benson and Cook, 205; House et al., 12291.

216. Benson and Cook, 205. Capillary forces, surface tension, and other electromagnetic forces can enable fingering within small spaces. This is coincident with residual

trapping. Residual trapping in certain conditions can reach up to 15 to 25 percent of the injected CO_2. See also Hellevang, 596.

217. Benson and Cook, 205. See also House et al., 12291.

218. Benson and Cook, 206. See also House et al., 12291.

219. Benson and Cook, 205.

220. Ibid.

221. Ibid.

222. Benson and Cook, 206; Hellevang, 594.

223. Benson and Cook, 206. Dissolution of CO_2 into water volumes is reduced as pressure increases, as temperatures decrease, and as salinity rises.

224. Ibid. Hellevang, 594.

225. Benson and Cook, 206. The underlying reference was to "millimeters to centimeters per year," clearly implying that it might take several centuries to reach that first meter of travel.

226. Ibid.

227. Ide et al., 1; Smit et al., 365.

228. Ide et al., 1; Kharaka et al., 414.

229. Examples are SO_x, NO_x, or H_2S. See Ariel A. Chialvo, Lukas Vlcek, and David R. Cole, "Acid Gases in CO_2-Rich Subsurface Geologic Environments," *Reviews in Mineralogy and Geochemistry* 77, no. 1 (2013): 362–363.

230. Benson and Cook, 220. See also Mahmut Sengul, "Reservoir-Well Integrity Aspects of Carbon Capture and Storage," in *Carbon Capture and Storage: R&D Technologies for Sustainable Energy Future,* ed. Malti Goel, Baleshwar Kumar, and S. Nirmal Charan (Oxford: Alpha Science, 2008), 192. Sengul suggests that while some policymakers might want to co-inject H_2S with CO_2, the combination could prove more corrosive than CO_2 alone.

231. Because this study does not engage in the liability of transport or other upstream items, we do not discuss the additional concerns raised by impure CO_2 at those earlier stages.

232. Benson and Cook, 220. See also Chialvo et al., 362, who note concerns related to pipeline corrosion during transport to the injection site.

233. Benson and Cook, 220; Chialvo et al., 363.

234. Benson and Cook, 220.

235. Ibid.

236. Brent Miyazaki, "Well Integrity: An Overlooked Source of Risk and Liability for Underground Natural Gas Storage. Lessons Learned from Incidents in the U.S.A.," in *Underground Gas Storage: Worldwide Experiences and Future Development in the UK and Europe*, ed. D. J. Evans and R. A. Chadwick (London: British Geological Society, Keyworth, 2009): 167; Kharaka et al., 414; Ide et al., 2. Idal et al., 2, 3, also point to broader issues in play, as the technology in plugging has not substantially changed in the past forty to fifty years. While certain chemical additives have improved the overall viability of plugs, the remainders of technological improvement are primarily centered on the size of plugs and how many plugs might be needed per well.

237. Ide et al., 2. See also Sengul "Reservoir-Well Integrity," 191.

238. Miyazaki, 167; Ide et al., 1. Here, the notion of being improperly sealed denotes a technological character of the plugging efforts and not a regard for whether regulations on plugging and abandoning were properly followed. Indeed, for many earlier wells, they were plugged without the benefit of regulatory guidance. A larger concern exists with deeper wells from later periods that were improperly plugged due to technologically inferior cements that might weaken or fail when exposed to injected CO_2 volumes or carbonic acid. Many wells were plugged with no expectation of needing to resist such injected fluids. Indeed, records also demonstrate that much earlier wells were plugged with whatever was literally on hand at the time, including, according to Ide et al., 2, such plugging materials as "tree stumps, logs, animal carcasses, and mud." See also Sengul, 190–191.

239. Kharaka et al., 414; Ide et al., 2. Regulation of plugging and abandonment began in California as early as 1915 and in Texas in 1919. The oil industry had of course begun much earlier (e.g., in 1866 in Texas), and scores of thousands of wells were drilled prior to those early regulatory efforts. And in some states, regulatory guidance arrived much later (e.g., in 1949 for Indiana), which had begun oil production back in 1876. Ibid.

240. First adopted in 1952, updated repeatedly since. Ide et al., 3. See also J. William Carey, "Geochemistry of Wellbore Integrity in CO_2 Sequestration: Portland Cement-Steel-Brine-CO_2 Interactions," *Reviews in Mineralogy and Geochemistry* 77, no. 1 (2013): 509.

241. Ide et al., 3; Carey, 509. See also Miyazaki, 167–168, who note that improper casings and poor completions still occur despite modern regulations. Ibid.

242. Ide et al., 3. "In a properly constructed well, there is little potential for CO_2 bearing fluids in the reservoir to chemically impact well integrity" (Carey, 509).

243. Elizabeth Aldrich, Cassandra Koerner, and David Solan, "Analysis of Liability Regimes for Carbon Capture and Sequestration: A Review for Policymakers," *Energy Policy Institute* (2011): 7.

244. Ibid., 7.

245. Ingelson et al., 437.

246. Ibid., 437–438.

247. Ibid., 438.

248. Ibid.

249. Ibid., 437, citing Benson and Cook, 245, 247.

250. Aines et al., 2422.

251. Ibid., 2422–2423.

252. Ibid., 2422.

253. Aines et al., 2423, explain that much of their model was based on publicly available modeling data on CO_2 behavior; for example, the National Institute of Standards and Technology provides a set of tables that detail CO_2 behavior for modelers.

254. Ibid. During its free-flow period, the Sheep Mountain well vented approximately 7,000 to 11,000 tons of CO_2 per day. Wilson, Friedmann, and Pollak, 5948.

255. Ibid., 5946, citing Curtis M. Oldenburg and André J. A. Unger, "On Leakage and Seepage from Geologic Carbon Sequestration Sites," *Vadose Zone Journal* 2, no. 3 (2003). See also Stephen M. Testa and James A. Jacobs, *Oil Spills and Gas Leaks: Environmental Response, Prevention, and Cost Recovery* (New York: McGraw-Hill, 2014), 194, where they apply similar language to natural gas storage issues.

256. Benson and Cook, 244.

257. Donna M. Attanasio, "Surveying the Risks of Carbon Dioxide: Geological Sequestration and Storage Projects in the United States," *Environmental Law Reporter News and Analysis* 39 (2009): 10386.

258. Wilson, Friedmann, and Pollak, 5946.

259. Benson and Cook, 246, 251. See also the discussion in this chapter in section 2.3.

260. Ibid., 251.

261. Ingelson et al., 436. Wilson, Klass, and Bergan point to the fact that at very high concentrations (greater than 30 percent), CO_2 may cause immediate human death from asphyxiation. See also Alexandra B. Klass and Elizabeth J. Wilson, "Climate Change and Carbon Sequestration: Assessing a Liability Regime for Long-Term Storage of Carbon Dioxide," *Emory Law Journal* 58 (2008): 118.

262. NASA, for example, lists the components from larger to smaller: 78.08 percent nitrogen (N_2), 20.95 percent oxygen (O_2), argon (Ar) 9340 ppm; carbon dioxide (CO_2) 400 ppm, neon (Ne) 18.18 ppm; helium (He) 5.24 ppm; CH_4 1.7 ppm;

krypton (Kr) 1.14 ppm; and hydrogen (H$_2$) 0.55 ppm. See NASA, "Earth Fact Sheet," http://nssdc.gsfc.nasa.gov/planetary/factsheet/earthfact.html.

263. Victor Byers Flatt, "Paving the Legal Path for Carbon Sequestration," *Duke Environmental Law and Policy Forum* 19 (2009): 221.

264. Ibid.

265. Benson and Cook, 246; Kharaka et al., 415.

266. Flatt, 221.

267. Benson and Cook, 246; Kharaka et al., 415.

268. Ingelson, et al., 436.

269. See the discussion in this chapter in section 2.3. See also Aines et al., 2422.

270. TEEL-2 is the level that enables potential permanent damage, whereas TEEL-3 is the level that enables potential "life-threatening adverse health effects or death." The US Department of Energy Subcommittee on Consequence Assessment and Protective Actions (SCAPA) determines the definitions of TEEL settings for several thousand airborne hazards. See the US Department of Energy's public database on protective action criteria (PAC) at http://www.atlintl.com/DOE/teels/teel.html. The TEEL levels of CO$_2$ are continuous exposure for 60 minutes at 30,000 ppm for TEEL-2 and 50,000 ppm for TEEL-3. See also Kharaka et al., 415.

271. Ingelson et al., 436–437, quoting from Benson, "Carbon Dioxide." See also Smit et al., 495–496.

272. Benson and Cook, 247. Deaths have occurred at the Poggio dell'Ulivo fields in Italy, but those deaths stand in the lack of reported deaths from many other global sources of natural venting events.

273. Aines et al., 2423. Again, as part of their effort to demonstrate the reliability of models to predict impacts of CO$_2$ release events, they relied on the National Atmospheric Advisory Release Capacity (NARAC) based at the Lawrence Livermore National Laboratories. NARAC is designated by the US Homeland Security Council as the provider of the Interagency Modeling and Atmospheric Assessment Center, which is to provide a single source for federal atmospheric dispersion predictions during "Incidents of National Significance," and as such maintains models to predict the dispersion of various airborne threats. Ibid., 2424.

274. Ibid.

275. Ibid. A danger zone is the zone within which CO$_2$ exceeds TEEL-2 and TEEL-3. See the chapter discussion on TEEL levels in n. 269.

276. Aines et al., 2424. This distance reflects TEEL-2 levels; TEEL-3 would be more limited and remain closer to the wellhead. Similarly, it was reported that CO$_2$

concentrations could fall by a factor of 50 after 100 meters of aerial transport and by a factor of 125 after 200 meters; thus, the overall risks would be expected to be very localized to the leak. Smit et al., 496.

277. A standard medical test for chronic respiratory disease and heart failure is the six-minute walk: a healthy adult is expected to be able to walk 400 to 700 meters within six minutes. See Paul L. Enright, "The Six-Minute Walk Test," *Respiratory Care* 48, no. 8 (2003). It has been reported that the severe medical events, such as loss of consciousness, would likely occur within fifteen minutes. See Kharaka et al., 415.

278. Aines et al., 2424.

279. Ibid.

280. Ibid.

281. Ibid., citing Les Skinner, "CO_2 Blowouts: An Emerging Problem: Well Control and Intervention," *World Oil* 224, no. 1 (2003).

282. Aines et al., 2425.

283. Aldrich, et al., 4.

284. Ibid.; Adelman and Duncan, 8.

285. Aldrich et al., 4–5, citing Alexandra B. Klass and Elizabeth J. Wilson, "Carbon Capture and Sequestration: Identifying and Managing Risks," *Issues in Legal Scholarship* 8, no. 1 (2009). The IPCC clearly held that this release of CO_2 from Lake Nyos is "not representative of the seepage through wells or factures that may occur from underground geological storage sites." Benson and Cook, 211.

286. See Elizabeth J . Wilson, Timothy L. Johnson, and David W. Keith, "Regulating the Ultimate Sink: Managing the Risks of Geologic CO_2 Storage," *Environmental Science and Technology* 37, no. 16 (2003): 3477.

287. Aldrich et al., 5; Flatt, 221; Smit et al., 492. But see Smit et al.'s discussion on potential damage to the vadose zone, the shallow subsurface, and nearby plants, (492–493).

288. Ide et al., 1.

289. Wilson, Friedmann, and Pollak, 5946.

290. Benson and Cook, 251.

291. Ibid., 251–52.

292. Ibid., 252. Kharaka et al., 416.

293. Benson and Cook, 252. Benson and Cook actually called for "effective regulatory oversight," but it is unclear if their term of art refers to a broader sense of public

legal engagement or a more limited sense of public regulations. In an effort to maintain a minimum of normative calls, the paraphrase here relies on the broader notion of clear legal duties and expectations.

294. Ibid.

295. Although some researchers retain their overall doubts on the certainties of overall safety claimed in the majority literature, see John Pendergrass, "Long-Term Stewardship of Geological Sequestration of CO_2," *Environmental Law Reporter: News and Analysis* 43, no. 8, (2013): 10660.

296. Ian Havercroft and Richard Macrory, *Legal Liability and Carbon Capture and Storage: A Comparative Perspective* (London: Global CCS Institute, University College London, Faculty of Law, 2014), 11.

297. See Pendergrass, 10659, who disputes Adelman and Duncan's assertions that "experience in other fields suggests that uncertainty is a given and that technology and human endeavors do not always operate as expected or intended."

298. Benson and Cook, 224.

299. See ibid., listing thirteen items of care and five items of concern (225).

300. For example, in the EU, the CCS directive (Council Directive 2009/31/EC, Annex I) provides that each member state should direct its own requirements and procedures for site selection; site selection rules remain within the member state's competence. Avelien Haan-Kamminga, Martha Roggenkamp, and Edwin Woerdman, "Legal Uncertainties of Carbon Capture and Storage in the EU: The Netherlands as an Example," *Carbon and Climate Law Review* 4, no. 3 (2010): 245.

301. Benson and Cook, 224.

302. Ibid.

303. Ibid., 225.

304. Ibid. "Regional in nature" refers to the breadth and domain of the rock, suggesting that it is sufficient in size to cover both initial injections and later migrations of the fluids.

305. Here meaning faults, fractures, or well structures. See also the discussion in ibid., 228, emphasizing the difficulties of locating ancient wells or recent wells that went uncompleted for production purposes for their lack of metallic content susceptible to remote detection methods.

306. Ibid., 225, 227, discontinuously.

307. Ibid., 227.

308. Ibid.

309. Ibid.

310. Ibid.

311. Ibid.

312. Ibid., 227–28.

313. A. McGarr et al., "Coping with Earthquakes Induced by Fluid Injection." *Science* 347, no. 6224 (2015), 830; Smit et al., 492 ; Kharaka et al., 415.

314. McGarr et al., 830 ; Kharaka et al., 415.

315. Jonas J. Monast, Brooks R. Pearson, and Lincoln F. Pratson, "A Cooperative Federalism Framework for CCS Regulation," *Environmental and Energy Law and Policy Journal* 7 (2012): 37.

316. McGarr et al., 830. See also Smit et al., 489, which explains the Mohr-Coulomb analysis of expected shear stress as it relates to induced seismicity.

317. McGarr et al., 830. See also Smit et al., 488, on the likelihood of induced seismic events being detectable by humans at the surface.

318. McGarr et al., 830.

319. Ibid. Current models are built on a one-year basis, which can be iterated on declining levels of reliability.

320. Ibid., 831. One of the critical issues is that after the CO_2 reaches a certain scale and radial distance from the wellbore, the plume would be harder to affect with mitigation efforts. Another issue is that careful initiation of the injection procedures could prevent accidental reopening of faults or enable earlier efforts to mitigate such concerns prior to irreversible damage to the subsurface, enabling larger volumes to be injected later in time.

321. Ibid.

322. Ibid.

323. Ibid.

324. Ibid.

325. Ibid. See Julian J. Bommer et al., "Control of Hazard Due to Seismicity Induced by a Hot Fractured Rock Geothermal Project." *Engineering Geology* 83, no. 4 (2006): 295, reporting on a similar seismic "traffic light" network installed in El Salvador, such as a dashboard of green-yellow-red indicators of safe and permitted, warning but permitted, and unpermitted periods of injection. Similar methods are widely used in the oil and gas industry for both risk management and financial controls because they enable managers to observe a variety of data inputs and quickly identify areas of concern without needing to process numbers visually. Only the data for critical concerns are presented after a primary screening.

326. Ibid., 297–298.

327. Ibid.

328. Ibid., 293.

329. Ibid.

330. Ibid., 291–93.

331. Ibid., 292.

332. Ibid., 292–93. They provide reference to the US Army Engineering Manual EM 1110–2-3800, which regulates ground shocks from blasting events, the Barneich system for traffic-induced vibrations and shocks, and the Athanaso-Poulos and Pelekis system for addressing construction sites piling shocks (292, 292, and 293, respectively).

333. Wilson, Klass, and Bergan, 4576.

334. Ibid. See also Smit et al., 492: "In well-chosen sites, the main impact is nuisance: earthquakes that might be felt but that usually do not do any measurable damage."

335. Flatt, 221.

336. Adelman and Duncan, 11; Smit et al., 487. See also Hellevang, 593.

337. The term *buoyant* as used here refers to the gravity effects on the CO_2, which would provide motive energy for the CO_2 to move vertically toward the surface if otherwise unimpeded in its movements.

338. Flatt, 221.

339. Ibid.

340. Benson and Cook, 256. The original citation is to a lateral spread of 100 square kilometers., which equates to a radial distance as already given.

341. As CO_2 encounters water (H_2O), carbonic acid (H_2CO_3) is created. Carbonic acid is well known to dissolve limestone; cave interiors often feature stalactites and stalagmites resulting from long-term carbonic acid action on limestone formations. See also Monast et al., 37; Smit et al., 487.

342. Wilson, Friedmann, and Pollak, 5946. See also Monast et al., 37.

343. Wilson, Friedmann, and Pollak, 5946; Benson and Cook, 248. See also Monast et al., 37.

344. Benson and Cook, 247. See also Monast et al., 37; Smit et al., 487.

345. For the centrality of the discussions on needs for refinements to both subterranean property law, tort law, and of liable parties (see the discussion in this chapter

in section 2.3), are primarily drawn in response to needs to determine legal rights affecting those subterranean water bodies. This is not unique to CCS activities, as nonconventional fracturing technologies have stimulated similar concerns for contamination of water supplies. See David A. Dana and Hannah J. Wiseman, "A Market Approach to Regulating the Energy Revolution: Assurance Bonds, Insurance, and the Certain and Uncertain Risks of Hydraulic Fracturing," *Iowa Law Review* 99 (2014): 1543.

346. Attanasio, 10386.

347. Wilson, Friedmann, and Pollak, 5949.

348. Ibid.

349. Ibid.

350. Ibid.

351. Aldrich et al., 5.

352. Adelman and Duncan, 4.

353. Ibid., 5, 15.

354. Wilson, Friedmann, and Pollak, 5946.

355. Benson and Cook, 250.

356. See the discussion on induced seismicity in this chapter, section B. See also Aldrich et al., 5.

357. Aldrich et al., 5, citing Sminchak et al., 2002.

358. Directive 2009/31/EC of the European Parliament and of the Council of April 23, 2009 on the geological storage of carbon dioxide and amending Council Directive 85/337/EEC, European Parliament and Council Directives 2000/60/EC, 2001/80/EC, 2004/35/EC, 2006/12/EC, 2008/1/EC and Regulation (EC) No. 1013/2006 (Text with EEA relevance); OJ L 140, 5.6.2009, 114–135.

359. See CCS Directive, ch. 3, art. 9, sec. 2–6, and ch. 4, art. 13, sec. 1(a)–(g). See also the scientific requirements at CCS Directive, annex I.

360. The selection of storage sites is included within the various articles that address the issuance of a storage permit, but most on point are the requirements set out at CCS Directive, chap. 2, art. 4 and at the directive's annex I.

361. See also chapter discussion in sections 2.3A, B, and C. In review, these requirements are sufficient injectivity conditions, indications of sufficient storage capacity, and a tight (secure) cap rock system. Hellevang, 601. Alternatively, the conditions have been listed by Smit et al., 365, as sufficiently deep storage to ensure supercritical fluid CO_2 and a secure cap rock.

362. See Adelman and Duncan, 18–19.

363. Trabucchi and Patton, 11.

364. Ibid., 10.

365. Ibid., 12.

366. James J. Dooley, Chiara Trabucchi, and Lindene Patton, "Design Considerations for Financing a National Trust to Advance the Deployment of Geologic CO_2 Storage and Motivate Best Practices," *International Journal of Greenhouse Gas Control* 4, no. 2 (2010): 2.

367. See Trabucchi et al.

368. See Ian J. Duncan, Jean-Philippe Nicot, and Jong-Won Choi, "Risk Assessment for Future CO_2 Sequestration Projects Based CO_2 Enhanced Oil Recovery in the US," *Energy Procedia* 1, no. 1 (2009): 20–37.

369. Trabucchi et al., 393.

370. But again, the existence of remedial technologies and means of restoration further enables the effectiveness of providing the very damages that civil liability rules most need in order to be functional. See the theoretical discussion in chapter 3, section 3.

371. For an overview of the potential risks coming from CCS, see Wilson, Johnson, and Keith, 3477.

372. As was already noted, the risks involved in a CCS project to some extent depend on the life cycle of the CCS project. Trabucchi and Patton, 8, distinguish risks in different phases. First, they distinguish among the capture, transport, and sequestration (storage), and within the sequestration, they subsequently distinguish siting/design, operation (CO_2 injection), closure and postclosure, and, finally, the long-term stewardship. Risks can of course emerge at the point of capture of the CO_2 where it is generated. There could be an improper capture or leakage at the capture point. In the second phase, that of the transport, risks could also emerge given the nature of carbon dioxide (9). However, in the first two phases, the risks are not particularly difficult to handle from a liability and insurance perspective since those risks are either not novel or are comparable to existing risks already addressed in the chemical or transportation industries. Insurance policies and liability rules for such concerns are foreseeable extensions of those in place for those respective industries.

373. Melisa F. Pollak and Elizabeth J. Wilson, "Regulating Geologic Sequestration in the United States: Early Rules Take Divergent Approaches," *Environmental Science and Technology* 43, no. 9 (2009): 3035. Their terms are "site selection and characterization," "operational injection of CO_2," "post-injection monitoring period," and "final stewardship phase."

374. According to Dana and Wiseman (1542), the short-term framework for fracturing accidents involves the several days surrounding and following a spill or injection event. The medium term refers to several years thereafter on the detection of contact with water resources and the resultant injuries and remediation thereof (1542–1543). Medium-term accidents are expected to retain high-quality information with regard to tortfeasor identity and chains of causation (1543). The long-term framework does not expect to retain such quality information and may face challenges in determining the who and how of the injuries to the water resources (1528). These concerns have been referred to as the concept of the long-tail risk factor: "There is the long-term risk, a highly uncertain risk—often referred to as `the long-tail risk'—that once all the unconventional development is done, we will discover that this activity degraded the environment and endangered public health in ways that cannot be linked to specific, identified accidents at active well operations" (1528). However, the concepts of long-term risk for CCS, measured in centuries and millennia, and that of fracturing related risks, measured in decades, are thusly differentiable and should be handled with some care in a comparative context. Liabilities solutions that might be feasible for one technology group might not be efficient for the other, given the difference in time scales.

375. Pollak and Wilson, 3035.

376. Brendan Beck et al., "Development and Distribution of the IEA CCS Model Regulatory Framework," *Energy Procedia* 4 (2011): 5938.

377. Ibid.

378. Havercroft and Macrory, 11.

379. Ingelson et al., 435.

380. Ibid., 435–436.

381. Ibid., 436.

382. Ibid.

383. Aldrich et al., 6.

384. Ibid.

385. Ibid., 7. Those liabilities included (1) injury to workers, (2) nuisance, trespass, or other preventions of access to storage rights, (3) poor well construction, (4) poorly sited wells, (5) surface and subsurface property damage, (6) induced seismicity, (7) groundwater contamination, (8) environmental damage, (9) damage to the confinement area, and (10) atmospheric release of CO_2.

386. Aldrich et al., 7. Those concerns included (1) integrity of the wells and their control apparatus, (2) integrity of the storage formation and its affiliated reservoirs, (3) fractures or faults adjacent to the well or storage formations, (4) migration of the

CO_2 plume, (5) migration or displacement of aquifers, including briny aquifers, and (6) transport of certain minerals into aquifers relied on for drinking water supplies.

387. Ibid.

388. Benson and Cook, 252.

389. A vadose zone is a "region from the water table to the ground surface, also called the unsaturated zone because it is partially water-saturated." Ibid., 253, with reference to the glossary definition in annex II.

390. Smit et al., 498; Benson and Cook, 252. Benson and Cook also list accumulation of CO_2 within closed structures, but as that requires some form of surface flux or other venting mechanism, it is implied within the incidents listed here.

391. Benson and Cook, 234; Hellevang, 598; R. A. Chadwick et al., "Review of Monitoring Issues and Technologies Associated with the Long-Term Underground Storage of Carbon Dioxide," *Geological Society, London, Special Publications* 313, no. 1 (2009): 257; Charles Jenkins, Andy Chadwick, and Susan D. Hovorka, "The State of the Art in Monitoring and Verification—Ten Years On," *International Journal of Greenhouse Gas Control* 40 (2015): 342. Jenkins et al. found that technological innovation in detection and surveillance technologies was driven by two different clusters of interest. Technology to monitor gas movements deep within the earth was driven by investment from the oil and gas companies, motivated by a "commercial need to improve time-lapse monitoring of producing fields." However, they found that technological development for shallow soil and near-surface monitoring were driven by the environmental science community and were motivated by "vague concerns and unease" (343).

392. Benson and Cook, 235.

393. Ibid.

394. Ibid.; Jenkins et al., 336.

395. Benson and Cook, 235; Jenkins et al., 336.

396. Benson and Cook, 235.

397. Smit et al., 498; Benson and Cook, 235.

398. Benson and Cook, 235; Jenkins et al., 330, 335–336.

399. Benson and Cook, 236. See the discussion on related techniques as applied at Weyburn. Chadwick et al., 269–270.

400. Benson and Cook, 236. See also Jenkins et al.

401. Benson and Cook, 237; Hellevang, 598–600; Smit et al., 500. For an update on the technologies discussed within Benson and Cook, see Jenkins et al., 333–338.

402. Benson and Cook, 237. For a discussion on three-dimensional surveillance in the field, that is, slices of four-dimensional surveillance, See Hellevang, 599. See also Chadwick et al., 259–260 and at 264. See also Jenkins, Chadwick, and Hovorka, 333–334.

403. Benson and Cook, 237. For a discussion on hot spots as CO_2 approaches the surface, see Chadwick et al., 259.

404. Benson and Cook, 237. For a discussion on updates to near-well geophysical measurement technology, see Jenkins et al., 337.

405. Benson and Cook, 237; Chadwick et al., 261.

406. Benson and Cook, 237; Chadwick et al., 261–262; Smit et al., 506.

407. Benson and Cook, 253.

408. Ibid., 237; Jenkins et al., 337.

409. Benson and Cook, 237. A noise log is the result of sonic tests that creates a "record of the sound measured at different positions in the borehole. Since fluid turbulence generates sound, high noise amplitudes indicate locations of greater turbulence such as leaks, channels and perforations. Noise logging is used primarily for channel detection, but has also been used to measure flow rates, identify open perforations, detect sand production and locate gas-liquid interfaces." See Schlumberger, "Oil Field Glossary," http://www.glossary.oilfield.slb.com/en/Terms/n/noise_log.aspx.

410. Benson and Cook, 253. See also De Figueiredo, 266, for a discussion on the applicability of plugging and abandoning rules to AGI injection sites in Texas.

411. Benson and Cook, 253.

412. Ibid. See Richard D. Lynch et al., "Dynamic Kill of an Uncontrolled CO_2 Well," *Journal of Petroleum Technology* 37, no. 8 (1985), for a historical account of a well kill performed on the Sheep Mountain Unit CO_2 well.

413. Benson and Cook, 252. See Lynch et al.

414. Benson and Cook, 253; Smit et al., 411. For a slide presentation that provides clear graphical illustrations of various remediation technologies, see Jean-Charles Manceau et al., "Methodologies and Technologies for Mitigation of Undesired CO_2 Migration in the Subsurface," in *The Seventh Trondheim CCS Conference* (2013).

415. J. R. Brydie et al., "The Development of a Leak Remediation Technology for Potential Non-Wellbore Related Leaks from CO_2 Storage Sites," *Energy Procedia* 63 (2014): 4603. See also Vello A. Kuuskraa, "Overview of Mitigation and Remediation Options for Geological Storage of CO_2" (staff workshop on technical papers for AB1925, Report to the California Legislature, 2007).

416. See Emily Rochon et al., *False Hope: Why Carbon Capture and Storage Won't Save the Climate* (Amsterdam: Greenpeace International, 2008).

Chapter 3

1. Brendan Beck et al., "Development and Distribution of the IEA CCS Model Regulatory Framework," *Energy Procedia* 4 (2011): 5934. They identified "twenty-nine key CCS regulatory issues" (5936). However, many of those issues were previously outstanding in energy law or environmental law more broadly.

2. David P. Flynn and Susan M. Marriott, "Carbon Sequestration: A Liability Pathway to Commercial Viability," *Natural Resources and Environment* 24 (2009): 37. They follow Klass and Wilson in finding three major types of postclosure liability: (1) groundwater contamination, (2) induced seismic events, and (3) surface releases to the atmosphere. Ibid., 37–38. Alexandra B. Klass and Elizabeth J. Wilson, "Climate Change and Carbon Sequestration: Assessing a Liability Regime for Long-Term Storage of Carbon Dioxide," *Emory Law Journal* 58 (2008): 103–179

3. See the discussion on James McLaren and James Fahey's seminal CCS liability paper: "Key Legal and Regulatory Considerations for the Geosequestration of Carbon Dioxide in Australia," *Australian Resources and Energy Law Journal* 24 (2005): 45–73.

4. The section attempts to provide this review primarily in chronological order, but some of the literature is discussed adjacent to other closely related pieces. Readers will likely detect several threads operating somewhat in parallel: those associated the passage of the EU's CCS Directive, those associated with efforts to pass federal and state-level CCS regulations within the United States, and legal literature drawn primarily from the Commonwealth areas such as New Zealand and Australia. While these threads do share many common concerns, it is noteworthy that each is drawn from legal systems of their own unique character and do diverge in thought at times, as might be expected.

5. Similar issues have been raised with regard for the need of legal clarity from oil and gas operations and the imminent increased implementation of fracturing technologies. Concerns center on the historical record; there have been a massive number of orphaned operations from traditional oil and gas producers and marketers. David A. Dana and Hannah J. Wiseman, "A Market Approach to Regulating the Energy Revolution: Assurance Bonds, Insurance, and the Certain and Uncertain Risks of Hydraulic Fracturing," *Iowa Law Review* 99 (2014): 1561. They count 190,000 orphaned underground petroleum tanks, 57,000 orphaned but unplugged oil or gas wells, and an estimate of 557,000 abandoned mining locations within the United States alone. As such, there are reasonable concerns to be raised that liability rules might benefit from improvement on historical rules.

6. Flynn and Marriott, 38.

7. Jonas J. Monast, Brooks R. Pearson, and Lincoln F. Pratson, "A Cooperative Federalism Framework for CCS Regulation," *Environmental and Energy Law and Policy Journal* 7 (2012): 38.

8. McLaren and Fahey, 63.

9. Ibid. They state that most cases of trespass, negligence, and nuisance face limitation periods of six years. Under Australian tort rules, the clock begins to run as soon as the "the relevant cause of action accrues," not from when the victim becomes aware of the tortious act or injury.

10. Ibid.

11. Ibid., 62–63.

12. M. J. Mace, Chris Hendriks, and Rogier Coenraads, "Regulatory Challenges to the Implementation of Carbon Capture and Geological Storage within the European Union under EU and International Law," *International Journal of Greenhouse Gas Control* 1, no. 2 (2007): 255.

13. Their list of phrases: *pollution, land-based pollution, wastes, hazardous wastes, industrial wastes, liquid wastes, harmful substances, dangerous substances, dangerous activities, operator, ship, sea, dumping, disposal, placement,* and *storage.*

14. Ibid.

15. Melisa F. Pollak and Elizabeth J. Wilson, "Regulating Geologic Sequestration in the United States: Early Rules Take Divergent Approaches," *Environmental Science and Technology* 43, no. 9 (2009): 3036.

16. Ibid., 3039.

17. Ibid., 3040.

18. Ibid., 3040–41.

19. Chiara Trabucchi and L. Patton, "Storing Carbon: Options for Liability Risk Management, Financial Responsibility," World Climate Change Report 170 (Washington, DC: Bureau of National Affairs, 2008), 9.

20. See ibid., 11, arguing that if the risk were to be transferred to the public, this would proffer a competitive disadvantage to environmentally superior operations.

21. Ibid., 13.

22. Directive 2009/31/EC of the European Parliament and of the Council of April 23, 2009, on the geological storage of carbon dioxide and amending Council Directive 85/337/EEC, European Parliament and Council Directives 2000/60/EC, 2001/80/EC, 2004/35/EC, 2006/12/EC, 2008/1/EC and Regulation (EC) No 1013/2006 (Text with EEA relevance); OJ L 140, 5.6.2009, 114–135.

23. CCS Directive, Preamble, para. 30. See also the discussion in chapter 5, section 5.2.2, on the inheritance of the CCS Directive from previous EU regulations.

24. See CCS Directive, ch. 4, art. 16, sec. 1, and see art. 17, sec. 5. See also ch. 3, art. 6, 8; chap. 4, art. 18; and the preamble, para. 34.

25. Victor Byers Flatt, "Paving the Legal Path for Carbon Sequestration," *Duke Environmental Law and Policy Forum* 19 (2009): 215.

26. Ibid., 215–216.

27. Ibid., 220.

28. Ibid., 220, 221, respectively.

29. Flatt also raises concerns that preexisting case history from EOR cases could provide conflicting case law for CCS projects. He connects those issues with needs for unitization to provide legal alignment with the diverse potential rights owners present in EOR settings. He found that states that did not require unitization prior to the onset of EOR injections did hold the operator of the EOR operations to be liable for mineral losses due to nuisance or trespass. Ibid., 231.

30. Ibid., 232.

31. Within the United States, the question of federal preemption over state government powers should be addressed. Ibid., 216–220.

32. Adam Gardner Rankin, "Geologic Sequestration of CO_2: How EPA's Proposal Falls Short," *Natural Resources Journal* 49 (2009): 915–916.

33. Ibid. The first option is discussed in this book in chapter 5, the second in chapter 7, and the third within proposals for mixed or tiered solutions in chapter 9.

34. Rankin, 921.

35. Ibid., 932.

36. Ibid.

37. Ibid., 938.

38. Allan Ingelson, Anne Kleffner, and Norma Nielson, "Long-Term Liability for Carbon Capture and Storage in Depleted North American Oil and Gas Reservoirs: A Comparative Analysis," *Energy Law Journal* 31 (2010), 434, citing Mark Anthony De Figueiredo, D. M. Reiner, and H. J. Herzog, "Framing the Long-Term in Situ Liability Issue for Geologic Carbon Storage in the United States," *Mitigation and Adaptation Strategies for Global Change* 10, no. 4 (2005): 647.

39. Ibid., 435.

40. Ibid., citing John C. Reitz, "How to Do Comparative Law," *American Journal of Comparative Law* 46, no. 4 (1998): 617–636.

41. Ingelson et al., 463–464.

42. Ibid., 463.

43. Ibid., 464.

44. Ibid.

45. Ibid.

46. Ibid.

47. Ibid.

48. See ibid., 466–469.

49. Ibid., 467.

50. Katherine Abend, "Geological Sequestration of Carbon Dioxide: Legal Issues and Recommendations for Regulators." *Appalachian Natural Resources Law Journal* 5 (2010): 16.

51. Ibid. Abend does propose that the burden should be placed on the operator to demonstrate a lack of substantial risk to water sources, because water resources will be even more critical with global warming.

52. David E. Adelman and Ian J. Duncan, "The Limits of Liability in Promoting Safe Geologic Sequestration of CO_2," *Duke Environmental Law and Policy Forum* 22 (2011): 5.

53. Ibid., 22.

54. Nicola Swayne and Angela Phillips, "Legal Liability for Carbon Capture and Storage in Australia: Where Should the Losses Fall?" *Environmental and Planning Law Journal* 29 (3) (2012).

55. Lincoln L. Davies, Kirsten Uchitel, and John Ruple, "Understanding Barriers to Commercial-Scale Carbon Capture and Sequestration in the United States: An Empirical Assessment," *Energy Policy* 59 (2013): 747. They polled 501 potential respondents and received 229 responses. They provide details of their survey methodology and the questions asked in a supplemental annex (760).

56. Ibid., 748. The other item in the top four was the high cost of CCS implementation.

57. Jacobs separated her analysis of carbon capture from carbon sequestration. In her terminology, *capture* referred to events upstream of the injection facility, and *sequestration* referred to events at and below the injection facility. Wendy B. Jacobs, "Carbon Capture and Sequestration," in *Global Climate Change and U.S. Law*, 2nd ed., ed. Jody Freeman and Michael Gerrard (Chicago: ABA, 2014).

58. Ibid., 16, 27.

59. Ibid., 15.

60. Ibid., 16.

61. Ibid.

62. Ibid., 26.

63. Adelman and Duncan, 31. They present an argument that traditional tort liability may provide meaningful deterrence against poor site selection and operation.

64. Ibid., 30–31.

65. See above in this section.

66. See Chiara Trabucchi, Michael Donlan, and Sarah Wade, "A Multi-Disciplinary Framework to Monetize Financial Consequences Arising from CCS Projects and Motivate Effective Financial Responsibility," *International Journal of Greenhouse Gas Control* 4, no. 2 (2010): 388.

67. Elizabeth J. Wilson, Alexandra B. Klass, and Sara Bergan, "Assessing a Liability Regime for Carbon Capture and Storage," *Energy Procedia* 1 (2009): 4575.

68. See Mark Anthony De Figueiredo, D. M. Reiner, and H. J. Herzog, "Framing the Long-Term in Situ Liability Issue for Geologic Carbon Storage in the United States," *Mitigation and Adaptation Strategies for Global Change* 10, no. 4 (2005): 655.

69. Adelman and Duncan, 31.

70. Similarly, rules of civil liability were found to be more effective for resolving fracturing disputes for two main reasons. Dana and Wiseman, 1556–1557. First, courts can be more responsive to the quickly evolving technology and can thus adapt to the needs of each case. Second, courts might be more resistant than public regulatory bodies to "capture by either industry or environmental interest groups."

71. John Pendergrass, "Long-Term Stewardship of Geological Sequestration of CO_2," *Environmental Law Reporter: News and Analysis* 43, no. 8 (2013): 10659. Pendergrass asserts that even if it were a given that appropriate legislation to support a regulatory system could be passed, which he disputed, drafting an "'effective performance-based regulations' is a difficult and uncertain task."

72. See Klass and Wilson, "Climate Change and Carbon Sequestration," 109; see also chap. 6.

73. See the survey drawn from the literature of law and economics on tort and accident law in section 3.3.

74. See section 3.2.

75. See the analyses on the rule of strict liability in section 3.4 and the analyses on the rule of negligence in section 3.5.

76. See the conclusions in section 3.6.

77. See chapter 4.

78. Avelien Haan-Kamminga, "Long-Term Liability for Geological Carbon Storage in the European Union," *Journal of Energy and Natural Resources Law* 29, no. 3 (2011): 315.

79. Ibid., 314.

80. Ibid.

81. Ibid., 314–315.

82. Because of the mathematical design of law and economic models for accident law, the common law perspective is shared with law and economic models; the models account for activity levels and levels of due care. Because strict liability, nuisance, and trespass all lack due care factors, they model similarly based on incentivizing on optimal activity levels and expected damages. (Negligence models include consideration of due care impacts on expected damages in addition to activity levels.) Thus, trespass and nuisance will be assumed for the purposes of this book to be sufficiently similar to strict liability that they do not require separate analyses from strict liability.

83. Ingelson et al. found that both Canada and the United States offered a short menu of civil liability options to govern long-term liabilities resultant from CCS storage accidents (438–441). Canada presented nuisance, trespass, negligence, strict liability, and claims arising from riparian injuries (439). The United States similarly provided the same but without riparian injuries (441), albeit they found that such might arise depending on further clarifications of water rights law with regard to CCS activities by individual states. See the discussion on the need for clarification on US water rights in chapter 4, section 4.2.2 of this book. Flatt had previously found that negligence, trespass, nuisance, and strict liability were available in the United States (222–223). Australia is said to have fewer options before it—that trespass, nuisance, and negligence avail. See Swayne and Phillips.

84. For a recent summary of the literature on the choice between strict liability and negligence, see also Hans-Bernd Schäfer and Frank Müller-Langer, "Strict Liability versus Negligence," in *Tort Law and Economics*, ed. Michael G. Faure, 3–45 (Cheltenham: Edward Elgar, 2009).

85. Not to suggest that there is complete alignment on this issue, as will be seen in the subsequent sections (3.2.2, 3.2.3, and 3.2.4) on trespass, nuisance, and negligence.

86. In the United States, the common references for strict liability are at Restatement (Second) Torts, sec. 519, 520. Although the Restatement itself is not law, the notes to it list the cases for each state on point.

87. Flatt, 223. In recent years, Canadian courts have applied strict liability for pipeline failures that led to leakage and for subsurface leakage from storage tanks. Ingelson et al., 440. The referenced cases were decided in 2003 and 1987, respectively.

88. Hannah Coman, "Balancing the Need for Energy and Clean Water: The Case for Applying Strict Liability in Hydraulic Fracturing Suits," *Boston College Environmental Affairs Law Review* 39 (2012): 149, citing Restatement (Second) Torts, sec. 520, cmt. e (1977).

89. See Lucas Bergkamp and Barbara Goldsmith, *The EU Environmental Liability Directive: A Commentary* (Oxford: Oxford University Press, 2013); K. De Smedt, *Environmental Liability in a Federal System: A Law and Economics Analysis* (Antwerp: Intersentia, 2007); Michael G. Faure and Jing Liu, "New Models for the Compensation of Natural Resources Damage," *Kentucky Journal of Equine, Agricultural and Natural Resources Law* 4 (2011): 261–314; Jing Liu, *Compensating Ecological Damage: Comparative and Economic Observations* (Antwerp: Intersentia, 2013); and Barbara Pozzo Zanchetta, "The Liability Problem in Modern Environmental Statutes," *European Review of Private Law* 4, no. 2 (1996): 112–129.

90. For an overview of the justifications for strict liability, see Lucas Bergkamp, *Liability and Environment: Private and Public Law Aspects of Civil Liability for Environmental Harm in an International Context* (The Hague: Kluwer Law International, 2001), 119–150.

91. See Benjamin J. Richardson, "Financial Institutions for Sustainability" *Environmental Liability* 8, no. 2 (2000): 165, See also M. Wilde, *Civil Liability for Environmental Damage: Comparative Analysis of Law and Policy in Europe and the US*, 2nd ed. (Alphen aan den Rhijn: Kluwer Law International, 2013), 210–220.

92. Ingelson et al., 467.

93. Darlene A. Cypser and Scott D. Davis, "Liability for Induced Earthquakes," *Journal Environmental Law and Litigation* 9 (1994): 551.

94. Ibid., 558. They describe a variety of injection activities that could encompass both CCS injections and hydraulic fracturing.

95. Ibid., 560, 563–564, respectively. They relied heavily on the finding from *Green v. General Petroleum* that while natural forces eventually became the final cause in fact, the previous human-sourced oil drilling activity was sufficiently proximate to sustain liability. See *Green v. General Petroleum Corp.*, 270 P.952 (Cal. 1928).

96. Cypser and Davis, "Liability for Induced Earthquakes," 569. They specifically cited a case of hydraulic fracturing: *Greg v. Delhi-Taylor Oil Corp.*, 344 S.W.2d 411 (Tex. 1961). They did warn that the tort of "vibrational trespass" had been subsumed under forms of strict liability in certain jurisdictions.

97. Ibid., 570.

98. Ibid., 572–573. See Restatement (First) of Torts, sec. 520 (1938).

99. Cypser and Davis, 573. See Restatement (First) of Torts, sec. 520 (1977). They found three of the six factors suggestive of a finding for abnormally hazardous activity.

100. Cypser and Davis, 575.

101. Ibid., 576–581, 581–586, respectively.

102. Haan-Kamminga, 323.

103. Ibid.

104. Christopher Bidlack, "Regulating the Inevitable: Understanding the Legal Consequences of and Providing for the Regulation of the Geologic Sequestration of Carbon Dioxide," *Journal of Land Resources and Environmental Law* 30 (2010): 210.

105. Ibid.

106. For a review of American law on point, see De Figueiredo et al., 649–652; see also Nathan R. Hoffman, "The Feasibility of Applying Strict-Liability Principles to Carbon Capture and Storage," *Washburn Law Journal* 49 (2009): 539–544.

107. For a review of American law on point, see De Figueiredo et al., 649–652; see also Hoffman, 539–544.

108. For a summary of the liability regime for long-term liability for stored CO_2 in various OECD countries, see Garrett and Beck (2011).

109. Adelman and Duncan, 41.

110. Ibid., 43–44.

111. Joe Schremmer, "Avoidable Fraccident: An Argument against Strict Liability for Hydraulic Fracturing," *University of Kansas Law Review* 60 (2011): 1253. See "Environmental Impact of Fracking," 1220–1224, for a discussion on the risks to potable water supplies.

112. Joe Schremmer, "Avoidable Fraccident: An Argument against Strict Liability for Hydraulic Fracturing," *University of Kansas Law Review* 60 (2011): 1254.

113. Ibid., 1237.

114. Ibid., 1239.

115. Ibid., 1249.

116. Ibid., 1252.

117. Ibid., 1253.

118. Ibid., 1242.

119. Ibid., 1253.

120. See the policy recommendation for a rule of strict liability in this book in section 9.1.2; the analyses supporting that policy are examined in sections 3.4 and 3.5 and concluded in section 3.6.

121. See also the lengthier discussion on this point in section 3.2.4.

122. Restatement (Second) on Torts, sec. 157. See also Flatt, 223.

123. Trespass arises if an actor intentionally "(a) enters land in the possession of the other, or *causes a thing* or a third person to do so, or (b) remains on the land, or (c) fails to remove from the land a thing which he is under a duty to remove [italics added]." Restatement (Second) on Torts, sec. 158. Additionally, the ancient writ of trespass was in strict liability, an awareness of which aids the understanding that the complexity of negligence is not present within this particular tort.

124. Dan B. Dobbs, *The Law of Torts* (Minneapolis: West Group, 2000), 95, 98. See also Restatement (Second) on Torts, sec. 158.

125. Dobbs, 95, 100. See also Restatement (Second) on Torts, sec. 158.

126. Dobbs, 95, 204–8. See also Restatement (Second) on Torts, sec. 158.

127. Dobbs, 95, 113–14. While some form of injury is required, the injury might be loss of exclusive usufruct (e.g., rental value or unjust enrichment from being on the premises), emotional distress, and for damages to the notions of "exclusive possession." See also Restatement (Second) on Torts, sec. 158.

128. Ingelson et al., 439, citing Alastair R. Lucas, William A. Tilleman, and Elaine Lois Hughes, *Environmental Law and Policy* (Toronto: Emond Montgomery Publication, 2003), 93.

129. Restatement (Second) on Torts, sec. 157.

130. Dobbs, 101.

131. As such, the notions of trespass will for the most part match our ultimate policy recommendation for strict liability. See the policy recommendation in chapter 9, section 9.1.2.

132. Barry Barton, Kimberley Jordan, and Greg Severinsen, *Carbon Capture and Storage: Designing the Legal and Regulatory Framework for New Zealand* (Hamilton: Centre for Environmental, Resources and Energy Law, 2013), 220. See also McLaren and Fahey, 60. They highlight that acts that substantiate torts of negligence or intentional torts have often been found to simultaneously be torts of trespass within oil and gas cases.

133. Ingelson et al., 438, citing Hughes, Lucas, and Tilleman, *Environmental Law and Policy*, 108. See also Ian Havercroft and Richard Macrory, *Legal Liability and Carbon Capture and Storage: A Comparative Perspective* (London: Global CCS Institute, University College London, Faculty of Law, 2014), 13.

134. Havercroft and Macrory, 13.

135. Ingelson et al., 438. See *Gregg v. Delhi-Taylor Oil Corp.*, 344 S.W.2d 411 (Tex. 1961); *Railroad Commission of Texas v. Manziel*, 361 S.W.2d 560 (Tex. 1962); *Geo Viking Inc. v. Tex-Lee Operating Co.*, 817 S.W.2d 357 (Tex. App. 1991); *Mission Res. v. Garza Energy Trust*, 166 S.W.3d 301 (Tex. App. 2005).

136. Ingelson et al., 438. See *ANR Pipeline Co. v. 60 Acres of Land*, 418 F. Supp. 2d 933 (W.D. Mich. 2006).

137. Ingelson et al., 438. See *Chance v. BP Chem, Inc.*, 670 N.E.2d 985 (Ohio 1996).

138. Ingelson et al., 439. While not attempting to store gas volumes underground, the process antecedent to producing from shale rocks involves the injection of highly compressed gases along with microscopic particles, which results in subterranean cracks permanently held open by the injected particles. The compression and injection of the propellant gases are comparable to the compression and injection of CO_2 plumes and could result in similar case facts.

139. Dobbs, 106. However, Dobbs notes that in such cases, the courts have blurred the rules of nuisance and negligence with the acts and elements of trespass; nevertheless, the ultimate judgments found liability for the incursion of the subterranean liquids.

140. Ibid., 112.

141. Havercroft and Macrory, 13. This distinction is not universal; it appears to be stronger in the United Kingdom than in the United States (14).

142. Restatement (Second) on Torts, sec. 158. The Restatement provides that the "the air space above" the land is identically protected to the land itself.

143. Flatt, 224.

144. Ibid.

145. Also, given its analytical similarity to strict liability, the rule of trespass will not be separately analyzed for its robustness.

146. Barton et al., 219.

147. Havercroft and Macrory, 14.

148. Ibid. Thus, factors such as recklessness, carelessness, intentionality, or accidentalness do not factor into findings of nuisance.

149. McLaren and Fahey, 60–61. Nuisance, similar to trespass, lacks a requirement to observe any duty of care and thus is functionally similar to strict liability.

150. Ibid., 61.

151. Barton et al., 219.

152. Ibid. An EOR-related saline flood that permanently damaged a water well used to provide freshwater for cattle was found to be an act of nuisance in *Gulf Oil Corp. v. Hughes*, 371 P.2d 81 (Okla. 1962).

153. The linguistic character of the primary element of nuisance displays some uncertainty. Dobbs, 1330. The Restatement (Second) of Torts, sec. 822, refers to nuisances as "invasions," whereas courts have more often described nuisances as "interferences," given that no physical trespass is required. Dobbs, 1330.

154. Ingelson et al., 439; Flatt, 223.

155. Dobbs, 1321.

156. Ibid., 1334. See also Restatement (Second) of Torts, sec. 821B.

157. Dobbs, 1321–1322.

158. Ibid., 1324.

159. Ingelson et al., 439. Barton et al., 218.

160. Havercroft and Macrory, 16.

161. Ibid.

162. Ibid.

163. The element of a prescribed duty of care sets negligence apart from strict liability, trespass, nuisance, and intentional torts in general.

164. Dobbs, 269. See also Flatt, 222, listing duty, breach, causation, and damages as the four elements of negligence. Havercroft and Macrory list four traditional elements for negligence: a duty of care, a breach of that duty, causation, and damage. Havercroft and Macrory, 17. For a similar set of elements for Australia in application to CCS, see McLaren and Fahey, 61.

165. Dobbs, 334.

166. Flatt, 222.

167. There are clearly cases from other forms of gas injection or from subterranean storage, but those cases would lack the public policy climate change benefit derived from the assumption of risk; thus those earlier cases might be distinguishable from CCS-based cases. To the extent that those earlier cases can be relied on, see the discussion attached to n. 174.

168. Flatt, 224.

169. Ibid.

170. Ibid.

171. Ibid., 222.

172. Bidlack, 210.

173. Ibid.

174. Ingelson et al., 440. They list four cases that involve both oil and majors (Imperial Oil of ExxonMobil) and independents (Irving Oil).

175. Barton et al., 217.

176. Ibid.

177. Ibid., 217–18.

178. Ibid., 218.

179. Ibid. It is worth noting that New Zealand's accident compensation scheme reduces the economic importance of a specific rule for CCS, as the Accident Compensation Act of 2001 already provides a no-fault scheme for all injuries derived from accidents both at work and otherwise. Suing for personal injury or wrongful death is barred in most events. Thus, the rules of civil liability apply only to property and consequential loss. As a broader result, the functional absence of a rule of strict liability is of no difference for personal injuries or death from CCS-related accidents.

180. Havercroft and Macrory, 17.

181. Ibid. They do mention that the application of the precautionary principle would depend on the balance of public actor versus private actor roles in operating the CCS facilities. The precautionary principle itself can be found within the Rio Declaration, "Principle 15: In order to protect the environment, the precautionary approach shall be widely applied by States according to their capabilities. Where there are threats of serious or irreversible damage, lack of full scientific certainty shall not be used as a reason for postponing cost-effective measures to prevent environmental degradation." Rio Declaration on Environment and Development, United Nations Environment Programme. Report of the United Nations Conference on the Human Environment, Stockholm, 5-16 June 1972; (United Nations publication, Sales No. E.73.II.A.14 and corrigendum), chap. I.

182. Dobbs, 303.

183. When the duty of care is set higher, the delta of difference between the expected costs to the tortfeasor under strict liability and negligence decreases; in extreme cases, the delta might become negligible. If the delta of expected costs

becomes negligible, the tortfeasor would react similarly with regard to activity level and precautionary levels.

184. See policy recommendations in chapter 9, section 9.1.2.

185. See the discussions on the theoretical approach of the school of law and economics on tort theory in section 3.3, the efficient applications of strict liability in section 3.4, and the efficient application of negligence in section 3.5.

186. See Alessio M. Pacces and Louis T. Visscher, "Methodology of Law and Economics," in *Law and Method: Interdisciplinary Research into Law*, ed. Bart van Klink and Sanne Taekema (Tübingen: Möhr Siebeck, 2011), 85.

187. Benjamin Franklin in the *Pennsylvania Gazette*, February 4, 1735 The full quote was indeed in reference to safety considerations and the prevention of home-based fires, and thus is quite fitting for the purposes of modern accident law: "In the first Place, as an Ounce of Prevention is worth a Pound of Cure, I would advise 'em to take care how they suffer living Coals in a full Shovel, to be carried out of one Room into another, or up or down Stairs, unless in a Warming pan shut; for Scraps of Fire may fall into Chinks and make no Appearance until Midnight; when your Stairs being in Flames, you may be forced, (as I once was) to leap out of your Windows, and hazard your Necks to avoid being oven-roasted."

188. See Gary T. Schwartz, "Mixed Theories of Tort Law: Affirming Both Deterrence and Corrective Justice," *Texas Law Review* 75 (1996).

189. The preventive effect of liability rules was, for example, explicitly stressed in the so-called EU White Paper on Environmental Liability, which preceded the European Environmental Liability Directive. EU Commission, White Paper on Environmental Liability, Brussels, 2000 (COM) 2000 66 final. For comments on this white paper, see Michael G. Faure, *Deterrence, Insurability, and Compensation in Environmental Liability: Future Developments in the European Union* (Vienna: Springer, 2003).

190. For an overview of empirical evidence concerning the effects of liability rules, see B. Van Velthoven, "Empirics of Tort," in *Tort Law and Economics*, ed. Michael G. Faure (Cheltenham: Edward Elgar, 2010). For an overview of empirical evidence with respect to the deterrent effect of environmental liability rules, see Michael G. Faure, "Effectiveness of Environmental Law: What Does the Evidence Tell Us?" *William and Mary Environmental Law and Policy Review* 36, no. 2 (2012), 301–305.

191. Michael G. Faure, "Designing Incentives Regulation for the Environment," Maastricht Faculty of Law Working Paper 2008–7 (2008), 32. See also Faure, "Environmental Liability," citing Schwartz, 1801.

192. Giuseppe Dari-Mattiacci, "Tort Law and Economics," in *Economic Analysis of Law: A European Perspective*, ed. Aristides Hatzis (Cheltenham: Edward Elgar 2006), 3; Faure, "Environmental Liability," 249.

193. Dari-Mattiacci, 3; see also Steven Shavell's treatise on accident law: *Economic Analysis of Accident Law* (Cambridge, MA: Harvard University Press, 1987), 263.

194. Dari-Mattiacci, 2. This is not to suggest that the literature has gone against the traditional perspective of the classical *lex Aquilia*, which required tortfeasors to be responsible for the damages they caused to the property of other persons, but rather provided an investigation into how best to minimize the combined losses to the victims by reducing the accidents incurred, thus reducing resultant damages and the sum amount of the damages from those accidents that do occur. In that sense, the law and economics literature has expanded on the traditional perspective.

195. Coase presented an argument that if the costs to negotiate were zero, then initial allocations of rights, such as property rights, could be readily renegotiated to attain the efficient allocation of those right. See Ronald H. Coase, *The Problem of Social Cost* (London: Palgrave Macmillan, 1960), 1. More critically, he observed that such was not what was generally observed: that the costs of negotiations were non-zero and initial allocations did in fact limit the renegotiations of those allocations. Furthermore, Pigou's externalities were in fact overlapping allocations of property rights. By combining these observations, Coase came to the insight that one role of law was to determine how to best allocate rights to enable real-world actors to best enable market forces and negotiations to attain efficient allocations.

196. For a general survey of the economic analysis of environmental liability and the role of tort law in that respect, see Lucas Bergkamp, *Liability and Environment* (The Hague: Kluwer Law International, 2001), 67–119; K. De Smedt, *Environmental Liability in a Federal System: A Law and Economics Analysis* (Antwerp: Intersentia, 2007), 28–64; M. Wilde, *Civil Liability for Environmental Damage: Comparative Analysis of Law and Policy in Europe and the US*, 2nd ed. (Alphen aan den Rhijn: Kluwer Law International, 2013), 138–148.

197. Pigou is the economist most commonly given credit for popularizing the concept of externalities. He advocated that adverse externalities could be offset by imposing a tax on the actor creating the externality and thus forcing that actor to include, or internalize, those costs otherwise imposed on third parties. See Arthur Cecil Pigou, *The Economics of Welfare* (London: Macmillan, 1920).

198. See ibid. For an applied discussion on optimal enforcement policies to address existing externalities, see R. Van den Bergh and L. T. Visscher, "Optimal Enforcement of Safety Law," in *Mitigating Risk in the Context of Safety and Security. How Relevant Is a Rational Approach?* ed. R.V. de Mulder (Rotterdam: Erasmus University Rotterdam, 2008), 29.

199. As we will argue, obviously many instruments other than liability rules could cure externalities. One of them is ex ante regulation by government, which we discuss in chapter 6. Yet other solutions to cure externalities are market-based instruments, such as emission trading, which has become popular as an instrument to

mitigate climate change. See Faure, "Effectiveness of Environmental Law," on the different instruments to remedy environmental pollution as well as on their optimal combination. In a related area, nuclear liability, international conventions provide low limits, so-called liability caps, on the liability of the nuclear operator. It has been held that those limits on liability constitute a subsidy that leads to an insufficient internalization of the externality caused by the nuclear risk. For estimates of those subsidies, see Jeffrey A. Dubin and Geoffrey S. Rothwell, "Subsidy to Nuclear Power through Price-Anderson Liability Limit," *Contemporary Economic Policy* 8, no. 3 (1990); Michael G. Faure and Karine Fiore, "An Economic Analysis of the Nuclear Liability Subsidy," *Pace Environmental Law Review* 26 (2009); and Anthony G. Heyes and Catherine Liston-Heyes, "Subsidy to Nuclear Power through Price-Anderson Liability Limit: Comment," *Contemporary Economic Policy* 16, no. 1 (1998).

200. Economists stress that the exposure to liability of a potential injurer will provide substantial incentives for accident prevention. A classic contribution in this respect is from Steven Shavell, "Strict Liability versus Negligence," *Journal of Legal Studies* 9, no. 1 (1980); see also Shavell, *Economic Analysis of Accident Law.*

201. It was on this basis that Shavell developed a model explaining the optimal level of accident avoidance. See Shavell, "Strict Liability versus Negligence," 1–25.

202. Dari-Mattiacci, 3.

203. Although here the preventive function of liability rules is stressed, Calabresi has pointed at the fact that liability rules may equally aim at loss spreading, which he refers to as a reduction of secondary costs, and at a reduction of administrative costs. Guido Calabresi, *The Costs of Accidents: A Legal and Economic Analysis* (New Haven: Yale University Press, 1970).

204. Ibid., 17.

205. Hans-Bernd Schäfer and Andreas Schönenberger, "Strict Liability versus Negligence," in *Encyclopedia of Law and Economics*, ed. B. Bouckaert and G. De Geest (Cheltenham: Edward Elgar, 2000), 598. They cite Calabresi.

206. Ibid.

207. Ibid., 602. They cite Jörg Finsinger and Mark V. Pauly, "The Double Liability Rule," *Geneva Papers on Risk and Insurance Theory* 15, no. 2 (1990): 159.

208. Dari-Mattiacci, 4.

209. Ibid., 4–5. See also Shavell, "Strict Liability versus Negligence."

210. Matthieu Glachant, "The Use of Regulatory Mechanism Design in Environmental Policy: A Theoretical Critique," in *Sustainability and Firms: Technological Change and the Changing Regulatory Environment* (Cheltenham: Edward Elgar, 1998), 181.

211. See Steven Shavell, "On Moral Hazard and Insurance," *Quarterly Journal of Economics* 93, no. 4 (1979): 541.

212. Jacob Nussim and Avraham D. Tabbach, "A Revised Model of Unilateral Accidents," *International Review of Law and Economics* 29, no. 2 (2009): 169. See also ibid., n. 2. See also Calabresi, "The Costs of Accidents"; Shavell, "Strict Liability versus Negligence"; Schäfer and Schönenberger; Schäfer and Müller-Langer, "Strict Liability versus Negligence," in *Tort Law and Economics*, ed. Michael G. Faure, 3–45 (Cheltenham: Edward Elgar, 2009).

213. See Shavell, "Strict Liability versus Negligence."

214. See Schäfer and Schönenberger.

215. See also the chapter discussion in section 3.4.1.

216. See the chapter discussion in section 3.4.2. Generally the requirement for unilateral accidents requires the necessary information to determine hazard conditions to be held by the tortfeasor to the functional exclusion of the victims.

217. See the chapter discussion in section 3.4.3.

218. See the chapter discussion in section 3.4.4.

219. See the chapter discussion in section 3.4.5.

220. Greater detail on each point here is explored in the following sections; this current exposition is but an introduction to those discussions.

221. Dobbs, 941. That earlier case law was limited in the United States by the Massachusetts case of *Brown v. Kendall*, 60 Mass. 292 (1850), which began a transition toward what became negligence law in the United States.

222. Dobbs, 947, 945, respectively. See also Roy Andrew Partain, "Moerman versus Pierson: The Nexus of Occupancy in Animals Ferae Naturae and Liability in Tort," *Soongsil Law Review* 28 (2012), for a comprehensive review of the development of liability rules associated with animals *ferae naturae* from classical to modern times.

223. *Rylands v. Fletcher*, L.R. 3 H.L. 330 (1868). L.R. 1 Ex. 265 (Eng.); [1868] UKHL 1 (17 July 1868), http://www.bailii.org/uk/cases/UKHL/1868/1.html.

224. Dobbs details the application of the early common law writ of trespass when applied to cattle trespassing on another's land: "The writ of trespass carried strict liability with it, so that with some exceptions, the owner of wandering cattle was strictly liable, just as he was strictly liable in other cases where the writ of Trespass was used. ... The rule of strict liability applied to barnyard animals generally. It did not apply to pets like dogs and cats, although the keeper of such animals might be liable for negligently or intentionally causing them to enter the land" (942–943). Similarly, "The English rule held that keepers of wild animals were strictly liable for harm caused by such animals" (947). See also Partain, "Moerman versus Pierson,"

281–284, for a discussion on the historical evolution of liability to owners of animals *ferae naturae*, such as the fleeing bear mentioned in at J. Inst. 4.9.1, "Ceterum sciendum est aedilitio edicto prohiberi nos canem verrem aprum ursum leonem ibi habere, qua vulgo iter fit: et si adversus ea factum erit et nocitum homini libero esse dicetur, quod bonum et aequum iudici videtur, tanti dominus condemnetur, ceterarum rerum, quanti damnum datum sit, dupli."

225. *Rylands v. Fletcher*, (1868) L.R. 1 Ex. 265 (Eng.); [1868] UKHL 1 (July 17, 1868).

226. Cranworth refered to *Baird v. Williamson*, (1863) 15 C. B. (N. S.) 376; reference found within *Rylands v. Fletcher*, (1868) L.R. 1 Ex. 265 (Eng.); [1868] UKHL 1 (July 17, 1868).

227. *Rylands v. Fletcher*, (1868) L.R. 1 Ex. 265 (Eng.); [1868] UKHL 1 (July 17, 1868).

228. Restatement (Second) of Torts, sec. 520.

229. Coman, 149, 151–153.

230. Ibid., 150.

231. Ibid., 153.

232. Ibid., 151, 153.

233. Ibid., 153–154.

234. Shavell examines the ultrahazardous from a bilateral perspective in part because a unilateral accident was already established to be more efficiently addressed with strict liability, even those ultrahazardous in nature.

235. Shavell, "Strict Liability versus Negligence," 24. See "Concluding Comments. #4.

236. Ibid. Italics added.

237. "PROPOSITION 4. Suppose that the tortfeasor and victim are strangers. Then none of the normal liability rules is efficient. Strict liability with a defense of contributory negligence is superior to the negligence rule if it is sufficiently important to lower tortfeasor activity levels. Strict liability without the defense and no liability are each inferior to whichever rule is better: either strict liability with the defense or the negligence rule." Shavell, "Strict Liability versus Negligence," 19.

"PROPOSITION 6. Suppose that injurers are sellers and that victims are strangers. Then the results are as given in Propositions 4 and 5." Ibid., 20.

238. Shavell found that neither policy was expected to attain efficiency, but that a rule of strict liability was expected to be more robust. Due to the way the policy choices either increased the activity of the tortfeasor or the victim and under an assumption that the risky activity was socially beneficial, Shavell found two interesting social welfare results. A policy choice for a rule of strict liability could lead to

insufficient happenings of socially useful activities; that is, strict liability could lead to net positive welfare results, albeit below optimal levels. A policy choice for a rule of negligence could lead to overindulgence in risky activities; that is, negligence could lead to net negative social welfare results.

239. Schäfer and Müller-Langer, 10. See also Calabresi.

240. Shavell, "Strict Liability versus Negligence," 24. See "Concluding Comments," #4.

241. Ibid., 19–20. See proposition 5.

242. Ibid.

243. Schäfer and Schönenberger, 606, citing A. M. Polinsky, *Strict Liability versus Negligence in a Market Setting* (Cambridge, MA: National Bureau of Economic Research, 1980).

244. Ibid., citing Shavell, "Strict Liability versus Negligence,"19.

245. Strict liability requires the tortfeasor to bear all of the costs, so tortfeasors have stronger incentives to ensure the net positive worth of their activities. Under negligence, the tortfeasor will escape some of the consequences and costs of his actions so long as he meets his duty of care. Schäfer and Schönenberger, 606.

246. Rankin, 922.

247. Ibid.

248. Ibid.

249. Similarly, claims that a contributor to a landfill could contribute "cleanly" without fault of hazardous pollution but remain liable under CERCLA (Rankin, 922) are not present at CCS in that all of the injected fluids would be CO_2 and that there would likely be but one singular operator and owner of the injected fluids.

250. That is, under the Restatement (Second) of Torts, and not under the simpler rules of the Restatement (First) of Torts.

251. See the discussion in chapter 2, section 2.4.

252. Schäfer and Müller-Langer, 25, offer a recent demonstration of this result. Their paper documents the history of similar findings, including the seminal models of Shavell and the Landes-Posner systems.

253. Schäfer and Müller-Langer, 25.

254. Shavell, "Strict Liability versus Negligence," 11.

255. Ibid., 11–12 and equation 2.

256. Ibid., 11.

257. Ibid., 12.

258. Ibid.

259. Ibid.

260. "PROPOSITION 1. Suppose that injurers and victims are strangers. Then strict liability is efficient and is superior to the negligence rule, which is superior to having no liability at all." Shavell, "Strict Liability versus Negligence," 12. Note that Hylton's positive theory of strict liability model can be shown to replicate the basic tenet of the Shavell-Landes-Posner model: under unilateral accidents, both strict liability and negligence are efficient. See Keith N. Hylton, "A Positive Theory of Strict Liability," *Review of Law and Economics* 4, no. 1 (2008): 6. An identical result is reached for Shavell's seller-stranger scenario. Shavell, "Strict Liability versus Negligence," 14.

261. Shavell, "Strict Liability versus Negligence," 11–12. See equation 2.

262. Ibid., 11–12.

263. Hylton, "A Positive Theory of Strict Liability," 23.

264. Schäfer and Müller-Langer found that negligence would require an ex ante forecast of ex post duty of care, and that such forecasts would likely be inaccurate resulting in inefficiency of preventive care and thus also an inefficient activity level results. That said, Schäfer and Müller-Langer did offer the caveat that efficiency might be obtained under a negligence rule, just unreliably so. See Schäfer and Müller-Langer, 26.

265. Ibid.

266. Schäfer and Schönenberger, 605. Especially relevant for the case of complicated or exotic industrial technologies, strict liability imposes the research costs on the party mostly likely to find the answer and to find that answer at the lowest costs.

267. Coasian transaction costs have been noted as blocking negligence from operating better with this particular problem. See James R. Chelius, "Liability for Industrial Accidents: A Comparison of Negligence and Strict Liability Systems," *Journal of Legal Studies* 5, no. 2 (1976): 296–297. The very employment of historical norms in setting duty-of-care standards for negligence has also been cited as one of the downfalls of the negligence rule in its underperformance to provide proper incentives for innovations. See Gideon Parchomovsky and Alex Stein, "Torts and Innovation," *Michigan Law Review* 107, no. 285 (2008): 303–306.

268. Nussim and Tabbach, 170.

269. Ibid., 173.

270. Ibid., 172. In some sense, this is captured by the idea of jointly permitting certain activity levels and safety standards within an environmental regulatory setting;

as such, to the extent that regulatory means can better combine these two targets than civil liability might, regulatory means would be preferable.

271. Schäfer and Müller-Langer, 18.

272. See Daniel L. Rubinfeld, "The Efficiency of Comparative Negligence," *Journal of Legal Studies* 16, no. 2 (1987): 375; Oren Bar-Gill and Omri Ben-Shahar, "The Uneasy Case for Comparative Negligence," *American Law and Economics Review* 5, no. 2 (2003): 433; and Thomas J. Miceli, "On Negligence Rules and Self-Selection," *Review of Law and Economics* 2, no. 3 (2006): 349.

273. Schäfer and Müller-Langer, 17–18. See Robert Cooter and Thomas Ulen, *Law and Economics* (New York: Pearson, 2004), 388.

274. See Schäfer and Müller-Langer, 17–18; Cooter and Ulen, 388.

275. Schäfer and Müller-Langer, 18, citing Miceli, who in turn was citing Marcel Kahan, "Causation and Incentives to Take Care under the Negligence Rule," *Journal of Legal Studies* 18 (1989): 427.

276. Shavell, "Liability and Safety Regulation"; Schäfer and Schönenberger, 604.

277. Schäfer and Müller-Langer, 16; Schäfer and Schönenberger, 604.

278. Schäfer and Schönenberger, 604. Clearly both forms of adjudication would also require several findings, such as causation, but because those matters would be common to both, they would not provide for substantial cost differences, even if the nuances of the issues were distinguishable between the two rules.

279. See Faure, "Regulatory Strategies in Environmental Liability," 137. See also Faure, "Effectiveness of Environmental Law," 17.

280. Schäfer and Müller-Langer, 16.

281. Ibid. See also Schäfer and Schönenberger, 604. If it is cheaper because it is obvious, then rational litigators would also expect the courts to render foreseeable judgments and thus preempt the need for actual litigation, leading to pretrial settlements.

282. Coman, 147.

283. Ibid., 148.

284. See Hylton, "A Positive Theory of Strict Liability"; Nussim and Tabbach. S ee also Tim Friehe, "Precaution v. Avoidance: A Comparison of Liability Rules," *Economics Letters* 105, no. 3 (2009).

285. Martin Nell and Andreas Richter, "The Design of Liability Rules for Highly Risky Activities: Is Strict Liability Superior When Risk Allocation Matters?" *International Review of Law and Economics* 23, no. 1 (2003).

286. Friehe, "Precaution v. Avoidance."

287. Nell and Richter, 33.

288. Nell and Richter, ibid., provide a list of reasons that corporate entities might be risk averse: (1) corporate notions of risk aversion operate only for well-financed diversified portfolio holders, which is contrary to many investors both private and public; (2) even for such parties as well-diversified portfolio holders, they can achieve genuine risk neutrality only if there is no system risk component, which might not be true for certain highly risky (investment) activities; (3) there is much evidence of structural imperfections in the capital market, which could frustrate efforts to diversify risk; (4) transaction costs tend to prevent portfolios from being sufficiently diversified; (5) entrepreneurial decisions within firms are made by risk-averse humans who are guided by careful strategies to remain in employment and are often rewarded for conservative stewardship of capital; and (6) those same human managers will have the potential to display risk aversion or pessimism against the risk of large losses.

289. Ibid.

290. Ibid., 31.

291. Ibid., 42.

292. For a discussion on the connection between risk neutrality and the standard models, see Michael G. Faure, "Economic Analysis of Tort Law," in *Tort Law and Economics*, ed. Michael G. Faure (Cheltenham: Edward Elgar, 2010). See also Alfred Endres and Reimund Schwarze, *Allokationswirkungen einer Umwelthaftpflicht-versicherung* (Berlin: Springer, 1992).

293. Nell and Richter, 40.

294. Insurance usually comes with loading fees, which causes the price to be in excess of the offered insurance payment; thus, even if insurance were complete, it would cost more than the expected damages it insures and the tortfeasor would buy less than complete coverage.

295. Nell and Richter, 42.

296. Ibid., 31.

297. Ibid., 43.

298. Ibid., 42.

299. Ibid.

300. While CCS liability insurance is already on the market, there is not a wide variety of such products on the market at present. While the product range is per-haps not yet extensive, the actual CCS projects that would need it are also scarce, so perhaps the market is supported given the scarcity of customers. As the number of expected CCS projects is expected to increase, there is little evidence to suggest that

more insurance could not be brought to the market. There is also a decades-spanning history of gas injection and gas storage from which to base actuarial tables. There are also the historical accidents, such as at Crystal Geysers, which inform insurance companies of the potential risks of massive CO_2 leakages. Thus, there is a lack of evidence to suggest that CO_2 insurance could not be reasonably provided sufficient to match the needs of CCS operators given sufficient time for the market to react to CCS operator demand for such products. In conclusion, the arguments for application of a rule of negligence predicated on a market shortage of insurance products for risk-averse customers are potentially not strong ones.

301. Nell and Richter, 42. Friehe's research also found a rule of negligence with a "highest due care" standard would be more robust when the number of potential victims is large and insurance is provided. Friehe, "Precaution v. Avoidance."

302. Nussim and Tabbach, 175, citing Steven M. Shavell, "The Judgment Proof Problem," *International Review of Law and Economics* 6 (1986): 43.

303. Ibid.

304. Ibid., 176.

305. Ibid., 175.

306. Friehe, "Precaution v. Avoidance," 215, lemmas 1 and 2.

307. Ibid., 215–16. *Avoidance* is defined as the efforts made to reduce the likelihood of being held responsible, not the avoidance of an accident itself.

308. Schäfer and Müller-Langer, 8. Court errors do occur and must be taken into account. There are three primary listed sources for court errors: error in determinations in the level of efficient care, error in the assessments of a tortfeasor's actual rendered level of care, and the parties' own inabilities to monitor and render specific levels of care continuously.

309. Nussim and Tabbach, 173.

310. Ibid.

311. Schäfer and Müller-Langer, 9.

312. Schäfer and Schönenberger, 605.

313. Nussim and Tabbach, 174. An overestimate of damage costs reinforces the calculus to avoid damages by operating at the due care level.

314. Ibid., 174–175.

315. Ibid., 174.

316. See Schäfer and Müller-Langer, 9.

317. Nussim and Tabbach, 174–175.

318. Schäfer and Schönenberger, 605.

319. Robert Cooter, "Prices and Sanctions," *Columbia Law Review* 84, no. 6 (1984): 1523. See also Louis T. Visscher, "Tort Damages," in *Tort Law and Economics: Encyclopedia of Law and Economics*, edited by Michael G. Faure (Cheltenham: Edward Elgar, 2009), 153.

320. Schäfer and Müller-Langer, 26.

321. Ibid.

322. Ibid., 9.

Chapter 4

1. See the discussion in chapter 3, sections 3.1 and 3.3.

2. See Chiara Trabucchi and L. Patton, "Storing Carbon: Options for Liability Risk Management, Financial Responsibility," World Climate Change Report 170 (Washington, DC: Bureau of National Affairs, 2008), 3, n, 7. They argue that "moral hazard refers to the specific situation where the risks of an unplanned event increase, because the responsible party is (partially) insulated from being held fully liable for resulting harm. If CCS facilities are not held completely responsible for the consequences of their actions, arguably they will be less careful in their siting and operating decisions."

3. James J. Dooley, Chiara Trabucchi, and Lindene Patton, "Design Considerations for Financing a National Trust to Advance the Deployment of Geologic CO_2 Storage and Motivate Best Practices," *International Journal of Greenhouse Gas Control* 4, no. 2 (2010): 386.

4. David Hawkins, George Peridas, and John Steelman, "Twelve Years after Sleipner: Moving CCS from Hype to Pipe," *Energy Procedia* 1, no. 1 (2009): 4407.

5. David E. Adelman and Ian J. Duncan, "The Limits of Liability in Promoting Safe Geologic Sequestration of CO_2," *Duke Environmental Law and Policy Forum* 22 (2011): 20.

6. See the discussion in chapter 3, sections 3.1 and 3.3.

7. See Israel Gilead, "Tort Law and Internalization: The Gap between Private Loss and Social Cost," *International Review of Law and Economics* 17, no. 4 (1997).

8. See Michael J. Trebilcock, "The Social Insurance-Deterrence Dilemma of Modern North American Tort Law: A Canadian Perspective on the Liability Insurance Crisis," *San Diego Law Review* 24 (1987). It is partially for this reason that the literature and most of the enacted CCS regulations, such as within the EU's CCS Directive have encouraged and implemented liability limits for postclosure accidents.

9. To be discussed in chapters 7 and 8.

10. One reason that liability rules are not well suited for compensation is that empirical research shows that only a small fraction of accident victims are actually compensated through liability rules. See B. Van Velthoven, "Empirics of Tort," in *Tort Law and Economics*, ed. Michael G. Faure (Cheltenham: Edward Elgar, 2010), 463–468. Hence, such compensation-focused rules create substantial inequality. Insurance and other compensation models are much better able than liability rules to compensate accident victims.

11. Ingelson, Kleffner, and Nielson have called for the existing environmental rules within each jurisdiction to set the paradigms on who is the responsible actor and who bears the attendant liabilities. Allan Ingelson, Anne Kleffner, and Norma Nielson, "Long-Term Liability for Carbon Capture and Storage in Depleted North American Oil and Gas Reservoirs: A Comparative Analysis," *Energy Law Journal* 31 (2010): 467. They suggest that the EU's CCS Directive sets a standard with regard to its reliance on the existing EU environmental liability frameworks.

12. Trabucchi and Patton, 4.

13. Ibid., 5–6.

14. Ibid. An additional problem that may arise is that those who produced the CO_2 would not necessarily be the same as those who would be injecting the CO_2 into the geologic storage sites. More particularly, in the early years, the storage site operators might be oil and gas producers, since a large part of the expertise with respect to CO_2 injection lays with the oil and gas producers. Hawkins et al., 4406. The literature therefore argues that there is to some extent a mismatch since the expertise that is present with the oil and gas industry is not necessarily transferred to the power industry where CCS is most needed.

15. See also Justine Garrett and Brendan Beck, *Carbon Capture and Storage: Legal Regulatory Review Edition*, 2nd ed. (Paris: OECD/ International Energy Agency, 2011), 9.

16. Katherine Abend, "Geological Sequestration of Carbon Dioxide: Legal Issues and Recommendations for Regulators," *Appalachian Natural Resources Law Journal* 5 (2010):18.

17. Ibid., 18–19.

18. Ibid., 18. Abend refers to governments having the ability to match the storage needs of CCS, which are in the "hundreds or thousands of years" time frame. She does not provide any evidence of historical governments that could have met such time frame demands.

19. Ibid., 19.

20. Alexandra B. Klass and Elizabeth J. Wilson, "Climate Change and Carbon Sequestration: Assessing a Liability Regime for Long-Term Storage of Carbon

Dioxide," *Emory Law Journal* 58 (2008): 109. Some scholars have argued that it is not the fear of unlimited liability that would constitute an important barrier for the development of CCS projects, but of technological uncertainties. Therefore they hold that capping liability is both useless and unnecessary. Adelman and Duncan, 64. It would be unnecessary because the potential harm resulting from demonstration projects would be so modest that a financial limit on liability would not be needed. It is therefore held that a cap on liability will probably have little more than a symbolic value.

21. Klass and Wilson, "Climate Change," 171.

22. Other studies point to the importance of first mover advantages, but do not necessarily call for a limitation on liability. Rather, they argue that legal systems with a stable regulatory system in place to encourage CCS would provide a first mover advantage to the industry within the particular legal system. Z. Makuch, S. Georgieva, and B. Oraee-Mirzamani, *Carbon Capture and Storage: Regulating Long-Term Liability* (London: Centre for Environmental Policy, Imperial College London, 2011), 16–17. Awarding benefits to first movers, particularly in the form of liability limits, is, however, a disputed issue.

23. While this study assumes that CO_2 will continue to be undesirable for the long term, it has previously seen economic value in EOR activities. It is possible that in some future scenario, efforts to find alternative means of CO_2 storage and usage might provide a superior market value for the previously stored CO_2. If so, the future question of ownership might take on additional importance. Thus, the future ownership of the gas once injected should be clarified before its storage.

24. Victor Byers Flatt, "Paving the Legal Path for Carbon Sequestration," *Duke Environmental Law and Policy Forum* 19 (2009): 233, citing Elizabeth J. Wilson and Mark A. de Figueiredo, "Geologic Carbon Dioxide Sequestration: An Analysis of Subsurface Property Law," *Environmental Law Reporter News and Analysis* 36, no. 2 (2006): 10121.

25. Abend, 14. The intent of this rule is that once the hydrocarbons are removed, the land no longer contains hydrocarbons to have a right to.

26. Ibid.

27. Flatt, 233. The phrase *surface rights* does not limit the owner to the surface, as might first appear. Surface rights within the American tradition of property law refer to the concept of landownership as predicated on surface-level descriptions of the land, with the traditional inclusion of the subsurface unless otherwise separately held, as can be done with regards to mineral rights.

28. See Ronald H. Coase, "The Federal Communications Commission," *Journal of Law and Economics* 2 (1959). See also a much more recent review of similar topics:

Ronald Coase, William Meckling, and Jora Minasian, *Problems of Radio Frequency Allocation* (Santa Monica, CA: Rand Corporation, 1995).

29. See Coase, "The Federal Communications Commission," 27.

30. See Ronald H. Coase, *The Problem of Social Cost* (Basingstoke, UK: Palgrave Macmillan, 1960), 1, n. 1.

31. Flatt, 233.

32. Ibid.

33. Ibid., 234.

34. Ibid., 233–34, relying on the analysis provided by Wilson and De Figueiredo, 10123.

35. Nicola Swayne and Angela Phillips, "Legal Liability for Carbon Capture and Storage in Australia: Where Should the Losses Fall?" *Environmental and Planning Law Journal* 29 (3) (2012).

36. Flatt, 235.

37. Ibid., citing Wilson and De Figueiredo, 10117.

38. Ibid.

39. Ibid.

40. Ibid., citing Restatement (Second) of Torts, sec. 858.

41. Ibid., citing Wilson and De Figueiredo, 10117.

42. For a more complete discussion on the complicated intellectual history underlying *Pierson v. Post*, see Roy Andrew Partain, "Moerman versus Pierson: The Nexus of Occupancy in Animals Ferae Naturae and Liability in Tort," *Soonngsil Law Review* 28 (2012).

43. Ian Havercroft and Richard Macrory, *Legal Liability and Carbon Capture and Storage: A Comparative Perspective* (London: Global CCS Institute, University College London: Faculty of Law, 2014), 18.

44. Ibid., 17, citing *Holland v. Hodgson* (1872) LR 7 CP 328 per Blackburn J., 335.

45. Havercroft and Macrory, 18.

46. Equally, see Bergkamp, *Liability and Environment,* 53, who also holds that foreseeability should be an important requirement for liability. In that perspective, unforeseeability would hence exclude liability.

47. The Paris Convention establishes a system of absolute liability. Classical exonerations such as force majeure, acts of God, or intervening acts of third persons are no longer applicable. The operator is only not liable for damage caused by a grave

natural disaster of an exceptional character, unless the legislation of the contracting party in whose territory his nuclear installation is situated, provides to the contrary. Similar stipulations can be found under the Vienna Convention, which is equally regulating nuclear liability. For details, see Jing Liu, *Compensating Ecological Damage: Comparative and Economic Observations* (Antwerp: Intersentia, 2013), 210–212.

48. If the nuclear damage is caused by a "grave natural disaster of an exceptional character or by an insurrection," the nuclear operator can be exonerated from liability on the basis of section 3 of the act on compensation for nuclear damage.

49. The government of Japan did not admit the earthquake and tsunami to be disasters of an "exceptional character." One does wonder what types of standards might be invoked to determine which types of CO_2 leakage are of exceptional character and which are not.

50. See Michael G. Faure and Jing Liu, "The Tsunami of March 2011 and the Subsequent Nuclear Incident at Fukushima: Who Compensates the Victims?" *William and Mary Environmental Law and Policy Review* 37 (2012), 192.

51. Makuch et al., 10.

52. Ibid., 15.

53. This is, moreover, in line with US case law concerning the question of whether natural disasters constitute an excuse from environmental liability. For a detailed discussion, see A. Wibisana, "The Myths of Environmental Compensation in Indonesia: Lessons from the Sidoarjo Mudflow," in *Regulating Disasters, Climate Change and Environmental Harm. Lessons from the Indonesian Experience*, ed. Michael G. Faure and A. Wibisana (Cheltenham: Edward Elgar, 2013).

54. See the discussions in chapter 2.

55. See the discussions in chapter 4, section 4.2.

56. For the historical sources of channeling in the 1969 Convention on Civil Liability for Oil Pollution Damage, see Hui Wang, *Civil Liability for Marine Oil Pollution Damage: A Comparative and Economic Study of the International, US and Chinese Compensation Regimes* (Alphen aan den Rijn: Kluwer Law International, 2011), 82–88. And for the origins of the liability channeling in the conventions on nuclear liability, see Liu, 212.

57. See Michael G. Faure and T. Hartlief, *Insurance and Expanding Systemic Risks* (Paris: OECD, 2003), 127–128. And see Michael Trebilcock and Ralph A. Winter, "The Economics of Nuclear Accident Law," *International Review of Law and Economics* 17, no. 2 (1997): 232–235.

58. Adelman and Duncan, 20.

59. See Makuch et al., 11–12.

60. A much broader review of the potential for regulations to complement and supplement rules of civil liability is provided in chapter 6, section 6.3.

61. See the discussions in chapter 6.

62. For example, in the debate preceding the Environmental Liability Directive, member states could not reach agreement on this point. As a result, it is left to the member states whether to include such a "compliance with permit defence." For a more detailed legal and economic analysis of the history of the regulatory compliance defense in the Environmental Liability Directive, see K. De Smedt, *Environmental Liability in a Federal System: A Law and Economics Analysis* (Antwerp: Intersentia, 2007), 225–231.

63. Michael G. Faure and Roger Van den Bergh, "Negligence, Strict Liability and Regulation of Safety under Belgian Law: An Introductory Economic Analysis," *Geneva Papers on Risk and Insurance* 12, no. 43 (1987), 110; see also Steven Shavell, "Liability for Harm versus Regulation of Safety," *Journal of Legal Studies* 13, no. 2 (1984): 365.

64. Paul Burrows, "Combining Regulation and Legal Liability for the Control of External Costs," *International Review of Law and Economics* 19, no. 2 (1999). Schwartz added to the debate by discussing whether compliance with federal safety statutes should have a justificative effect in state tort cases. See Alan Schwartz, "Statutory Interpretation, Capture, and Tort Law: The Regulatory Compliance Defense," *American Law and Economics Review* 2, no. 1 (2000).

65. This result holds under both a negligence rule and a rule of strict liability.

66. To efficiently set the prevention and activity levels.

67. Charles D. Kolstad, Thomas S. Ulen, and Gary V. Johnson, "Ex Post Liability for Harm vs. Ex Ante Safety Regulation: Substitutes or Complements?" *American Economic Review* 80 (1990): 888–901.

68. Burrows, "Combining Regulation."

69. Some of those problems are discussed in chapter 3, section 3.4.5, and in chapter 6, sections 6.1.7 and 6.2.2.B.

70. Michael G. Faure, Ingeborg M. Koopmans, and Johannes C. Oudijk, "Imposing Criminal Liability on Government Officials under Environmental Law: A Legal and Economic Analysis," *Loyola Los Angeles International and Comparative Law Journal* 18 (1995).

71. Note, however, that industry argues against such a liability of the licensor, claiming that this may entail the risk that licensors would be too reluctant in allowing emission if this could give rise to their liability. G. J . Niezen, "Aansprakelijkheid voor milieuschade in de Europese Unie," in *Ongebonden Recht Bedrijven* (Berlin: Kluwer, 2000), 171.

72. Susan Rose-Ackerman, *Rethinking the Progressive Agenda: The Reform of the American Regulatory State* (New York: Free Press, 1992): 123. See also Alessandra Arcuri, "Controlling Environmental Risk in Europe: The Complementary Role of an EC Environmental Liability Regime," *Tijdschrift voor Milieuaansprakelijkheid* 15, no. 2 (2001): 43–44.

73. See Bergkamp, *Liability and Environment*, 239–255.

74. See M. Wilde, *Civil Liability for Environmental Damage: Comparative Analysis of Law and Policy in Europe and the US*, 2nd ed. (Alphen aan den Rhijn: Kluwer Law International, 2013), 224–227.

75. See Adelman and Duncan, 46, who present an argument that liability can be a useful complement to traditional regulatory requirements.

76. Specifically for CCS-related risks, De Figueiredo also refers to the "conventional wisdom" that operators would be saved from liability by complying with all applicable regulations, but clearly holds that regulatory compliance is not a safe harbor for liability. See De Figueiredo, 378–383.

77. Flatt, 226.

78. Ibid.

79. Ibid.

80. See Jeffrey Trauberman, "Statutory Reform of Toxic Torts: Relieving Legal, Scientific, and Economic Burdens on the Chemical Victim," *Harvard Environmental Law Review* 7 (1983).

81. See Samuel D. Estep, "Radiation Injuries and Statistics: The Need for a New Approach to Injury Litigation," *Michigan Law Review* 59, no. 2 (1960).

82. Problems related to causal uncertainty can especially arise in cases of environmental liability; the solutions in many legal systems are not always very clear. For a more detailed discussion, see Wilde, 74–97, 235–249.

83. Those often amount to a situation whereby operators can never prove that their site did not cause a particular loss, as a result of which they may also be held liable for damage that never emerged from their site.

84. See also Bergkamp, "Liability and Environment," 287–291 and at 368–369.

85. On the potential damage, see chapter 2, section 2.4.

86. Joint and several liability was therefore meant to assist the judiciary in circumstances where it was impossible to reasonably ascertain the relative contributions of different parties. See Richardson, *Environmental Regulation through Financial Organisations*, 163–164.

87. See section 4.4.

88. See Tom H. Tietenberg, "Indivisible Toxic Torts: The Economics of Joint and Several Liability," *Land Economics* 65, no. 4 (1989).

89. For an analysis of joint and several liability under full solvency, see Lewis A. Kornhauser and Richard L. Revesz, "Sharing Damages among Multiple Tortfeasors," *Yale Law Journal* 98, no. 5 (1989); for the case of insolvency, see Lewis A. Kornhauser and Richard L. Revesz, "Apportioning Damages among Potentially Insolvent Actors," *Journal of Legal Studies* 19, no. 2 (1990).

90. For this reason, joint and several liability in case of environmental harm is, for example, opposed by Bergkamp. See Lucas Bergkamp, "The Proper Scope of Joint and Several Liability," *Tijdschrift voor Milieuschade en Aansprakelijkheidsrech* 14 (2000). See also Bergkamp, *Liability and Environment,* 300–303.

91. See David Gerard and Elizabeth J. Wilson, "Environmental Bonds and the Challenge of Long-Term Carbon Sequestration," *Journal of Environmental Management* 90, no. 2 (2009): 1099.

92. Ingelson et al., 467.

93. Ibid.

94. Adelman and Duncan, 42–43.

95. Ibid., 45–46.

96. For a similar conclusion with respect to environmental harm, especially in cases of causal uncertainty, see Bergkamp, *Liability and Environment,* 303–306.

97. A discussion of the potential damage is in chapter 2, section 2.4.

98. See Makuch et al., 10–11.

99. Hawkins et al., 4407.

100. Ibid.

101. For those evaluations, see Chiara Trabucchi, Michael Donlan, and Sarah Wade, "A Multi-Disciplinary Framework to Monetize Financial Consequences Arising from CCS Projects and Motivate Effective Financial Responsibility," *International Journal of Greenhouse Gas Control* 4, no. 2 (2010).

102. On the position of first movers and the policy to provide them with incentives, see the discussion in section 4.2.

103. De Figueiredo et al., 652.

104. Ibid. See also Ch. H. Haake and K. B. Marsh, "The Trouble with Angels: Carbon Capture and Storage Hurdles and Solutions," *World Climate Change Report* (Washington, DC: Bureau of National Affairs, 2009).

105. Trabucchi and Patton, 16. This would mean that operators would be responsible for consequences up to a dollar threshold per occurrence plus remediation costs.

106. For the compensation system under the Price-Anderson Act, see Michael G. Faure and Tom Vanden Borre, "Compensating Nuclear Damage: A Comparative Economic Analysis of the US and International Liability Schemes," *William and Mary Environmental Law and Policy Review* 33, no. 1 (2008): 240–245. See also Liu, 236–238.

107. On the nature of the liability regime and the limitation of liability in the US Oil Pollution Act, see Wang, 207–212.

108. Trabucchi and Patton, 17–21.

109. Elizabeth J. Wilson, Alexandra B. Klass, and Sara Bergan, "Assessing a Liability Regime for Carbon Capture and Storage," *Energy Procedia* 1, no. 1 (2009): 4581.

110. Klass and Wilson, "Climate Change and Carbon Sequestration," 164–65.

111. Ibid., 168.

112. Ibid., 171, 177.

113. Adelman and Duncan, 64.

114. Ibid.

115. See the discussion in chapter 3, section 3.3.

116. Shavell, "Strict Liability Versus Negligence," 8, 11.

117. See the discussion on strict liability in chapter 3, section 3.4.

118. See Richardson, *Environmental Regulation through Financial Organisations*, 366.

119. For a more detailed critical analysis of financial caps, especially in the context of marine oil pollution damage, see Wang, 311–323.

120. See Michael G. Faure, "Economic Models of Compensation for Damage Caused by Nuclear Accidents: Some Lessons for the Revision of the Paris and Vienna Conventions," *European Journal of Law and Economics* 2, no. 1 (1995): 21–43.

121. The reason for the underdeterrence is obviously the same as for the underdeterrence that results from the insolvency of the injurer. Underdeterrence arises because the injurer is not exposed to full liability as a result of his insolvency or a cap.

122. See also Marcus Radetzki and Marian Radetzki, "Private Arrangements to Cover Large-Scale Liabilities Caused by Nuclear and Other Industrial Catastrophes," *Geneva Papers on Risk and Insurance: Issues and Practice*, 25, no. 2 (2000).

123. See Anthony Heyes and Catherine Liston-Heyes, "Subsidy to Nuclear Power through Price-Anderson Liability Limit: Comment," *Contemporary Economic Policy* 16, no. 1 (1998); Anthony Heyes and Catherine Liston-Heyes, "Capping Environmental Liability: The Case of North American Nuclear Power," *Geneva Papers on Risk and Insurance: Issues and Practice* 25, no. 2 (2000).

124. Michael G. Faure and Karine Fiore, "An Economic Analysis of the Nuclear Liability Subsidy," *Pace Environmental Law Review* 26 (2009): 419–447.

125. For example, as far as nuclear liability is concerned, see Faure, "Economic Models of Compensation"; see also Trebilcock and Winter.

126. Adelman and Duncan, 64.

127. See Mark Anthony De Figueiredo, D. M. Reiner, and H. J. Herzog, "Framing the Long-Term in Situ Liability Issue for Geologic Carbon Storage in the United States," *Mitigation and Adaptation Strategies for Global Change* 10, no. 4 (2005): 652.

Chapter 5

1. Barry Barton, Kimberley Jordan, and Greg Severinsen, *Carbon Capture and Storage: Designing the Legal and Regulatory Framework for New Zealand* (Hamilton: Centre for Environmental, Resources and Energy Law, 2013), 223.

2. Ibid. They advocated for New Zealand to bifurcate these liabilities, following practice overseas. Havercroft and Macrory expanded on this by listing three questions: (1) What conditions would need to be met prior to transfer? (2) Which liabilities would be transferred and which not? and (3) Would there be any reopeners or clawbacks to restore liability to the former operators at some future time? Ian Havercroft and Richard Macrory, *Legal Liability and Carbon Capture and Storage: A Comparative Perspective* (London: Global CCS Institute, University College London: Faculty of Law, 2014), 37.

3. A first step is the process of site selection and its regulatory review. As often reviewed in the preceding chapter, the literature broadly agrees that the portfolio of long-term risks will depend on proper site selection and design to an important extent. Chiara Trabucchi and L. Patton, "Storing Carbon: Options for Liability Risk Management, Financial Responsibility," World Climate Change Report 170 (Washington, DC: Bureau of National Affairs, 2008), 9, mention that this first phase on the siting and design decision would normally take less than a year. Alexandra B. Klass and Elizabeth J. Wilson, "Climate Change and Carbon Sequestration: Assessing a Liability Regime for Long-Term Storage of Carbon Dioxide," *Emory Law Journal* 58 (2008): 115, however, mention that this phase could take one to ten years. The second step is the active operation of the site, that is, the period of active CO_2 injection. This period could last for several decades up to the point of closure. Trabucchi and Patton, 8. Klass and Wilson, "Climate Change and Carbon Sequestration," 115,

mention a period of twenty to thirty years. The third step is the period of injection site closure and its monitoring after closure. Authors mention different time periods for this closure monitoring. Trabucchi and Patton, 10, mention ten, twenty, or fifty years, whereas Klass and Wilson, "Climate Change and Carbon Sequestration," 115, refer to fifteen to thirty years. However, the important point from the perspective of industry is that this third step of monitoring after closure is also defined in time.

4. Klass and Wilson, "Climate Change and Carbon Sequestration," 115; Trabucchi and Patton, 8.

5. An interesting alternate problem within permanence is the coordination of offshore storage of CO_2 within UNCLOS. Avelien Haan-Kamminga, Martha Roggenkamp, and Edwin Woerdman, "Legal Uncertainties of Carbon Capture and Storage in the EU: The Netherlands as an Example," Carbon and Climate Law Review 4, no. 3 (2010): 245, citing UNCLOS, art. 60, para. 3. UNCLOS requires that all installations used for the exploitation of natural resources be removed from the offshore area once operational activities are completed to better secure the advantages of the high seas for all. When might that period end for offshore storage activities, necessitating removal of equipment (e.g., monitoring equipment might need to linger longer in place), and exactly which parts of the installation should be removed when? One wonders if ocean-surface awareness should be retained via buoys or such devices to enable vessels to be aware of potential venting hazards.

6. See Trabucchi and Patton, 10.

7. Justine Garrett and Brendan Beck, Carbon Capture and Storage: Legal Regulatory Review Edition, 2nd ed. (Paris: OECD/International Energy Agency, 2011), 9.

8. Ibid. It appears that a majority of states that have enacted CCS regulations have preferred to support a transfer of liability stewardship to public actors. Ibid. Yet the issue remains open in a majority of countries, and even in leading CCS countries such as the United States, regulations can be found on both sides of the issues.

9. See the chapter discussion in sections 5.2.1, 5.2.2, and 5.2.3.

10. See the chapter discussion in sections 5.4, 5.5, and 5.6.

11. See the chapter discussion in section 5.7; See also policy recommendations in chapter 9, section 9.1.6.

12. Garrett and Beck, 9.

13. Barton et al., 224.

14. Paul Bailey, Elizabeth McCullough, and Sonya Suter, "Can Governments Ensure Adherence to the Polluter Pays Principle in the Long-Term CCS Liability Context?" Sustainable Development Law and Policy 12 (2011): 47. See the chapter discussion in sections 5.1 and 5.2 above.

15. Ibid. See the chapter discussion in sections 5.1 and 5. 2 above.

16. Ibid., 47–48. See the chapter discussion in sections 5.1 and 5. 2 above.

17. Ibid., 48. See the discussions in chapter 8, section 8.3.

18. Jonas J. Monast, Brooks R. Pearson, and Lincoln F. Pratson, "A Cooperative Federalism Framework for CCS Regulation," *Environmental and Energy Law and Policy Journal* 7 (2012): 41.

19. Ibid.

20. Avelien Haan-Kamminga, "Long-Term Liability for Geological Carbon Storage in the European Union," *Journal of Energy and Natural Resources Law* 29, no. 3 (2011): 327.

21. Ibid., 328.

22. Allan Ingelson, Anne Kleffner, and Norma Nielson, "Long-Term Liability for Carbon Capture and Storage in Depleted North American Oil and Gas Reservoirs: A Comparative Analysis," *Energy Law Journal* 31 (2010): 468.

23. Ibid. They included Kansas, Wyoming, and Victoria as examples of such states.

24. Ibid., 468. They included Montana and member states of the EU as examples of such states.

25. To the profitable benefit of the corporation-cum-operator.

26. To the nuisance-avoiding benefit of each citizen to the extent that harms from climate change are averted.

27. Wendy B. Jacobs, "Carbon Capture and Sequestration," in *Global Climate Change and U.S. Law*, 2nd ed., ed. Jody Freeman and Michael Gerrard (Chicago: ABA, 2014), 27. Jacobs did not provide any reference as to whom she meant by "many legal commentators."

28. Ibid.

29. Ibid.

30. Ibid., 28.

31. Trabucchi and Patton, 15.

32. Klass and Wilson, "Climate Change and Carbon Sequestration," 174–175.

33. Ibid., 174. See also Elizabeth J. Wilson, Alexandra B. Klass, and Sara Bergan, "Assessing a Liability Regime for Carbon Capture and Storage," *Energy Procedia* 1, no. 1 (2009): 4581.

34. Klass and Wilson, "Climate Change and Carbon Sequestration," 174.

35. David E. Adelman and Ian J. Duncan, "The Limits of Liability in Promoting Safe Geologic Sequestration of CO_2," *Duke Environmental Law and Policy Forum* 22 (2011): 20–23.

36. Arguably, it would appear that damages, if sufficiently removed in time from the ex ante decision-making moment, would suffer from discounting under the time value of money. The net present value calculation creates a discount from future period economic events, for both revenues and costs, by application of the discount $(1 - r)^{-n}$ where n is the number of years in the future from which the discount applies to bring the economic event to present value. In this case, the present value of future liabilities would be [Ex ante cost of liability] = [Future value of liability] * $[(1 - r)^{-n}]$, where r is the applied interest rate. When one considers time frames of centuries or millennia, the effects of discounting extremely large values of n could be severe on the impact of whatever incentives might have in fact been generated by those extremely latent liabilities. For example, given an annual interest rate of 5 percent and a future injury resulting in $1 million of damages, the data of ex ante discounted impact is only $5,921 from an event one hundred years in the future and only $1 for an event approximately 270 years in the future. This comment is not to argue that such analysis is the only appropriate one, but for operators responsive to traditional financial and accounting models, it is an important element of their decision making toolset and demonstrates their potential accounting for future liabilities at the time of ex ante decision making.

37. Adelman and Duncan.

38. In the sense of not clearly distinguishing between the different phases in the CCS project life cycle.

39. Adelman and Duncan, 6.

40. Ibid., 20.

41. Ibid., 28–29.

42. Ibid.

43. Ibid., 46.

44. See Mark Anthony De Figueiredo, "The Liability of Carbon Dioxide Storage" (PhD diss., MIT, 2007), 383–386. De Figueiredo has suggested a liability time limit for the operator of ten years from the end of CO_2 injection. Ibid., 396.

45. Trabucchi and Patton,15.

46. Adelman and Duncan, 59.

47. Trabucchi and Patton, 8.

48. Adelman and Duncan, 59.

49. According to the literature, this period can potentially be very long and even last 2,000 years. See David Gerard and Elizabeth J. Wilson, "Environmental Bonds and the Challenge of Long-Term Carbon Sequestration," *Journal of Environmental Management* 90 (2009): 1099.

50. Wilson et al., 4581.

51. For example, in the European CCS Directive, liability can be passed to the state after the closure of a site when particular strict conditions have been met, including the provision of a financial contribution for the posttransfer period covering at least the costs for monitoring for the next thirty years. Z. Makuch, S. Georgieva, and B. Oraee-Mirzamani, *Carbon Capture and Storage: Regulating Long-Term Liability* (London: Centre for Environmental Policy, Imperial College London, 2011), 7–9. Also according to the IEA document, liability could be transferred to the operator only when there is evidence that there is no significant risk of physical leakage or seepage, a minimum time period has elapsed from the cessation of the injection, and a financial contribution is provided for the long-term stewardship of the site. Garrett and Beck, 9.

52. Especially strongly by Adelman and Duncan, 6.

53. Bailey et al., 47.

54. Ibid., citing Robin Warner, *Protecting the Oceans beyond National Jurisdiction: Strengthening the International Law Framework* (Leiden: Brill, 2009), 49.

55. Ibid., 47. Flynn and Marriott advocate the application of polluter pays in conjunction with a transfer of liabilities (or provision of indemnifications) to "ensure that owner/operators properly characterize potential GS [CCS] sites and safely and properly operate GS facilities" David P. Flynn and Susan M. Marriott, "Carbon Sequestration: A Liability Pathway to Commercial Viability," *Natural Resources and Environment* 24 (2009): 39.

56. See its enactment within the EU at article 191(2) of the Treaty on the Functioning of the European Union: "Union policy on the environment shall aim at a high level of protection taking into account the diversity of situations in the various regions of the Union. It shall be based on the precautionary principle and on the principles that preventive action should be taken, that environmental damage should as a priority be rectified at source and that *the polluter should pay* [italics added]." See also principle 16 of the Rio Declaration of 1992, where the economic principle of internalizing a costly externality is encouraged but not required: "National authorities should endeavour to promote the *internalization of environmental costs* and the use of economic instruments, taking into account the approach that the *polluter should, in principle, bear the cost of pollution*, with due regard to the public interest and without distorting international trade and investment [italics added]."

57. For example, Dana and Wiseman employed the polluter-pays principle without need of explication in their advocacy of implementing performance bonds and mandatory insurance requirements: "Assurance bonds and mandatory insurance, even when they do nothing to alter the conduct of industry actors, generate a pool of money that can be used for the remediation of the environmental harms that the actors knowingly or (more often) unknowingly created. Reserving this pool of money is critical because, absent such funds, there is a high likelihood that operators or public actors will never undertake environmental remediation." David A. Dana and Hannah J. Wiseman, "A Market Approach to Regulating the Energy Revolution: Assurance Bonds, Insurance, and the Certain and Uncertain Risks of Hydraulic Fracturing," *Iowa Law Review* 99 (2014): 1530.

58. Bailey et al., 47.

59. Ibid., 48.

60. Ingelson et al., 461.

61. Ibid.

62. Ibid.

63. Ibid.

64. Ibid. The reference is to the Price-Anderson Nuclear Industries Indemnity Act of 1957, 42 U.S.C. § 2210 et seq.

65. Adam Gardner Rankin, "Geologic Sequestration of CO_2: How EPA's Proposal Falls Short," *Natural Resources Journal* 49 (2009): 916.

66. Ibid., 917.

67. Ibid., 917–918.

68. Ibid.

69. Kim Gittleson, "Can a Company Live Forever?" *BBC News*, January19, 2012, http://www.bbc.com/news/business-16611040; "The Business of Survival: What Is the Secret of Corporate Longevity?" *Economist*, December 16, 2004.

70. Ibid.

71. Ibid. The *Economist* refers to the research of Leslie Hannah as the source of the seventy-five-years rule for the half-life of the world's top one hundred corporations as ranked by market capitalization. Although the *Economist* provided no citation to Hannah's research, two articles provide the data for that rule: Leslie Hannah, "Survival and Size Mobility among the World's Largest 100 Industrial Corporations, 1912–1995," *American Economic Review* 88, no. 2 (1998), and Hannah, "Marshall's Trees and the Global Forest: Were Giant Redwoods Different?" in *Learning by Doing in Markets, Firms, and Countries*, ed. Naomi R. Lamoreaux, Daniel M. G. Raff, and

Peter Temin (Chicago: University of Chicago Press, 1999). A similar rule is found in the same article that "one-third of the firms in the Fortune 500 in 1970 no longer existed in 1983, killed by merger, acquisition, bankruptcy or break-up," which yields a half-life for those Fortune 500 of eight years, two months. The BBC cites Richard Foster's research as supporting a finding of decreasing half-lives: "The average lifespan of a company listed in the SandP 500 index of leading US companies has decreased by more than 50 years in the last century, from 67 years in the 1920s *to just 15 years today* [italics added]." Gittleson.

72. This estimate of 1 percent is based on the seventy-five-year half-life of the top one hundred corporations taken to five hundred years. A substantially more dramatic case could be made on Foster's rule: it would take less than a century to reach those few survivors. Neither time frame is sufficient if the CCS is to be securely stored to prevent greenhouse gas hazards to the stability of the global climate.

73. Price-Anderson Nuclear Industries Indemnity Act of 1957, 42 U.S.C. § 2210 et seq.

74. Elizabeth Aldrich, Cassandra Koerner, and David Solan, *Analysis of Liability Regimes for Carbon Capture and Sequestration: A Review for Policymakers* (Chicago: Energy Policy Institute, 2011), 13.

75. Will Reisinger, Nolan Moser, and Trent A. Dougherty, "Reconciling King Coal and Climate Change: A Regulatory Framework for Carbon Capture and Storage," *Vermont Journal of Environmental Law* 11 (2009): 39, with reference to 42 U.S.C. § 2212.

76. Ibid. That the act succeeded in sponsoring the development of nuclear energy was of course not universally welcomed (40). Not only environmentalists but also conservative bodies such as the Cato Institute have raised concerns that the provision of indemnifications increased the risks of negligence behaviors. However, such concerns might be counterbalanced by the provision of efficient regulations to reduce the moral hazard risk of indemnifying such parties. See the discussion in chapter 6 in this book. Also, to the extent that the provision of CCS is effectively a low-risk public good with substantial positive externalities, the risk of negligence might be outweighed, and the provision of the indemnifications could be transaction cost reducing in the provision of that desirable public good.

77. Ibid., 39–40.

78. Bailey et al., 48. See 42 U.S.C.A. § 7901 (West 2014).

79. Bailey et al., 48. The fee was established in 1978 at $250,000; the value is routinely adjusted for inflationary purposes by the DOE.

80. Ibid., 48.

81. Ingelson et al., 460. Their references are to the National Childhood Vaccine Injury Act of 1986, 42 U.S.C. § 300aa-1 to § 300aa-34, and to the Support

Anti-Terrorism by Fostering Effective Technologies Act of 2002 (SAFETY Act), 6 U.S.C. § 441–44. The Terrorism Risk Insurance Act used a liability cap, limiting private exposure to $100 billion before the overflow would be assumed by the federal government. Monast et al., 40.

82. Ingelson et al., 460.

83. Bailey et al., 48.

84. Ibid.

85. Ibid., 50.

86. Ibid.

87. Ibid.

88. Ibid. As such, the application of the SWDA to CCS activities is in alignment with the polluter-pays principle. See the chapter discussion in section 5.2.1.

89. Bailey et al., 50.

90. Flatt, 239.

91. Ibid., 240.

92. Fred Eames and Scott Anderson, "The Layered Approach to Liability for Geological Sequestration of CO_2," *Environmental Law Reporter News and Analysis* 43 (2013): 10653. They also refer to the problems created by the site selection procedures.

93. Christopher Bidlack, "Regulating the Inevitable: Understanding the Legal Consequences of and Providing for the Regulation of the Geologic Sequestration of Carbon Dioxide," *Journal of Land Resources and Environmental Law* 30 (2010): 200.

94. Ibid.

95. Flynn and Marriott, 38.

96. Ibid.

97. Ibid.

98. Haan-Kamminga et al., 247.

99. Ibid.

100. Haan-Kamminga, 325.

101. Ibid.

102. Ibid.

103. Ibid.

104. Monast et al., 40.

105. See the policy recommendations in chapter 9, sections 9.1.5. and 9.1.6.

106. See the chapter discussions in sections 5.4, 5.5, and 5.6, respectively.

107. See the chapter policy discussions in section 5.7.

108. Garrett and Beck, 9.

109. Ibid., 11.

110. Ibid.

111. Ibid.

112. Ibid., 12.

113. Directive 2009/31/EC of the European Parliament and of the Council of April 23, 2009, on the geologic storage of carbon dioxide and amending Council Directive 85/337/EEC, European Parliament and Council Directives 2000/60/EC, 2001/80/EC, 2004/35/EC, 2006/12/EC, 2008/1/EC and Regulation (EC) No 1013/2006 (Text with EEA relevance); OJ L 140, 5.6.2009, 114–135.

114. CCS Directive, ch. 4, art. 18. See also CCS Directive, Preamble, para. 32, 33. Barton et al., 225.

115. CCS Directive, ch. 4, art. 18, sec. 1; Barton et al., 225.

116. CCS Directive, ch. 4, art. 18, sec. 1(a); Barton et al., 225.

117. CCS Directive, ch. 4, art. 18, sec. 1(b); Barton et al., 225.

118. CCS Directive, ch. 4, art. 18, sec. 1(c). See also ch. 4, art. 20, sec. 1.

119. CCS Directive, ch. 4, art. 18, sec. 1(d). Neither *injection facilities* nor *sealed* are defined terms under CCS Directive, ch. 1, art. 3. One assumes that such terms would bear interpretations based on preexisting use with regard to injection facilities already existing within oil and gas and other hazardous material injection facilities.

120. CCS Directive ch. 4, art. 18, sec. 7. However, the CCS Directive does provide a carve-out for operators who acted with such faults as negligence, willful deceit, or failure to operate with due diligence that such operators would not be indemnified under the transfer of liability. Ibid. See also CCS Directive, Preamble, para. 35.

121. See CCS Directive, ch. 4, art. 18, sec. 3, 4, 5 for details on the required reports prior to the closure and transfer of liabilities and ch. 4, art. 19 and 20, for the provisions on the operator's provision of financial security and of the financial mechanisms available to the member states. See also the CCS Directive, Preamble, para. 36, 37.

122. Bailey et al., 48.

123. Ibid.

124. Semere Solomon et al., "A Proposal of Regulatory Framework for Carbon Dioxide Storage in Geological Formations," International Risk Governance Council Workshop (2007), 7–8. Norway developed its laws to address the abandonment of mines and oil fields that would have similar long-term risks of seepage or of corrosive damages that could occur over long time periods akin to those of CCS storage operations (7).

125. Ibid., 7–8.

126. Ibid., 8.

127. Bailey et al., 48.

128. Ibid.

129. Ibid.

130. Ibid.

131. Ibid.

132. Ibid.

133. For purposes of this study, North America includes Panama and all nations north of it within Central and North America. States and provinces of the United States and Canada are listed separately from their federal governments.

134. Aldrich et al., 20.

135. Havercroft and Macrory, 42.

136. Bailey et al., 49. See 31 U.S.C.A. § 1341 (West 2014).

137. Bailey et al., 49. The Carbon Storage Stewardship Trust Fund Act of 2009 would have also required private liability insurance that would be superseded in case of excessive damages by funds accumulated by the DOE at the time of the postclosure transfers. Arnold W. Reitze Jr. and Marie Bradshaw Durrant, "State and Regional Control of Geological Carbon Sequestration," *Environmental Law Reporter News and Analysis* 41 (2011): 10368.

138. Peter Folger, *Carbon Capture and Sequestration: Research, Development, and Demonstration at the US Department of Energy* (Washington, DC: Congressional Research Service, 2012), 5.

139. Flynn and Marriott, 39.

140. Ibid., 38–39.

141. Ibid., 39.

142. Ibid., 39, citing 73 C.F.R. § 43,520. "USDWs" is an EPA-used phrase for "underground sources of drinking waters."

143. Reitze and Durant, 10367.

144. Ibid.

145. Ibid.

146. Ibid.

147. Ibid.

148. Ibid.

149. Monast et al., 13.

150. Ibid.

151. Matthew J. Lepore and Derek L. Turner, "Legislating Carbon Sequestration: Pore Space Ownership and Other Policy Considerations," *Colorado Lawyer* 40, no. 10 (2011): 63. Implied in the legislators' concerns appears to have been a logic that if other states did not provide such postclosure transfers of liability, then operators would be incentivized to store more in the areas that did so provide and thus potentially increase the CCS-related burdens for local communities burdened with both increased local storage risks of foreign CO_2 volumes and potential liability-based calls for future taxation.

152. Ibid.

153. Ibid.

154. Lincoln Davies et al., *Carbon Capture and Sequestration: A Regulatory Gap Assessment* (Salt Lake City: University of Utah, 2012), 29.

155. Aldrich et al., 20.

156. Ibid., 16; Louisiana HB § 661, 2009.

157. Ibid., 16.

158. Ibid.

159. Ibid.; Louisiana HB § 661, 1109(A)(2), 2009.

160. Aldrich et al., 17.

161. Davies et al., 28.

162. Mont. Code Ann. § 82–11–183(7); Aldrich et al., 15.

163. Davies et al., 28.

164. This assumption of liability needs to be coordinated with the operator's compliance to provide the necessary contributions to the fund. See Mont. Code Ann. § 82–11–183(9)(a) for when the operator retains long-term liability and § 82–11–181(1) for the liability can be transferred. See also Ingelson et al., 445–446.

165. Montana SB 498 § 4, 2009, See also Mont. Code Ann. § 82–11–183(8)(c). Aldrich et al., 15. See also Ingelson et al., 445–446.

166. Mont. Code Ann. § 82–11–183(8)(a) and (b). Ingelson et al., 445–446.

167. Aldrich et al., 14–15.

168. Ibid., 15. See also North Dakota SB 2095 § 38–22–17(6), 2009.

169. Davies et al., 28.

170. Aldrich et al., 15.

171. Ibid., 19–20.

172. Ibid.

173. Ibid., 18. Wyoming HB 17, § 1(d), 2010.

174. Delissa Hayano, "Guarding the Viability of Coal and Coal-Fired Power Plants: A Road Map for Wyoming's Cradle to grave Regulation of Geologic CO_2 Sequestration," *Wyoming Law Review* 9 (2009): 160.

175. Ibid.

176. Ibid.

177. Aldrich et al., 18; Texas SB 1387 § 10(c)(5)(B).

178. Bailey et al., 49.

179. Ibid.

180. Ibid.

181. Ibid.

182. Ibid.

183. Ibid.

184. Ingelson et al., 444; Havercroft and Macrory, 43.

185. Greenhouse Gas Geological Sequestration Act 2008 (Vict.).

186. Ingelson et al., 445, citing GGGSA § 7(a), § 7(d).

187. Barton et al., 225.

188. By no means is this to imply any policy guidance contrary to those listed in chapter 9, sections 9.1.5 and 9.1.6, and for the need of public planning and provision for the likely extinction of operators prior to the secure sun-setting of all potential long-term liabilities from CCS storage activities. The comment here, to coordinate with the polluter-pays principle, is to be taken *in pari materia*, and encouraged to the extent feasible within our policy proposals.

Chapter 6

1. See also the chapter discussion in sections 6.1 and 6.2.

2. See also the chapter discussion in sections 6.3.1 and 6.3.2.

3. Michael G. Faure and Stefan E. Weishaar, "The Role of Environmental Taxation: Economics and the Law," in *Handbook of Research on Environmental Taxation* (Cheltenham: Edward Elgar, 2012), 404–406.

4. In reference to public regulations in particular. Ibid., 404, citing Lucas Bergkamp, *Liability and Environment: Private and Public Law Aspects of Civil Liability for Environmental Harm in an International Context* (The Hague: Kluwer Law International, 2001).

5. Generally most modern public states develop regulations within democratic or at least publicly deliberative processes, so that the nature and character of the regulations are coordinated with social awareness.

6. See R. Van den Bergh and L. T. Visscher, "Optimal Enforcement of Safety Law," in *Mitigating Risk in the Context of Safety and Security: How Relevant Is a Rational Approach?* ed. R. V. de Mulder (Rotterdam: Erasmus University Rotterdam, 2008), 29: "Fines can be attached to norm breaking behaviour, irrespective of whether losses have occurred, and/or harmful behaviour."

7. For example, an injunction could be used to cease efforts prior to the irreversible moment. Van den Bergh and Visscher, 29, set out a temporally framed set of enforcement measures: (1) preclusionary measures, (2) act-based sanctions, and (3) harm-based sanctions. They demonstrated that regulations could provide policy tools at each temporal stage, while rules of civil liability would be primarily limited to harm-based sanctions with some access to preclusionary measures through injunction-type petitions.

8. Keith N. Hylton, "When Should We Prefer Tort Law to Environmental Regulation?" *Washburn Law Journal* 41 (2001): 4. Hylton used the term *private regime* to indicate rules of civil liability and not private or self-regulation.

9. Arguably, certain forms of civil liability can enable some forms of broader engagement within the public community, such as the role of amicus curiae. However, the

general models from accident law do not usually incorporate the potential strategies of these types of interactions of nonstanding parties to a dispute.

10. See Neil Gunningham, Martin Phillipson, and Peter Grabosky, "Harnessing Third Parties as Surrogate Regulators: Achieving Environmental Outcomes by Alternative Means," *Business Strategy and the Environment* 8, no. 4 (1999).

11. Steven Shavell, "Liability for Harm"; Steven Shavell, "A Model of the Optimal Use"; Steven Shavell, *Economic Analysis of Accident Law* (Cambridge: Harvard University Press, 1987), 277–290. For an application to environmental harm, see Alessandra Arcuri, "Controlling Environmental Risk in Europe: The Complementary Role of an EC Environmental Liability Regime," *Tijdschrift voot Milieuaansprakelijkheid* 15, no. 2 (2001): 39–40. See also M. Boyer, Y. Hiriart, and D. Martimort, *Frontiers in the Economics of Environmental Regulation and Liability* (Aldershot: Ashgate, 2006).

12. See Bergkamp, *Liability and Environment,* 213–216; see also Jing Liu, *Compensating Ecological Damage: Comparative and Economic Observations* (Antwerp: Intersentia, 2013), 47–50.

13. See Shavell, "Liability for Harm," 357, and Shavell, "A Model of the Optimal Use," 271.

14. Hylton, "A Positive Theory of Strict Liability," 12.

15. Ibid., 4.

16. David E. Adelman and Ian J. Duncan, "The Limits of Liability in Promoting Safe Geologic Sequestration of CO_2," *Duke Environmental Law and Policy Forum* 22 (2011): 22.

17. Ibid.

18. Ibid., 25.

19. Ibid.

20. This is referred to as the rational apathy problem. See Hans-Bernd Schäfer, "The Bundling of Similar Interests in Litigation: The Incentives for Class Actions and Legal Actions Taken by Associations," *European Journal of Law and Economics* 9 (2000): 183–223.

21. See S. Rose-Ackerman and M. Geisteld, "The Diversions between Social and Private Incentives to Sue: A Comment on Shavell, Menell and Kaplow," *Journal of Legal Studies* 16 (1987): 483–491. See also Steven Shavell, "The Social versus the Private Incentive to Bring Suit in a Costly Legal System," *Journal of Legal Studies* 11, no. 2 (1982): 333–339.

22. Adelman and Duncan, 61.

23. Alexandra B. Klass and Elizabeth J. Wilson, "Climate Change and Carbon Sequestration: Assessing a Liability Regime for Long-Term Storage of Carbon Dioxide," *Emory Law Journal* 58 (2008): 159, argue that tort liability fails to detect and assign blame for harm, may lead to an insolvency problem with operators, and is unable to provide a remedy in case of long-tail risks. See also David Gerard and Elizabeth J. Wilson, "Environmental Bonds and the Challenge of Long-Term Carbon Sequestration," *Journal of Environmental Management* 90 (2009): 1099.

24. Ibid.

25. See Chiara Trabucchi and L. Patton, "Storing Carbon: Options for Liability Risk Management, Financial Responsibility," World Climate Change Report 170 (Washington, DC: Bureau of National Affairs, 2008), 8–9.

26. Ibid., 14–16.

27. In the words of Adelman and Duncan, 29: "The failure of common law regimes to deter latent harms also places a high premium on effective ex ante regulation."

28. Shavell, "A Model of the Optimal Use"; see also Shavell, "Liability for Harm." See Michael G. Faure, "Regulatory Strategies in Environmental Liability," in *The Regulatory Function of European Private Law*, ed. F. Cafaggi, H. Watt, and H. Muir (Cheltenham: Edward Elgar, 2011), 140.

29. Shavell, "Strict Liability versus Negligence."

30. The theoretical idea is that the asymmetrical information presents a market failure to the involved parties and the regulatory body could solve that failure by collecting the necessary information and finding ways to correct those informational deficits. See George J. Stigler, "The Economics of Information," *Journal of Political Economics* 69, no. 3 (1961): 231. See also Alan Schwartz and Louis L. Wilde, "Intervening in Markets on the Basis of Imperfect Information: A Legal and Economic Analysis," *University of Pennsylvania Law Review* 127, no. 3 (1979): 630; Ejan Mackaay, *Economics of Information and Law* (Paris: Groupe de recherche en consommation, 1980).

31. For a seminal article, see Stigler, 213. See also Schwartz and Wilde.

32. Shavell, "Liability for Harm," 359.

33. Hylton, "A Positive Theory of Strict Liability," 3.

34. Arguably, certain forms of civil liability can enable some forms of broader engagement within the public community, such as the role of amicus curiae. However, the general models from accident law do not usually incorporate the potential strategies of these types of interactions of nonstanding parties to a dispute.

35. Glachant, 3.

36. Ibid. Glachant states that the determination between regulation, rules of civil liability, or indeed of no governance other than the marketplace might be intractable without additional but missing information.

37. Ibid.

38. Ibid., 9–10.

39. Ibid., 7.

40. Ibid., 3.

41. Ibid., 7. Therein is a detailed explanation of the critical steps of the information-revealing mechanism; its mathematical depth exceeds the purview of this book.

42. However, once this information is thus obtained, it could be reintroduced into the market to provide potential coordination of regulations with rules of civil liability to reinforce both systems.

43. Shavell demonstrated that when the expected costs from judgments would exceed the assets of tortfeasors, then public regulations should be more robust than either a rule of strict liability or a rule of negligence. See also Shavell, "Liability for Harm."

44. See the discussions on expected company longevity in chapter 5, section 5.4.

45. Shavell, "Liability for Harm," 360.

46. Steven Shavell, "The Judgment Proof Problem."

47. See Richardson, *Environmental Regulation through Financial Organisations*, 169.

48. If insurance came into the picture, it could overcome the problems of underdeterrence, provided that the moral hazard problem, caused by insurance, can be cured. We will discuss the functioning of insurance in chapter 7, section 7.2.

49. Steven Shavell, "Criminal Law and the Optimal Use of Nonmonetary Sanctions as a Deterrent," *Columbia Law Review* 85, no. 6 (1985): 1236–1237. See also Steven Shavell, "The Optimal Use of Nonmonetary Sanctions as a Deterrent," *American Economic Review* 77, no. 4 (1987): 590.

50. Ibid.

51. Steven M. Shavell, "The Judgment Proof Problem," *International Review of Lw and Economics* 6 (1986): 43.

52. Ibid.

53. Gerard and Wilson, 1099.

54. See Shavell, "Liability for Harm," 363. See also William M. Landes and Richard A. Posner, "Tort Law as a Regulatory Regime for Catastrophic Personal Injuries," *Journal of Legal Studies* 13, no. 3 (1984): 417.

55. Shavell, "A Model of the Optimal Use," with reference to Shavell, "Liability for Harm."

56. Ibid.

57. Ibid.

58. See Landes and Posner, "Tort Law." See also Howard C. Kunreuther and Paul K. Freeman, "Insurability, Environmental Risks and the Law," in *The Law and Economics of the Environment*, ed. Anthony Heyes (Cheltenham: Edward Elgar, 2001), 302.

59. Shavell, "A Model of the Optimal Use," with reference to Shavell, "Liability for Harm."

60. Hans-Bernd Schäfer and Andreas Schönenberger, "Strict Liability versus Negligence," in *Encyclopedia of Law and Economics*, ed. B. Bouckaert and G. De Geest (Cheltenham: Edward Elgar, 2000), 605.

61. Ibid.

62. Shavell, "Liability for Harm," 363.

63. More particularly with respect to ecological damage, Liu, 86, argues that the probability of being sued is low, which could lead to underdeterrence.

64. See Landes and Posner, 417.

65. For alternatives to liability suits, see H. Bocken, "Alternatives to Liability and Liability Insurance for the Compensation of Pollution Damages," *Tijdschrift voor Milieuaansprakelijkheid* 1 (1987).

66. Michael G. Faure, "Compensation for Non-Pecuniary Losses from an Economic Perspective," in *European Tort Law: Liber Amicorum for Helmut Koziol*, ed. U. Magnus and J. Spier (Frankfurt: Peter Lang, 2000). For the economic assessment of pain and suffering, see Vaia Karapanou and Louis Visscher, "Towards a Better Assessment of Pain and Suffering Damages," *Journal of European Tort Law* 1, no. 1 (2010): 48.

67. According to Cooter, this is the most important difference between tort law and criminal law: tort law simply puts a price on activities in the form of compensation due for harmful acts, whereas criminal law simply prohibits some activities from being undertaken by imposing a sanction on the one who nevertheless engages in this activity. Robert Cooter, "Prices and Sanctions," *Columbia Law Review* 84 (1984): 1523.

68. See the discussion on risks in chapter 2, section 2.4.

69. See the discussions on long-term risks in chapter 2, sections 2.4 and 2.5.

70. Gerard and Wilson, 1099.

71. Hylton, "When Should We Prefer Tort Law?" 12.

72. Ibid., 4.

73. See Michael G. Faure and S. Ubachs, "Comparative Benefits and Optimal Use of Environmental Taxes," *Critical Issues in Environmental Taxation* 1 (2003).

74. Also, this follows a pattern from criminal law, in that the regulatory body focused on the prevention of acts that hurt the public welfare instead of focusing on how to improve it.

75. Thomas H. Tietenberg and Lynne Lewis, *Environmental and Natural Resource Economics* (Reading, MA: Addison-Wesley, 2000).

76. Ibid.

77. Ibid.

78. Ibid.

79. Michael G. Faure, "Designing Incentives Regulation for the Environment," Maastricht Faculty of Law Working Paper 2008–7 (2008).

80. Ibid., 26–27. See also Tietenberg and Lewis.

81. Faure, "Designing Incentives."

82. Shavell 1976. Stickiness is related to a variety of phenomena, primarily the complex interactions of various transaction costs that prevent more continuous adjustments to pricing and cost data over time.

83. For a seminal paper on path dependency on the effects of technological choice, see W. Brian Arthur, "Competing Technologies, Increasing Returns, and Lock-In by Historical Events," *Economic Journal* 99, no. 394 (1989): 116.

84. In some literature, it is also referred to as private regulation.

85. See Anthony Ogus, "Self-Regulation," in *Encyclopedia of Law and Economics*, ed. B. Bouckaert and G. De Geest (Cheltenham: Edward Elgar, 2000). See also Niels Philipsen. *Regulation of and by Pharmacists in the Netherlands and Belgium: An Economic Approach* (Cambridge: Intersentia, 2003), 35–41.

86. Anthony Ogus," Rethinking Self-Regulation," *Oxford Journal of Legal Studies* 15, no. 1 (1995): 97–98; Roger Van den Bergh, "De maatschappelijke wenselijkheid van gedragscodes vanuit rechtseconomisch perspectief," *Weekblad voor Privaatrecht, Notariaat en Registratie* 6772 (2008): 793.

87. Anthony Ogus, "The Paradoxes of Legal Paternalism and How to Resolve Them," *Legal Studies* 30, no. 1 (2010): 61–73. See also Ronald Coase, *The Problem of Social Cost* (Basingstoke, UK: Palgrave Macmillan, 1960).

88. Ogus, "Paradoxes of Legal Paternalism."

89. In order to do so, the regulatory authority must be able to identify instances of market failures and select those regulatory measures that predictably correct the market failure at the least cost.

90. Ibid.

91. The argument is not made here that such parties would be the best informed, only that such parties ought to be reasonably knowledgeable about such concerns. Due to the potential advancement of technology and related matters, and their likely involved role in that development, they might also be in possession of relevant information in advance of other parties such as regulators.

92. Anthony Ogus, "Regulation Revisited," *Public Law,* no. 2 (2009): 232.

93. Ogus, "Paradoxes of Legal Paternalism."

94. When that group is primarily composed of potential tortfeasors, that regulation can also be called self-regulation; Ogus, "Paradoxes of Legal Paternalism." See Ogus, "Rethinking Self-Regulation," and Anthony Ogus, "Self-Regulation," in *Encyclopedia of Law and Economics,* ed. B. Bouckaert and G. De Geest (Cheltenham: Edward Elgar, 2000).

95. Neil Gunningham, Peter N. Grabosky, and Darren Sinclair, *Smart Regulation: Designing Environmental Policy* (Oxford: Clarendon Press, 1998).

96. Ibid., 212.

97. Ibid., 214, 216.

98. Ibid.

99. Ibid., 217.

100. Ibid., 218.

101. Ibid., 212. See the discussion in this chapter, section 6.2.3, on the potential use of private insurance as surrogate regulators for CCS activities.

102. For example, the Institute of Electrical and Electronics Engineers Standards Association provides a wide variety of industrial standards across many sections of the economy. http://standards.ieee.org.

103. See the ethical rules adopted by the American Medical Association: http://www.ama-assn.org/ama/pub/physician-resources/medical-ethics/code-medical-ethics.page.

104. For example, the American Petroleum Institute maintains an "inventory of over six hundred standards and recommended practices." See "Publications, Standards, and Statistics Overview," http://www.api.org/publications-standards-and-statistics.

105. Furthermore, private regulation need not stand alone, outside the public authority. When private regulation and public regulation have been integrated together, the result been called integrated regulatory design. Gunningham, Phillipson, and Grabosky, 220.

106. Private regulatory efforts are robust when (1) internal monitoring costs of members' rule compliance are low, (2) the number of local markets is small, and (3) customers are relatively mobile between suppliers. Anthony Ogus, "Regulation Revisited," Public Law 2 (2009): 233. In the case of CCS, these terms might be interpreted as (1) operators could readily monitor each other's performances, (2) the overall number of injection sites is not so large as to make recording keeping difficult, and (3) the operators could be swapped in and out of injection facilities as needed by local governance or authorities. One would assume that given the long-established track record of compliance on API standards, the first requirement could be readily accomplished. The second requirement would also not likely pose a substantial problem, in that both computer-based accounting and the supplemental efforts of green public interest groups are likely to reduce the overall transaction costs of that effort. Finally, the third requirement could be met by ensuring that permits to inject require operational capacity to replace operators while injection and storage operations are underway; this is routinely done within the oil and gas industry and should pose no special problem for CCS efforts.

107. Ogus, "Regulation Revisited," 233. If trust persists between the principal authority and the private agency, the internal costs of coordination between the two parties will be reduced in contrast to a command and control model of regulation. Also, since the private agency bears its own regulatory costs, there is an efficiency gain from avoiding the need to tax them to in turn govern them.

108. Ibid.

109. Ibid., 231.

110. Ibid., 234, citing Peter Grajzl and Peter Murrell, "Allocating Lawmaking Powers: Self-Regulation vs. Government Regulation," Journal of Comparative Economics 35 (2007).

111. James C. Miller III, "The FTC and Voluntary Standards: Maximizing the Net Benefits of Self-Regulation," Cato Journal 4 (1984). See also Ogus, "Rethinking Self-Regulation," 98.

112. Ogus, "Regulation Revisited," 235.

113. Ibid., citing Peter Grajzl and Andrzej Baniak, "Industry Self-Regulation, Subversion of Public Institutions, and Social Control of Torts," International Review of Law and Economics 29, no. 4 (2009).

114. Ibid., 234, citing Grajzl and Murrell.

115. Miller.

116. Ogus, "Rethinking Self-Regulation," 98. See also Van den Bergh, 794.

117. For a critical analysis of self-regulation in environmental law, see Richardson, *Environmental Regulation through Financial Organisations*, 142–161.

118. See Ogus, "Rethinking Self-Regulation"; see also Ogus, "Self-Regulation."

119. The logic here does assume that damages could and would be accurately brought to bear on the operator, thus providing the operator with the necessary information for this balanced decision-making process.

120. See Ogus, "Rethinking Self-Regulation"; see also Ogus, "Self-Regulation."

121. David A. Dana and Hannah J. Wiseman, "A Market Approach to Regulating the Energy Revolution: Assurance Bonds, Insurance, and the Certain and Uncertain Risks of Hydraulic Fracturing," *Iowa Law Review* 99 (2014): 1546.

122. See Omri Ben-Shahar and Kyle D. Logue, "Outsourcing Regulation: How Insurance Reduces Moral Hazard," *Michigan Law Review* 111, no. 2 (2012).

123. Dana and Wiseman, 1565. "Insurers also have an incentive to gather information regarding safety that will be relevant to setting the next premium. Moreover, an insured in a regime where an entity can only operate if it is able to acquire insurance has a strong incentive to cooperate in producing information lest the insured be denied coverage and thus unable to operate. 'Insurers' are thus 'strategically well placed to gather information and engage in risk management, and reflect these costs through premium differentiation.'" Ibid., citing Richardson, *Environmental Regulation through Financial Organisations*, 363.

124. Dana and Wiseman, 1566.

125. Ibid., 1564.

126. Ibid., 1565–1566.

127. Ibid., 1546.

128. Their review applied to the application of such policies to shale fracturing technologies, but the argument would appear to be equally applicable to CCS storage technologies. Ibid.

129. Ibid., 1529. This is because the obtenation of data can be drafted into the very policy that provides the insurance (1568).

130. Ibid., 1569. They did assume in their analysis that insurance policies would be sufficiently viable to pay for long-term concerns when such long-tail liabilities came due.

131. Ibid., 1570. They describe this logic pattern as "engaged in continual, co-adaptive evolution" (1569).

132. For an overview, see T. Heremans, *Professional Services in the EU Internal Market: Quality Regulation and Self-Regulation* (Oxford: Hart, 2012), 23–86.

133. Neal Shover, "Zelfregulering door ondernemingen: Ontwikkeling, beoordeling en bange voorgevoelens," *Tijdschrift voor Criminologie* 51, no. 2 (2008).

134. See Angela J. Campbell, "Self-Regulation and the Media," *Federal Communications Law Journal* 51, no. 3 (1999): 717.

135. Ibid., "Self-regulators often combine—and sometimes confuse—self-regulation with self-service."

136. See Thomas W. Reader, "Is Self-Regulation the Best Option for the Advertising Industry in the European Union: An Argument for the Harmonization of Advertising Laws through the Continued Use of Directives," *University of Pennsylvania Journal of International Business Law* 16 (1995): 182, 210.

137. See Cary Coglianese and David Lazer, "Management-Based Regulation: Prescribing Private Management to Achieve Public Goals," *Law and Society Review* 37, no. 4 (2003); see also Roderick Arthur William Rhodes, "The New Governance: Governing without Government," *Political Studies* 44, no. 4 (1996). On the concept of transnational private regulation, see Fabrizio Cafaggi, "New Foundations of Transnational Private Regulation," *Revista de Derecho Privado* 26 (2014); see also Colin Scott, Fabrizio Cafaggi, and Linda Senden, "The Conceptual and Constitutional Challenge of Transnational Private Regulation," *Journal of Law and Society* 38, no. 1 (2011).

138. See Barbara Baarsma et al., *Zelf doen? Inventarisatiestudie van zelfreguleringsinstrumenten* (Amsterdam: SEO, 2003), 8.

139. See Charles D. Kolstad, Thomas S. Ulen, and Gary V. Johnson, "Ex Post Liability For Harm vs. Ex Ante Safety Regulation: Substitutes or Complements?" *American Economic Review* 80 (1990): 888. See also Susan Rose-Ackerman, *Rethinking the Progressive Agenda: The Reform of the American Regulatory State* (New York: Free Press, 1992), and Susan Rose-Ackerman, "Public Law versus Private Law in Environmental Regulation: European Union Proposals in the Light of United States Experience," *Review of European Community and International Environmental Law* 4, no. 4 (1995): 312. See also Michael G. Faure, and Marieke Ruegg, "Environmental Standard Setting through General Principles of Environmental Law," in *Environmental Standards in the European Union in an Interdisciplinary Framework* (Antwerp: Maklu, 1994), 39–60; Paul Burrows, "Combining Regulation and Legal Liability for the Control of External Costs," *International Review of Law and Economics* 19 (1999); Arcuri, 39; Faure, "Regulatory Strategies in Environmental Liability," 143.

140. This perspective is not universally held within the literature. Wilson, Klass and Bergan, for example, have advocated that liability rules could still play a role, but more modest as a "secondary backstop" behind a future comprehensive regulatory framework aiming at the prevention of CCS-related risks (4576). In either case, however, it is agreed that both rules of civil liability and public regulations need to be implemented to manage the overall risks of CCS-related activities.

141. On those interdependencies between regulation and liability, see also Bergkamp, *Liability and Environment*, 233–236.

142. The literature generally concludes, also with respect to environmental risks, that liability and regulation should be combined in order to reach optimal deterrence. See P. W. Schmitz, "On the Joint Use of Liability and Safety Regulation," *International Review of Law and Economics* 20 (2000), and for empirical evidence with respect to environmental liability in France, see Pierre Bentata and Michael G. Faure, "The Role of Environmental Civil Liability: An Economic Analysis of the French Legal System," *Environmental Liability, Law, Policy and Practice* 20, no. 4 (2012).

143. See Adelman and Duncan, 46.

144. Ibid., 53, although they qualify the additional incentive provided by liability in this case as "limited."

145. See chapter 2, section 2.5.2.

146. On those pilots, see the discussion on CCS technologies in chapter 2, section 2.2.

147. In the sense of Neil Gunningham, Peter N. Grabosky, and Darren Sinclair, *Smart Regulation: Designing Environmental Policy* (Oxford: Clarendon Press, 1998).

148. See chapter 4, section 4.5.

149. Neil Gunningham and Darren Sinclair, "Regulatory Pluralism: Designing Policy Mixes for Environmental Protection," *Law and Policy* 21, no. 1 (1999): 50.

150. Faure and Weishaar, 405–406. While the bulk of the American response to regulating CCS activities, particularly for the protection of subterranean freshwater supplies, has relied on the Safe Water Drinking Act, the historical troubles of that act suggest that a purely regulatory method might be shortsighted. John Pendergrass, "Long-Term Stewardship of Geological Sequestration of CO_2," *Environmental Law Reporter: News and Analysis* 43 (2013): 10660. Pendergrass notes that the SWDA has suffered from inconsistencies in implementation from state to state under the delegable powers assigned to them under the act, varying levels of monitoring and enforcement due to local politics and budgets, and tension between federal oversight and the policies of state agencies. Pendergrass says "Effective long-term stewardship must be multi-layered because experience has shown that no single measure

is sufficient to protect against risks that have long latency periods and are not easily observed" (10661).

151. Faure, "Regulatory Strategies in Environmental Liability," 143.

152. The ability to bring compliance via private litigation for injuries received may reduce the value of coercion or agency capture efforts. See ibid.

153. See Van den Bergh and Visscher for a discussion of the Nyborg and Telle problem within their discussion of compliance strategies as an alternative to deterrence strategies.

154. Rose-Ackerman 1992, *supra* at note 139.

155. Bruno S. Frey, "Morality and Rationality in Environmental Policy," *Journal of Consumer Policy* 22, no. 4 (1999): 395.

156. Ibid.

157. As Frey stated, even the movement toward market-based incentives to reinforce regulatory frameworks is very much akin to selling indulgences; it provides the wrong message that environmental error can be washed clean with cash when in fact much of that damage cannot.

Chapter 7

1. Avelien Haan-Kamminga, "Long-Term Liability for Geological Carbon Storage in the European Union," *Journal of Energy and Natural Resources Law* 29 (2011): 313.

2. Ibid.

3. Richard A. Posner, *Economic Analysis of Law*, 6th ed. (New York: Aspen, 2003), 167–177.

4. See George L. Priest, "The Government, the Market, and the Problem of Catastrophic Loss," *Journal of Risk and Uncertainty* 12, no. 2–3 (1996): 235. Moreover, leaving victims of catastrophes without any relief would probably be incompatible with the concept of the welfare state, at least as this is conceived within most EU member states. See Reimund Schwarze and Gert G. Wagner, "In the Aftermath of Dresden: New Directions in German Flood Insurance," *Geneva Papers on Risk and Insurance: Issues and Practice* 29, no. 2 (2004). To the extent that damage resulting from CCS could be considered a disaster, a similar reasoning could be applied as well.

5. See Richard A. Posner, *Catastrophe: Risk and Response* (Oxford: Oxford University Press, 2004).

6. Some have indeed argued that it may be problematic if compensation would be guaranteed only for victims of catastrophes; for victims of other accidents, this

generosity would not apply. See U. Magnus, "Germany," in *Compensation for Victims of Catastrophes: A Comparative Legal Approach*, ed. Michael G. Faure and T. Hartlief (Vienna: Springer, 2006); See also Richard Zeckhauser, "The Economics of Catastrophes," *Journal of Risk and Uncertainty* 12, no. 2–3 (1996): 114.

7. Elizabeth Aldrich, Cassandra Koerner, and David Solan, *Analysis of Liability Regimes for Carbon Capture and Sequestration: A Review for Policymakers* (Chicago: Energy Policy Institute, 2011), 10.

8. Ibid., 11.

9. David A. Dana and Hannah J. Wiseman, "A Market Approach to Regulating the Energy Revolution: Assurance Bonds, Insurance, and the Certain and Uncertain Risks of Hydraulic Fracturing," *Iowa Law Review* 99 (2014): 1572.

10. Ibid.

11. Ibid.

12. Ibid., 1573. They write, "Despite the predictions made in the face of mandatory insurance proposals, insurance markets have consistently produced adequate insurance capacity once a mandate was enacted."

13. Ibid., 1523–1524.

14. Donna M. Attanasio, "Surveying the Risks of Carbon Dioxide: Geological Sequestration and Storage Projects in the United States," *Environmental Law Reporter: News and Analysis* 39 (2009): 10387.

15. Lincoln Davies et al., *Carbon Capture and Sequestration: A Regulatory Gap Assessment* (Salt Lake City: University of Utah, 2012), 29.

16. See Kenneth J. Arrow, "Aspects of the Theory of Risk-Bearing (Yrjo Jahnsson Lectures)" (Helsinki: Yrjo Jahnssonin Saatio, 1965). See also Karl Borch, "The Utility Concept Applied to the Theory of Insurance," *Astin Bulletin* 1, no. 5 (1961): 245–255.

17. See Steven Shavell, *Economic Analysis of Accident Law* (Cambridge: Harvard University Press, 1987) 190.

18. See Jing Liu, *Compensating Edological Damage: Comparative and Economic Observations* (Antwerp: Intersentia, 2013), 103–105.

19. See Benjamin Richardson, *Environmental Regulation through Financial Organisations: Comparative Perspectives on the Industrialized Nations* (London: Kluwer Law International, 2002), 329.

20. See Göran Skogh, "The Transactions Cost Theory of Insurance: Contracting Impediments and Costs," *Journal of Risk and Insurance* 56, no. 4 (1989): 726–732.

21. Richardson, *Environmental Regulation through Financial Organisations*, 328–329.

22. To be discussed in this and the next chapter.

23. On those different type of insurances for environmental liability, see Michael G. Faure, "Environmental Damage Insurance in the Netherlands," *Environmental Liability* 10, no. 1 (2002): 31–41.

24. Given that it concerns relatively new risks, one cannot see why insurers would explicitly exclude CCS-related risks from coverage. On first-party victim insurance, See Richardson, *Environmental Regulation through Financial Organisations*, 371.

25. Ibid., 329.

26. See Alberto Monti, "Environmental Risk: A Comparative Law and Economics Approach to Liability and Insurance," *European Review of Private Law* 9, no. 1 (2001).

27. Flatt raised concerns that there remains insufficient historical experience from which to build accurate actuarial packages on which to set accurate premium prices for insurance policies and thus that insurance would either be unavailable or inefficiently priced. Victor Byers Flatt, "Paving the Legal Path for Carbon Sequestration," *Duke Environmental Law and Policy Forum* 19 (2009): 225. In chapter 2, we reviewed the varieties of analogous historical experiences to CCS operations. It is likely that insurers with previous experience with the industries discussed would have sufficient actuarial data for their purposes.

28. Zurich began offering two different CCS-related insurance policies beginning in 2009. Swiss Reinsurance Company and ACE are listed as having been in the development of similar insurance products. Aldrich et al., 10.

29. Howard C. Kunreuther, R. Hogarth, and J. Meszaros, "Insurer Ambiguity and Market Failure," *Journal of Risk and Uncertainty* 7, no. 1 (1993). We will come back to the importance of insurer ambiguity when dealing with the insurability of retrospective liability.

30. As we will explain below, the advantage of a risk-sharing agreement is that operators themselves may, better than insurers, be able to obtain accurate information on the exposure to risk and thus cure the information asymmetry.

31. Alexandra B. Klass and Elizabeth J. Wilson, "Climate Change and Carbon Sequestration: Assessing a Liability Regime for Long-Term Storage of Carbon Dioxide," *Emory Law Journal* 58 (2008): 163.

32. Ibid., 164.

33. Chiara Trabucchi and L. Patton, "Storing Carbon: Options for Liability Risk Management, Financial Responsibility," World Climate Change Report 170 (Washington, DC: Bureau of National Affairs, 2008), 8–9.

34. Ibid., 10.

35. See Mark Anthony De Figueiredo, D. M. Reiner, and H. J. Herzog, "Framing the Long-Term in Situ Liability Issue for Geologic Carbon Storage in the United States," *Mitigation and Adaptation Strategies for Global Change* 10, no. 4 (2005): 654.

36. David E. Adelman and Ian J. Duncan, "The Limits of Liability in Promoting Safe Geologic Sequestration of CO_2," *Duke Environmental Law and Policy Forum* 22 (2011): 17.

37. Trabucchi and Patton, 13.

38. Chiara Trabucchi, Michael Donlan, and Sarah Wade, "A Multi-Disciplinary Framework to Monetize Financial Consequences Arising from CCS Projects and Motivate Effective Financial Responsibility," *International Journal of Greenhouse Gas Control* 4 (2010): 389, 391.

39. David Hawkins, George Peridas, and John Steelman, "Twelve Years after Sleipner: Moving CCS from Hype to Pipe," *Energy Procedia* 1 (2009): 4407.

40. See IEA (2011), 12, arguing that insurance is generally considered to be appropriate only during the operation of the plant, but may not be applied indefinitely across the postclosure phase.

41. See Z. Makuch, S. Georgieva, and B. Oraee-Mirzamani, *Carbon Capture and Storage: Regulating Long-Term Liability* (London: Centre for Environmental Policy, Imperial College London, 2011), 7, 9.

42. The insurers refer more particularly to the fact that the ELD includes a duty to restore the environment back to baseline, which, they say, provides an indefinite open end to the insurance policy since it may not be known when the regulatory authority judges that this baseline has been reached. For a discussion of other reasons behind the reluctance of insurance to cover the liability regime of the Environmental Liability Directive, see L. Bergkamp, N. Herbatschek, and S. Jayanti, "Financial Security and Insurance," in *The EU Environmental Liability Directive: A Commentary*, ed. L. Bergkamp and B. J. Goldsmith (Oxford: Oxford University Press, 2013), 126–27. For more on the insurers' approach to environmental liability, see Christian Lahnstein, "A Market-Based Analysis of the Financial Assurance Issues of Environmental Liability, Taking Special Account of Germany, Austria, Italy and Spain," in *Deterrence, Insurability and Compensation in Environmental Liability: Future Developments in the European Union*, ed. Michael G. Faure (Vienna: Springer, 2003), 303–330.

43. On financial security and insurance under the Environmental Liability Directive, see Bergkamp et al., 118–138.

44. On the importance of this issue for environmental liability insurance, see Kenneth S. Abraham, "Environmental Liability and the Limits of Insurance," *Columbia Law Review* 88, no. 5 (1988): 970–972.

45. See H. Cousy, "Recent Developments in Environmental Insurance," in *Recent Economic and Legal Developments in European Environmental Policy*, ed. F. Abraham, K. Deketelaere, and T. Stuyck (Leuven: Leuven University Press, 1995), 237–239; Richardson, *Environmental Regulation through Financial Organisations*, 344.

46. See Jaap Spier and A. T. Bolt, "De uitdijende reikwijdte van het aansprakelijkheidsrecht?" *De uitdijende reikwijdte van de aansprakelijkheid uit onrechtmatige daad*, Preadvies Nederlandse Juristenvereniging (Deventer: Kluwer, 1996), 207–352; J. H. Wansink, *De Algemene Aansprakelijkheidsverzekering* (Zwolle: W. E. J. Tjeenk Willink, 1994), 110; and J. H. Wansink, "De aansprakelijkheidsverzekering en de dekking van long-tail-risico's," *Aansprakelijkheid en Verzekering* (1995): 1–5.

47. Richardson, *Environmental Regulation through Financial Organisations*, 345.

48. See F. Van Huizen, "Enkele Begrenzingen van de (Beroeps-) Aansprakelijkheidsverzekering," in *Drie Treden* (De Ruiter-bundel, 1995), 325–339.

49. See Wansink, 110–12. See also O. A. Haazen and J. Spier, "Feitelijke ontwikkelingen en verzekerbaarheid," in *De Uitdijende Reikwijdte van de Aansprakelijkheid uit Onrechtmatige Daad*, ed. A. T. Bolt and J. Spier (Preadvies Nederlandse JuristenVereniging 1996): 56–79.

50. Kenneth S. Abraham, "Cost Internalization, Insurance, and Toxic Tort Compensation Funds," *Virginia Journal of Natural Resources Law* 2 (1982): 123–131. See also Richardson, *Environmental Regulation through Financial Organisations*, 80.

51. Richardson, *Environmental Regulation through Financial Organisations*, 52–64.

52. Insurance companies would usually underwrite policies with maximum terms of five to ten years that can be renewed. See US EPA, Geological Sequestration of Carbon Dioxide, Underground Injection Control (UIC) Program, Class VI Financial Responsibility Guidance (Washington, DC: EPA, July 2011), 18.

53. These are, for example, laid down in the CCS directive.

54. See Makuch et al., 7–9. See also Trabucchi and Patton, 15.

55. See chapter 5, sections 5.1 and 5.2.

56. In simple words: insurers need to hold sufficient financial resources to meet the claims they can expect. See Richardson, *Environmental Regulation through Financial Organisations*, 332–333.

57. See sections 7.2 and 7.3 in this chapter.

58. For details, see Michael G. Faure and Roger Van den Bergh, "Liability for Nuclear Accidents in Belgium from an Interest Group Perspective," *International Review of Law and Economics* 10, no. 3 (1990): 250–251.

59. See Michael G. Faure and Roger Van den Bergh, "Competition on the European Market for Liability Insurance and Efficient Accident Law," *Maastricht Journal of European and Comparative Law* 9 (2002).

60. See Michael G. Faure and Tom Vanden Borre, "Compensating Nuclear Damage: A Comparative Economic Analysis of the US and International Liability Schemes," *William and Mary Environmental Law and Policy Review* 33, no. 1 (2008): 248–251.

61. See, in that respect, especially Commission Regulation (EU) No. 267/2010 of March 24, 2010, on the application of article 101(3) of the Treaty on the Functioning of the European Union to certain categories of agreements, decisions, and concerted practices in the insurance sector.

62. See Ernst and Young, Study on Co(Re)-Insurance Pools and on Ad-Hoc Co(Re)-Insurance Agreements on the Subscription Market, EC Commission, Luxembourg (February 2013), http://europa.eu/competition/sectors/financialservices/insurance.html.

63. See Trabucchi and Patton, 9.

64. Richardson, *Environmental Regulation through Financial Organisations*, 354. On moral hazard and adverse selection in insurance, see Fred Wagner, "Tort Law and Liability Insurance," in *Tort Law and Economics*, ed. Michael G. Faure (Cheltenham: Edward Elgar, 2010), 386–392.

65. See chapter 3, section 3.1.

66. See Kenneth J. Arrow, "Uncertainty and the Welfare Economics of Medical Care," *American Economic Review* 53, no. 5 (1963).

67. See Wagner, 389–392.

68. This is especially identified by Shavell. See Steven Shavell, "On Moral Hazard and Insurance," *Quarterly Journal of Economics* 93 (1979).

69. See J. C. Van Eijk-Graveland, *Verzekerbaarheid van Opzet in het Schadeverzekeringsrecht* (Zwolle: Tjeenk Willink, 1999).

70. John M. Marshall, "Moral Hazard," *American Economic Review* 66, no. 5 (1976), 880–890.

71. See George A. Akerlof, "The Market for 'Lemons': Quality Uncertainty and the Market Mechanism," *Quarterly Journal of Economics* 84, no. 3 (1970): 488–500.

72. George L. Priest, "The Current Insurance Crisis and Modern Tort Law," *Yale Law Journal* 96, no. 7 (1987): 1521–90. Priest has been criticized by Viscusi, who claims that there were other reasons for the product liability crisis in the United States than adverse selection on its own. See W. Kip Viscusi, "The Dimensions of the Product Liability Crisis," *Journal of Legal Studies* 20, no. 1 (1991): 147–177.

73. Abraham, "Environmental Liability," 949–951.

74. Generally the question is whether the benefits of particularization outweigh their costs, which has to be addressed in tort law when a standard of care is defined. Posner, *Economic Analysis of Law*, 187–189. It also has to be addressed when legal rules are made. Isaac Ehrlich and Richard A. Posner, "An Economic Analysis of Legal Rulemaking," *Journal of Legal Studies* 3, no. 1 (1974): 257. And it has to be addressed as well when standards are set. Anthony I. Ogus, "Quantitative Rules and Judicial Decision-Making," in *The Economic Approach to Law*, ed. P. Burrows and C. Veljanovski (London: Butterworths, 1981), 210–225, and Anthony I. Ogus, "Standard Setting for Environmental Protection: Principles and Processes," in *Environmental Standards in the European Union*, ed. Michael G. Faure, J. Vervaele, and A. Weale (Antwerp: Maklu, 1994), 25–37.

75. On the costs of risk differentiation, see Bohrenstein 1989, 25–39. See also Wouter P. J. Wils, "Insurance Risk Classifications in the EC: Regulatory Outlook," *Oxford Journal of Legal Studies*, 14, no. 3 (1994): 449–467.

76. For a detailed description of moral hazard considerations for the CCS-related risks, see Trabucchi and Patton, 3, footnote 7.

77. Makuch et al., 11.

78. See section 7.2.4.

79. See Adelman and Duncan, who indicate that risk differentiation is possible on the basis of the choice of the sequestration site, but also the care measures taken during the operation.

80. See ibid., 60. According to the authors, such analysis should be neither technically demanding nor prohibitive in cost.

81. See also Trabucchi and Patton, 14, who stated that a risk profile of the particular site and operator would need to be designed and that operational risk management systems need to be implemented.

82. Richardson, *Environmental Regulation through Financial Organisations*, 364.

83. We stressed the importance of regulation for CCS in chapter 6.

84. For a discussion on the role for a public-private CCS safety board, see Trabucchi and Patton, 16.

85. Richardson, *Environmental Regulation through Financial Organisations*, 360–366.

86. See Makuch et al., 13, and the discussion on uncertainties relating to CCS-related risks in chapter 2, sections 2.4 and 2.5.

87. See Implementation of Directive 2008/31/EC on the geological storage of carbon dioxide, Guidance Document 4, 6, where it is held "lack of sufficient knowledge about the behaviour of sequestrate CO_2 also will inhibit insurance offerings."

88. See Makuch et al., 13.

89. See the discussion in section 7.2.5.

90. Elizabeth J. Wilson, Alexandra B. Klass, and Sara Bergan, "Assessing a Liability Regime for Carbon Capture and Storage," *Energy Procedia* 1, no. 1 (2009): 4580. They refer more specifically to the "lack of experience with large-scale CCS and inherent geologic heterogeneity," which may create problems.

91. See Klass and Wilson, "Climate Change and Carbon Sequestration," 163.

92. Ibid.

93. See Implementation of Directive 2009/31/EC on the geological storage of carbon dioxide, Guidance Document 4, 7.

94. See Wilson, Klass, and Bergan, 4578.

95. Implementation of Directive 2009/31/EC on the geological storage of carbon dioxide, Guidance Document 4, 4–6, 26–28.

96. US EPA, Geological Sequestration of Carbon Dioxide, Underground Injection Control (UIC) program, 9.

97. See Faure and Hartlief, 144.

98. On those captives, see Paul A. Bawcutt, *Captive Insurance Companies: Establishment, Operation, and Management* (Cambridge: Woodhead-Faulkner, 1991). See also T. Dowding, *Global Developments in Captive Insurance* (London: FT Energy, Pearson Professional, 1997).

99. See Bergkamp et al., 128–129.

100. See Shavell, "On Moral Hazard and Insurance."

101. Richardson, *Environmental Regulation through Financial Organisations*, 373.

102. Financial tests would thus have to be developed to assess an operator's financial capability to face liability for CCS-related risks. Compare the financial security under the Environmental Liability Directive. Bergkamp et al., 128.

103. See implementation of Directive 2009/31/EC on the geological storage of carbon dioxide, Guidance Document 4, 27. "Certainty also depends on stringency of required financial tests."

104. In the definitions, it is held that "self-insurance allows the owner or operator to submit financial statements and other information to prove that they are likely to remain in operation, based on indicators of the economic health of the organization, and that they will be able to complete all required GS activities." US EPA, Geological Sequestration of Carbon Dioxide, Underground Injection Control (UIC) program.

105. Ibid., 8–9.

106. Ibid., 22.

107. Ibid., 36–38.

108. Available at http://www.OPOL.org.uk.

109. See Forms B and Fr-4 on the OPOL website; other conditions to qualify as self-insurer apply as well.

110. Richardson, *Environmental Regulation through Financial Organisations*, 373.

111. See W. E. Belser, "Über die Zweckmäßigkeit der Poolung von Atomrisiken," *Versicherungswirtschaft* 18 (1959); James C. Dow, *Nuclear Energy and Insurance* (Livingston, UK: Witherby, 1989); "Nuclear Power: Insurance and the Pooling System," *Nuclear Pools' Bulletin,* special ed. (1992); S. M. S. Reitsma, "Nuclear Insurance Pools: History and Development," in *Nuclear Accidents-Liabilities and Guarantees* (OECD-IAEA, 1993).

112. Most pools provide for coverage for third-party liability as well as damage to the operator itself. The (mandatory) liability insurance of the operator covers in general the compensatory consequences of extracontractual liability of the operator of a nuclear installation for damage resulting from a nuclear incident, even if the incident was directly due to a grave natural disaster. This policy should be clearly separated from the policy covering potential damage to the operator itself.

113. For a more detailed discussion of nuclear risk-sharing agreements in the United States, see Liu, 246.

114. See Richardson, *Environmental Regulation through Financial Organisations*, 373–374.

115. See the website of the International Group of P&I Clubs: http://www.igpandi.org.

116. For details, see Hui Wang, *Civil Liability for Marine Oil Pollution Damage: A Comparative and Economic Study of the International, US, and Chinese Compensation Regimes* (Alphen aan den Rijn: Kluwer Law International, 2011), 337–338.

117. Interestingly, the pooling in the Price-Anderson Act is also mentioned as an example in the CCS-related literature. See particularly De Figueiredo et al., 653, and Ch. H. Haake and K. B. Marsh, "The Trouble with Angels: Carbon Capture and Storage Hurdles and Solutions," World Climate Change Report (Washington, DC: Bureau of National Affairs, 2009), 5–6.

118. 42 U.S.C.A. § 2210.

119. The Price-Anderson Act was revised in 1967, 1975, 1988, and 2005.

120. Pub. L. No. 85–256, 71 Stat. (1957) 576, 577.

121. Faure and Vanden Borre, 243.

122. 10 C.F.R. § 140.11 (4); 75 FR 16646, April 2, 2010.

123. See A&I's website for more information: http://www.nuclearinsurance.com.

124. Faure and Vanden Borre, 260.

125. H.R. Rep. No. 94–648, 10.

126. This amount consists of the $375 million of the operators' liability + 104 (the total number of operators in the United States), multiplied by their contribution to the second tier of $121.255 million + 5 percent for legal expenses. The correct total amount since 2013 is therefore $13,616,046,000. The maximum contribution per reactor per calendar year to the second tier since 2013 is $18.963 million.

127. See chapter 8, section 8.1.

128. See De Figueiredo et al., 653. See also Trabucchi and Patton, 18.

129. Although Guidance Document 4 in Article 19, Financial Security, and Article 20, Financial Mechanisms of the European Commission, is remarkably silent on the potential of risk-sharing agreements.

130. While notionally insurance meets this definition, it is generally excluded from consideration as a guarantee, as insurance is so well identified in its own right.

131. Compared to insurance, one can hold that if insurers specialize in CCS-related risks, they may have more information for an appropriate risk differentiation and premium setting; as a result, the costs of insurance could be lower than the costs of a bank guarantee.

132. See Bergkamp et al., 127.

133. See De Figueiredo et al., 653.

134. See IEA 2011, 30.

135. See Implementation of Directive 2009/31/EC on the Geological Storage of Carbon Dioxide, Guidance Document 4, 26–27. It is remarkable that Guidance Document 4 considers the bank guarantee or the irrevocable standby letter of credit as providing excellent certainty and having low costs (at least for creditworthy parties).

136. See US EPA, Geological Sequestration of Carbon Dioxide. Underground Injection Control (UIC) program, 17.

137. Ibid., 45.

138. Ibid., 21.

139. The US EPA Guidance Document equally contains recommended specifications concerning the wording of the trust documents. US EPA, Geological Sequestration of Carbon Dioxide, Underground Injection Control (UIC) program, 26–27.

140. Aldrich et al., 10. Monast, Pearson, and Pratson warned that the use of performance bonds is effective when the time period is well defined but not when open ended. Jonas J. Monast, Brooks R. Pearson, and Lincoln F. Pratson, "A Cooperative Federalism Framework for CCS Regulation," *Environmental and Energy Law and Policy Journal* 7 (2012): 39.

141. Aldrich et al., 10.

142. Ibid.

143. Ibid.

144. See Wilson, Klass, and Bergan, 4580. For a further detailed analysis of the feasibility of bonding schemes for geological storage, see Gerard and Wilson, "Environmental Bonds."

145. For an explanation of the working of these catastrophe bonds, see Jean-Robert Tyran and Peter Zweifel, "Environmental Risk Internalization through Capital Markets (ERICAM): The Case of Nuclear Power," *International Review of Law and Economics* 13, no. 4 (1993).

146. Aldrich et al., 10.

147. Dana and Wiseman, 1529.

148. Ibid., 1562.

149. Ibid.

150. Ibid.

151. Performance bonds routinely require the bonded party to comply with regulations in exchange for return of capital. Flatt, 226.

152. Ibid., 227.

153. In essence, Abend has proposed a data-revealing mechanism role in performance bonds, although not enunciated in her paper as such. See Katherine Abend, "Geological Sequestration of Carbon Dioxide: Legal Issues and Recommendations for Regulators," *Appalachian Natural Resources Law Journal* 5 (2010): 19.

154. Dana and Wiseman, 1547.

155. Ibid.

156. US EPA, Geological Sequestration of Carbon Dioxide. Underground Injection Control (UIC) program, 11–12.

157. Ibid., 18, 21–22.

158. See Fred Wagner, "Risk Securitization: An Alternative of Risk Transfer of Insurance Companies," *Geneva Papers on Risk and Insurance: Issues and Practice* (1998).

159. See Richard E. Smith, Emily A. Canelo, and Anthony M. Di Dio, "Reinventing Reinsurance Using the Capital Markets," *Geneva Papers on Risk and Insurance: Issues and Practice*, 22, no. 82 (1997).

160. See ibid.

161. See Jürgen Zech, "Will the International Financial Markets Replace Traditional Insurance Products?" *Geneva Papers on Risk and Insurance: Issues and Practice* 23, no. 89 (1998).

162. For an overview, see Véronique Bruggeman, "Capital Market Instruments for Natural Catastrophe and Terrorism Risks: A Bright Future?" *Environmental Law Reporter* 40 (2010).

163. See Klass and Wilson, "Climate Change and Carbon Sequestration," 162, who argue: "Bonding works well for short timeframes, but over the fifteen to thirty years required for post-closure financial responsibility, bonding could tie up capital and prove less efficient than insurance-based instruments." See also Gerard and Wilson, 1100.

164. Will Reisinger, Nolan Moser, and Trent A. Dougherty, "Reconciling King Coal and Climate Change: A Regulatory Framework for Carbon Capture and Storage," *Vermont Journal of Environmental Law* 11 (2009): 37, in reference to the Surface Mining Control and Reclamation Act of 1977, 30 U.S.C. § 1253(a).

165. Ibid., 37.

166. Ibid., 38. Their proposal was for a federal requirement on performance bonds, and their proposal was normative, whereas here the comment has been limited to a positive legal option. They also recommended that the duration of the bonding be ten years, matching the IOGCC's earlier recommendations.

167. Ibid., 38.

168. Allan Ingelson, Anne Kleffner, and Norma Nielson, "Long-Term Liability for Carbon Capture and Storage in Depleted North American Oil and Gas Reservoirs: A Comparative Analysis," *Emergy Law Journal* 31 (2010): 443. See Kan. Stat. Ann. § 55–1637((3)(e) (2007).

169. Mont. Code Ann. § 82–11–123(1)(f); Ingelson et al., 445.

170. Mont. Code Ann. § 82–11–182(2); Ingelson et al., 445.

171. Mont. Code Ann. § 82–11–183(1)(f) ; Ingelson et al., 442.

172. Mont. Code Ann. § 82–11–183(8)(d).

173. Wyo. Stat. Ann. § 35–11–313(f)(ii)(K).

174. Wyo. Stat. Ann. § 35–11–313(f)(ii)(O).

175. Wyo. Stat. Ann. § 35–11–313(f)(vii).

176. Ingelson et al., 442.

177. Aldrich et al., 9. See also the Fed UIC 2010 and the TEXAS 2010.

178. Aldrich et al., 9.

179. Ibid., 9–10.

180. Although the trust funds are to be placed in third-party hands, the third party arguably might have efficient uses for the capital in advance of the need of paying damages and might also be able to focus the use of that capital in alignment with CCS capital needs generally; for example, the funds might be applied to investing in corporations developing next-generation CCS storage or safety technologies. In such cases, an argument might be made that the capital could be used in a second-best sense.

181. Aldrich et al., 9.

182. Ibid., 10.

183. Ibid.

184. Those are summarized in an excellent way in Guidance Document 4 on the implementation of Directive 2009/31/EC on the Geological Storage of Carbon Dioxide, 4–7, 26–28 and in the US EPA Financial Responsibility Guidance Document. See US EPA, Geological Sequestration of Carbon Dioxide, Underground Injection Control (UIC) program, 45.

185. For that reason, the US EPA Financial Responsibility Document explicitly recommends the use of multiple instruments rather than a single one to meet financial responsibility. See US EPA, Geological Sequestration of Carbon Dioxide. Underground Injection Control (UIC) program, 38–39.

186. Aldrich et al., 13. In particular, they advocated for the US federal government, as opposed to state governments, to provide the indemnification.

187. Aldrich et al., 20.

188. Ibid., 22.

189. See chapter 7, section 7. 4.

190. See chapter 5.

Chapter 8

1. Chiara Trabucchi and L. Patton, "Storing Carbon: Options for Liability Risk Management, Financial Responsibility," World Climate Change Report 170 (Washington, DC: Bureau of National Affairs, 2008), 2, who argue that no financial risk management framework should inappropriately subsidize or otherwise provide economic advantage for CCS over future, as yet undeveloped or improved, technologies designed to make coal a cleaner source of power.

2. See chapter 4, section 4.1.

3. See Z. Makuch, S. Georgieva, and B. Oraee-Mirzamani, *Carbon Capture and Storage: Regulating Long-Term Liability* (London: Centre for Environmental Policy, Imperial College London, 2011), 25, who argue that the aim of a liability framework "should be to create the market conditions in which private insurance products can be offered in the CCS market."

4. Primarily in chapter 6.

5. See chapter 7, section 7.2.7.

6. See chapter 7, section 7.5.2.

7. Such as in the case of the international conventions for nuclear liability.

8. For example, in the United States via the September 11th Victim Compensation Fund. See Véronique Bruggeman, *Compensating Catastrophe Victims: A Comparative Law and Economics Approach* (London: Kluwer Law International, 2010), 465–481.

9. See Trabucchi and Patton, 18–19.

10. Peter J. Jost, "Limited Liability and the Requirement to Purchase Insurance," *International Review of Law and Economics* 16, no. 2 (1996). Similar arguments have been formulated by Polborn and by Skogh. See Mattias K. Polborn, "Mandatory Insurance and the Judgment-Proof Problem," *International Review of Law and Economics* 18, no. 2 (1998). See also Göran Skogh, "Mandatory Insurance: Transaction Costs Analysis of Insurance," in *Encyclopedia of Law and Economics, II, Civil Law and Economics*, ed. B. Bouckaert and G. De Geest (Cheltenham: Edward Elgar, 2000). Skogh has also pointed out that compulsory insurance may save on transaction costs.

11. See Howard C. Kunreuther and Paul K. Freeman. "Insurability, Environmental Risks and the Law," in *The Law and Economics of the Environment*, ed. Anthony Heyes (Cheltenham: Edward Elgar, 2001).

12. Benjamin Richardson, *Environmental Regulation through Financial Organisations: Comparative Perspectives on the Industrial Nations* (London: Kluwer Law International, 2002), 360.

13. For a summary of those warnings, see Michael G. Faure, "Economic Criteria for Compulsory Insurance," *Geneva Papers on Risk and Insurance* 31 (2006). See also Richardson, *Environmental Regulation through Financial Organisations*, 360–366.

14. See Steven Shavell, "The Judgment Proof Problem," *International Review of Law and Economics* 6 (1986): 43–58.

15. See Alberto Monti, "Environmental Risk: A Comparative Law and Economics Approach to Liability and Insurance," *European Review of Private Law* 9, no. 1 (2001).

16. Richardson, *Environmental Regulation through Financial Organisations*, 362–363.

17. See Trabucchi and Patton, 14, arguing in favor of the mandatory purchase of third-party instruments or self-insurance. They also demand evidence of financial responsibility for the operating life of the facility (16).

18. See section 7.4.

19. Implementation of Directive 2008/31/EC on the geological storage of carbon dioxide, Guidance Document 4.

20. EC Guidance Document 4 on the implementation of Directive 2009/31/EC on the Geological Storage of Carbon Dioxide follows from Article 7(10) of the CCS Directive, which states that applications for storage permits must include proof that the financial security or other equivalent provision as required under article 19 of the directive will be valid and effective before the injection begins. This amounts to a financial security requirement as defined in article 19 of the directive that "Member States shall ensure the proof that adequate provisions can be established, by way of financial security or any other equivalent, on the basis of arrangements to be decided by the Member States, is presented by the potential operator as part of the application for a storage permit."

21. US EPA Geological Sequestration of Carbon Dioxide, Underground Injection Control (UIC) Program, Class VI Financial Responsibility Guidance (Washington, DC: EPA, July 2011).

22. For the owner or operator submission requirements and the director review responsibilities, see ibid., 6–7.

23. Ibid., 9.

24. There are many examples of ex post compensation by government to victims of catastrophes. It can take the form of a structural fund solution (e.g., in Austria or Belgium) or ad hoc relief (such as the well-known September 11th Victim Compensation Fund in the United States). See Véronique Bruggeman, Michael G. Faure, and Tobias Heldt, "Insurance against Catastrophe: Government Stimulation of Insurance Markets for Catastrophic Events," *Duke Environmental Law & Policy Forum* 23, no. 185 (2012): 190–192.

25. Precisely this redistributional character of compensation via government is debated since disaster expenditures, for example, by the Federal Emergency Management Agency (FEMA) in the United States, are often politically motivated. See T. Garret and R. Sobel, "The Political Economy of FEMA Disaster Payment," *Economic Inquiry* 41, no. 3 (2003).

26. More particularly, Epstein classifies this type of ex post government compensation as "catastrophic responses to catastrophic risks." See Richard A. Epstein, "Catastrophic Responses to Catastrophic Risks," *Journal of Risk and Uncertainty* 12, no. 2–3 (1996). Equally critical on ex post relief by government are Louis Kaplow, "Incentives and Government Relief for Risk," *Journal of Risk and Uncertainty* 4, no. 2 (1991), and George L. Priest, "The Government, the Market, and the Problem of Catastrophic Loss," *Journal of Risk and Uncertainty* 12 (2–3) (1996).

27. See Saul Levmore and Kyle D. Logue, "Insuring against Terrorism—and Crime," *Michigan Law Review* 102, no. 2 (2003): 310.

28. This argument has been strongly made by Howard C. Kunreuther and E. Michel-Kerjan, "Challenges for Terrorism Risk Insurance in the United States," *Journal of Economic Perspectives* 18 (2004).

29. For more on climate adaptation and compensation, see Michael G. Faure, "Climate Change Adaptation and Compensation," in *Research Handbook on Climate Change Adaptation Law*, ed. Jonathon Verschuuren (Cheltenham: Edward Elgar, 2013).

30. Levmore and Logue, 281; see also Kaplow.

31. A. Gron and A. O. Sykes, "A Role for Government," *Regulation* 25 (2002); A. Gron and A. O. Sykes, "Terrorism and Insurance Markets: A Role for the Government as Insurer?" *Indiana Law Review* 36 (2003); Alfred Endres, Cornelia Ohl, and Bianca Rundshagen, "'Land unter!' Ein institutionenökonomischer Zwischenruf," *ListForum* 29 (2003): 290.

32. C. Gollier, "Some Aspects of the Economics of Catastrophe Risk Insurance," *Catastrophic Risks and Insurance* 8 (2005): 25.

33. Stephen Coate, "Altruism, the Samaritan's Dilemma, and Government Transfer Policy," *American Economic Review* 85 (1995).

34. More particularly by Paul A. Raschky and Hannelore Weck-Hannemann, "Charity Hazard: A Real Hazard to Natural Disaster Insurance?" *Environmental Hazards* 7, no. 4 (2007).

35. Epstein.

36. See chapter 3, section 3.3.

37. See chapter 4, section 4.1.

38. See Trabucchi and Patton, 11.

39. Ibid., 14.

40. See Makuch et al., 7, 14.

41. Compare Alexandra B. Klass and Elizabeth J. Wilson, "Climate Change and Carbon Sequestration: Assessing a Liability Regime for Long-Term Storage of Carbon Dioxide," *Emory Law Journal* 58 (2008): 108.

42. See chapter 7, section 1.

43. David E. Adelman and Ian J. Duncan. "The Limits of Liability in Promoting Safe Geologic Sequestration of CO_2," *Duke Environmental Law and Policy Forum* 22 (2011): 20–21.

44. See the summary of Adelman and Duncan's position in chapter 4, section 4.1.

45. See chapter 4, section 4.8.

46. Adelman and Duncan.

47. See chapter 5.

48. Government intervention in this case thus creates a form of cross-time diversification that the private market could not achieve. See Kunreuther and Michel-Kerjan, 210.

49. See David Hawkins, George Peridas, and John Steelman, "Twelve Years after Sleipner: Moving CCS from Hype to Pipe," *Energy Procedia* 1 (2009): 4405, who hold, "The lack of a regulatory framework specifically for CCS increases uncertainty and complicates project uptake. Alongside unfavourable economics, it is the most often quoted barrier that stands in the way of CCS deployment." See also Chiara Trabucchi, Michael Donlan, and Sarah Wade, "A Multi-Disciplinary Framework to Monetize Financial Consequences Arising from CCS Projects and Motivate Effective Financial Responsibility," *International Journal of Greenhouse Gas Control* 4, no. 2 (2010): 388, who hold, "However, because it represents a new and relatively unproven technology, concern about potential liability associated with CCS often is cited as a significant barrier to project deployment."

50. Tom Vanden Borre, "Shifts in Governance in Compensation for Nuclear Damage, 20 Years after Chernobyl," in *Shifts in Compensation for Environmental Damage,* ed. Michael G. Faure and A. Verheij (Vienna: Springer, 2007).

51. However, the different roles that such a fund could play (facilitative or financing) and the different phases in which it could intervene are not always carefully distinguished in the literature. We endeavor to make it clearer in this chapter.

52. See the policy recommendations on private and public contributions in chapter 9, sections 9.1.8 and 9.1.9, respectively.

53. Ian Havercroft and Richard Macrory, *Legal Liability and Carbon Capture and Storage: A Comparative Perspective* (London: Global CCS Institute, University College London, Faculty of Law, 2014), 45.

54. Here Flatt was referring to the scheme proposed by De Figueiredo et al. to align the unique risk of each storage facility with unique contribution schedules. See De Figueiredo, "The Liability of Carbon Dioxide Storage."

55. Victor Byers Flatt, "Paving the Legal Path for Carbon Sequestration," *Duke Environmental Law and Policy Forum* 19 (2009): 228.

56. Ibid. Arguably, while not raised directly by Flatt, one might seriously consider how few human-engineered structures have survived centuries beyond original construction without constant maintenance and rebuilding. The potential to successfully differentiate individual storage sites over centuries of future operation and storage events to ascribe differentiated capital contributions would likely be challenged, one would assume, by those sites chosen for higher fees; it remains unclear exactly how an administrative agency could meet its burden of evidentiary proof to assert such differentiated rates.

57. Flatt, 228.

58. Avelien Haan-Kamminga, "Long-Term Liability for Geological Carbon Storage in the European Union," *Journal of Energy and Natural Resources Law* 29 (2011): 328.

59. Ibid.

60. Barry Barton, Kimberley Jordan, and Greg Severinsen, *Carbon Capture and Storage: Designing the Legal and Regulatory Framework for New Zealand* (Hamilton: Centre for Environmental, Resources and Energy Law, 2013), 227–228.

61. Ibid., 227.

62. Ibid., 228.

63. Ibid.

64. Ibid.

65. Ibid.

66. Flatt, 227–228.

67. Ibid., 228–229.

68. Wendy B. Jacobs, "Carbon Capture and Sequestration," in *Global Climate Change and U.S. Law*, 2nd ed., ed. Jody Freeman and Michael Gerrard (Chicago: ABA, 2014), 28.

69. For example, Dana and Wiseman held that funds needed to be collected upfront from the operators of fracturing technologies, primarily from performance bonds

and mandatory insurance requirements, so that funds would be available for the eventual injuries caused by the fracturing activities. David A. Dana and Hannah J. Wiseman, "A Market Approach to Regulating the Energy Revolution: Assurance Bonds, Insurance, and the Certain and Uncertain Risks of Hydraulic Fracturing," *Iowa Law Review* 99 (2014): 1530. They were deeply concerned about both the large numbers of unplugged yet orphaned well sites historically (1561) and of the risks of operator insolvency (1558).

70. Jacobs, 28.

71. This was discussed in section 7.5.1.

72. See Adelman and Duncan, 29; Klass and Wilson, "Climate Change and Carbon Sequestration," 165–68; Trabucchi and Patton, 17–18.

73. Flatt, 227. See also Flynn and Marriott, calling for an industry-funded pool that operators would pay into during a "pay-in period." David P. Flynn and Susan M. Marriott, "Carbon Sequestration: A Liability Pathway to Commercial Viability," *Natural Resources and Environment* 24 (2009): 39. They recommended that funds be available for both monitoring and corrective applications.

74. Flatt, 227.

75. Ibid.

76. Barton et al., 225.

77. Were markets to fail, either informationally or in product availability, public provision or management of compensation funds might be more warranted, but the concerns of insurance availability were previously addressed in chapter 7 and the potential of public reinsurance is discussed in chapter 8, section 8.4. Because of the conclusions found in those areas, this section broadly assumes that sufficient insurance products would be available.

78. Michael G. Faure and Roger Van den Bergh, "Negligence, Strict Liability and Regulation of Safety under Belgian Law: An Introductory Economic Analysis," *Geneva Papers on Risk and Insurance* 12, no. 43 (1987).

79. See Michael G. Faure and Göran Skogh, "Compensation for Damages Caused by Nuclear Accidents: A Convention as Insurance," *Geneva Papers on Risk and Insurance: Issues and Practice* 65 (1992); also see Michael G. Faure, "Economic Models of Compensation for Damage Caused by Nuclear Accidents: Some Lessons for the Revision of the Paris and Vienna Conventions," *European Journal of Law and Economics* 2 (1995). See also the discussion in section 7.5.1.

80. See T. G. Coghlin, "Protection and Indemnity Clubs," *Lloyd's Maritime and Commercial Law Quarterly* 11 (1984): 403–16. See also the discussion in section 7.5.1.

81. Gron and Sykes, "A Role for Government."

82. Endres et al., 290.

83. See Elizabeth J. Wilson, Alexandra B. Klass, and Sara Bergan, "Assessing a Liability Regime for Carbon Capture and Storage," *Energy Procedia* 1 (2009): 4580.

84. See Hui Wang, *Civil Liability for Marine Oil Poluution Damage: A Comparative and Economic Study of the International, US and Chinese Compensation Regimes* (Alphen aan den Rijn: Kluwer Law International, 2011), 214–19. See also Jing Liu, *Compensating Ecological Damage: Comparative and Economic Observations* (Antwerp: Intersentia, 2013), 266.

85. Ibid. See also Klass and Wilson, "Climate Change and Carbon Sequestration," 170–171; Trabucchi and Patton, 21.

86. Wilson, Klass and Bergan, 4580.

87. James J. Dooley, Chiara Trabucchi, and Lindene Patton, "Design Considerations for Financing a National Trust to Advance the Deployment of Geologic CO_2 Storage and Motivate Best Practices," *International Journal of Greenhouse Gas Control* 42 (2010): 6.

88. This is also suggested by De Figueiredo, 391–398.

89. See chapter 7, section 7.2.4.

90. For example, if operators would be unable or unwilling to differentiate risks between them.

91. See section 8.2.

92. A disadvantage would obviously be that it would amount to an immobilization of capital from the public budget that would probably be used for other causes.

93. See V. C. Ammerlaan and W. H. Van Boom, "De Nederlandse Herverzekerings-maatschappij voor Terrorismeschaden en de rol van de overheid bij het vergoeden van terreurschade," *Nederlands Juristenblad* (2003).

94. Gron and Sykes, "A Role for Government."

95. Discussed in chapter 8.3.

96. Discussed in detail in chapter 7, section 7.5.1.

97. This would be because the private insurance market could not provide adequate catastrophe insurance coverage.

98. For an overview of the different forms of public intervention in disaster insurance schemes, see Alberto Monti, "Public-Private Initiatives to Cover Extreme Events," in *Extreme Events and Insurance: 2011 Annus Horribilis*, ed. Chr. Courbage and W. R. Stahel (Geneva: Geneva Association, 2012).

99. See Bruggeman, *Compensating Catastrophe Victims*, 321–326.

100. See ibid., 376–381.

101. For further discussions on TRIA, see ibid., 438–459.

102. See Dwight M. Jaffee and Thomas Russell, "Should Governments Support the Private Terrorism Insurance Market?" (paper presented at WRIEC Conference, Salt Lake City, August 2005).

103. Allan Ingelson, Anne Kleffner, and Norma Nielson, "Long-Term Liability for Carbon Capture and Storage in Depleted North American Oil and Gas Reservoirs: A Comparative Analysis," *Energy Law Journal* 31 (2010): 460.

104. Ibid., 460–461. Primary insurers must exceed their loss retention levels, as determined under the program.

105. Pool Re, "How the Scheme Works," http://www.poolre.co.uk/what-we-do. See also Ingelson et al., 461.

106. For a further discussion and critical analysis, see Bruggeman, Faure, and Heldt, 227–29.

107. Ingelson et al., 462.

108. Ibid.

109. For an overview, see Joanne Linnerooth-Bayer and Reinhard Mechler, "Disaster Safety Nets for Developing Countries: Extending Public–Private Partnerships," *Environmental Hazards* 7, no. 1 (2007).

110. This is a point strongly made by S. E. Harrington, "Rethinking Disaster Policy," *Regulation* 23, no. 1 (2000); Howard Kunreuther, "Mitigating Disaster Losses through Insurance," *Journal of Risk and Uncertainty* 12, no. 2–3 (1996): 180–183. See also Reimund Schwarze and Gert G. Wagner, "In the Aftermath of Dresden: New Directions in German Flood Insurance," *Geneva Papers on Risk and Insureance: Issues and Practice* 29, no. 2 (2004): 154–168.

111. These public-private initiatives to cover extreme risks are also supported by OECD recommendations. See Monti, "Public-Private Initiatives."

112. Michael G. Faure, "Financial Compensation for Victims of Catastrophes: A Law and Economics Perspective," *Law and Policy* 29, no. 3 (2007): 358.

113. See chapter 7, section 2.

114. See Epstein, "Catastrophic Responses to Catastrophic Risks." See also Howard C. Kunreuther and Mark Pauly, "Rules Rather Than Discretion: Lessons from Hurricane Katrina," *Journal of Risk and Uncertainty* 33, no. 1–2 (2006): 113, arguing that this government's role in assisting the supply side allows avoiding the inefficiencies and inequities associated with disaster assistance.

115. Haan-Kamminga, 326–327.

116. Ibid., 327.

117. Ibid. Monast, Pearson, and Pratson warned that the use of private liability insurance is effective when the time period is well defined but not when open ended. Jonas J. Monast, Brooks R. Pearson, and Lincoln F. Pratson, "A Cooperative Federalism Framework for CCS Regulation," *Environmental and Energy Law and Policy Journal* 7 (2012): 39.

118. See Gron and Sykes, "A Role for Government"; see also Gron and Sykes, "Terrorism and Insurance Markets."

119. Levmore and Logue, 304, who argue that otherwise disaster insurance would still not be "available."

120. Ibid., 311.

121. Trabucchi and Patton, 18–19.

122. Makuch et al., 24.

123. For more, see Bruggeman, Faure, and Heldt.

124. See chapter 7, section 7.2.6.

125. See chapter 7, section 7.5.

126. See chapter 8, section 8.3.

127. See Véronique Bruggeman, Michael G. Faure, and Karine Fiore, "The Government as Reinsurer of Catastrophe Risks?" *Geneva Papers on Risk and Insurance-Issues and Practice* 35, no. 3 (2010); see also Bruggeman, Faure, and Heldt, 221–223.

128. See Bruggeman, Faure, and Fiore.

129. See Bruggeman, Faure, and Heldt, 223.

130. Will Reisinger, Nolan Moser, and Trent A. Dougherty, "Reconciling King Coal and Climate Change: A Regulatory Framework for Carbon Capture and Storage," *Vermont Journal of Environmental Law* 11 (2009): 41–42. Either the CCS storage site could be under the management of a public utility from the start, perhaps with contractual delegation to a storage services operator, or the public utility phase could be established at a later phase of storage operations after the operator transfers its activities and liabilities to the state. For the discussion here, the assumption is made that the decision to operate the storage site under a public utility implies that the utility is operation prior to the onset of injection.

131. Reisinger, Moser, and Dougherty, 41.

132. Ibid.

133. Ibid. It is unclear if their proposal would be implemented state-by-state or in a multistate mode. See ibid.

134. Ibid.

135. Ibid.

136. Ibid.

137. Ibid., 42.

138. These details are not discussed at all within their paper. See Reisinger et al., 42.

Chapter 9

1. See chapter 2, sections 2.4 and 2.5.

2. See chapter 4, section 4.8.

3. Those are the classic uncertainties with respect to CCS projects. See chapter 2, section 2.4.

4. This notion of climate change liability is often referred to in different senses. Some use *climate change liability* to refer to the possibilities of using tort law to force emitters toward mitigation measures aiming at reducing CO_2 emissions. In that respect, see Michael G. Faure and Marjan Peeters, *Climate Change Liability* (Cheltenham: Edward Elgar, 2011). However, in the context of this section, *climate change liability* is meant as the liability of operators that may arise in case of an escape of CO_2 from a storage site.

5. See chapter 7.

6. See chapter 8, section 8.2.

7. See chapter 8, section 8.4.

8. See Sally Benson and Peter Cook, "Underground Geological Storage," in *IPCC Special Report on Carbon Dioxide Capture and Storage*, ed. B. Metz, O. Davidson, H. De Coninck, M. Loos, and L. Meyer (Cambridge: Cambridge University Press, 2005), 14.

9. See Emily Rochon et al., *False Hope: Why Carbon Capture and Storage Won't Save the Climate* (Amsterdam: Greenpeace International, 2008).

Bibliography

Abend, Katherine. "Geological Sequestration of Carbon Dioxide: Legal Issues and Recommendations for Regulators." *Appalachian Natural Resources Law Journal* 5 (2010): 1–23.

Abraham, Kenneth S. "Cost Internalization, Insurance, and Toxic Tort Compensation Funds." *Virginia Journal of Natural Resources Law* 2 (1982): 123–148.

Abraham, Kenneth S. "Environmental Liability and the Limits of Insurance." *Columbia Law Review* 88 (5) (1988): 942–988.

Adelman, David E., and Ian J. Duncan. "The Limits of Liability in Promoting Safe Geologic Sequestration of CO_2." *Duke Environmental Law and Policy Forum* 22 (2011): 1–66.

Aines, Roger D., Martin J. Leach, Todd H. Weisgraber, Matthew D. Simpson, S. Julio Friedmann, and Carol J. Bruton. "Quantifying the Potential Exposure Hazard Due to Energetic Releases of CO_2 from a Failed Sequestration Well." *Energy Procedia* 1 (1) (2009): 2421–2429.

Akerlof, George A. "'The Market for Lemons': Quality Uncertainty and the Market Mechanism." *Quarterly Journal of Economics* 84 (3) (1970): 488–500.

Aldrich, Elizabeth, Cassandra Koerner, and David Solan. *Analysis of Liability Regimes for Carbon Capture and Sequestration: A Review for Policymakers.* Chicago: Energy Policy Institute, 2011.

Ammerlaan, V. C., and W. H. Van Boom. "De Nederlandse Herverzekerings-maatschappij voor Terrorismeschaden en de rol van de overheid bij het vergoeden van terreurschade." *Nederlands Juristenblad* 2003 (45/46): 2330–2337.

Annunziatellis, A., S. E. Beaubien, S. Bigi, G. Ciotoli, M. Coltella, and S. Lombardi. "Gas Migration along Fault Systems and through the Vadose Zone in the Latera Caldera (Central Italy): Implications for CO_2 Geological Storage." *International Journal of Greenhouse Gas Control* 2 (3) (2008): 353–372.

Arcuri, Alessandra. "Controlling Environmental Risk in Europe: The Complementary Role of an EC Environmental Liability Regime." *Tijdschrift voor Milieuaansprakelijkheid* 15 (2) (2001): 39–40.

Arrow, Kenneth J. "Uncertainty and the Welfare Economics of Medical Care." *American Economic Review* 53 (5) (1963): 941–973.

Arrow, Kenneth J. *Aspects of the Theory of Risk-Bearing (Yrjo Jahnsson Lectures)*. Helsinki: Yrjo Jahnssonin Saatio, 1965.

Arthur, W. Brian. "Competing Technologies, Increasing Returns, and Lock-In by Historical Events." *Economic Journal* 99 (394) (1989): 116–131.

Attanasio, Donna M. "Surveying the Risks of Carbon Dioxide: Geological Sequestration and Storage Projects in the United States." *Environmental Law Reporter News and Analysis* 39 (2009): 310–376.

Baarsma, Barbara, Flóra Felsö, Sjoerd van Geffen, José Mulder, and André Oostdijk. *Zelf doen? Inventarisatiestudie van zelfreguleringsinstrumenten. No. 664*. Amsterdam: SEO, 2003.

Bachu, Stefan. "Sequestration of CO_2 in Geological Media: Criteria and Approach for Site Selection in Response to Climate Change." *Energy Conversion and Management* 41 (9) (2000): 953–970.

Bailey, Paul, Elizabeth McCulough, and Sonya Suter. "Can Governments Ensure Adherence to the Polluter Pays Principle in the Long-Term CCS Liability Context?" *Sustainable Development Law and Policy* 12 (2011): 46–51.

Bankes, Nigel. "*The Legal and Regulatory Issues Associated with Carbon Capture and Storage in Arctic States*." Carbon and Climate Law Review 6 (2012): 21–32.

Bar-Gill, Oren, and Omri Ben-Shahar. "The Uneasy Case for Comparative Negligence." *American Law and Economics Review* 5 (2) (2003): 433–469.

Barton, Barry, Kimberley Jordan, and Greg Severinsen. *Carbon Capture and Storage: Designing the Legal and Regulatory Framework for New Zealand*. Hamilton: Centre for Environmental, Resources and Energy Law, 2013.

Bawcutt, Paul A. *Captive Insurance Companies: Establishment, Operation, and Management*. Cambridge: Woodhead-Faulkner, 1991.

Beck, Brendan, Justine Garrett, Ian Havercroft, David Wagner, and Paul Zakkour. "Development and Distribution of the IEA CCS Model Regulatory Framework." *Energy Procedia* 4 (2011): 5933–5940.

Belser, W. E. "Über die Zweckmäßigkeit der Poolung von Atomrisiken." *Versicherungswirtschaft* 18 (1959).

Ben-Shahar, Omri, and Kyle D. Logue. "Outsourcing Regulation: How Insurance Reduces Moral Hazard." *Michigan Law Review* 111 (2) (2012): 197–248.

Benson, Sally M. *Carbon Dioxide Capture and Storage in Underground Geologic Formations*. Berkeley, CA: Lawrence Berkeley National Laboratory, 2004.

Benson, Sally, and Peter Cook. "Underground Geological Storage." In *IPCC Special Report on Carbon Dioxide Capture and Storage*, ed. B. Metz, O. Davidson, H. De Coninck, M. Loos, and L. Meyer. Cambridge: Cambridge University Press, 2005.

Bentata, Pierre, and Michael G. Faure. "The Role of Environmental Civil Liability: An Economic Analysis of the French Legal System." *Environmental Liability, Law, Policy and Practice* 20 (4) (2012): 120–128.

Bentham, M., and M. Kirby. "CO_2 Storage in Saline Aquifers." *Oil and Gas Science and Technology* 60 (3) (2005): 559–567.

Bergkamp, Lucas. "The Proper Scope of Joint and Several Liability." *Tijdschrift voor Milieuschade en Aansprakelijkheidsrech* 14 (2000): 153–156.

Bergkamp, Lucas. *Liability and Environment: Private and Public Law Aspects of Civil Liability for Environmental Harm in an International Context*. The Hague: Kluwer Law International, 2001.

Bergkamp, Lucas, and Barbara Goldsmith. *The EU Environmental Liability Directive: A Commentary*. Oxford: Oxford University Press, 2013.

Bergkamp, L., N. Herbatschek, and S. Jayanti. "Financial Security and Insurance." In *The EU Environmental Liability Directive: A Commentary*, ed. L. Bergkamp and B. J. Goldsmith, 118–138. Oxford: Oxford University Press, 2013.

Bickle, Mike, Niko Kampman, and Max Wigley. "Natural Analogues." *Reviews in Mineralogy and Geochemistry* 77 (1) (2013): 15–71.

Bidlack, Christopher. "Regulating the Inevitable: Understanding the Legal Consequences of and Providing for the Regulation of the Geologic Sequestration of Carbon Dioxide." *Journal Land Resources and Environmental Law* 30 (2010): 199–227.

Bocken, H. "Alternatives to Liability and Liability Insurance for the Compensation of Pollution Damages." *Tijdschrift voor Milieuaansprakelijkheid* 1 (1987): 83–87.

Bode, Sven, and Martina Jung. "Carbon Dioxide Capture and Storage—Liability for Non-Permanence under the UNFCCC." *International Environmental Agreement: Politics, Law and Economics* 6 (2) (2006): 173–186.

Bommer, Julian J., Stephen Oates, José Mauricio Cepeda, Conrad Lindholm, Juliet Bird, Rodolfo Torres, Griselda Marroquín, and José Rivas. "Control of Hazard Due to Seismicity Induced by a Hot Fractured Rock Geothermal Project." *Engineering Geology* 83 (4) (2006): 287–306.

Borch, Karl. "The Utility Concept Applied to the Theory of Insurance." *ASTIN Bulletin* 1 (5) (1961): 245–255.

Borenstein, Severin. "The Economics of Costly Risk Sorting in Competitive Insurance Markets." International Review of Law and Economics 9 (1) (1989): 25–39.

Boyer, M., Y. Hiriart, and D. Martimort. Frontiers in the Economics of Environmental Regulation and Liability. Aldershot: Ashgate, 2006.

Bruggeman, Véronique. Compensating Catastrophe Victims: A Comparative Law and Economics Approach. The Hague: Kluwer Law International, 2010.

Bruggeman, Véronique. "Capital Market Instruments for Natural Catastrophe and Terrorism Risks: A Bright Future?" Environmental Law Reporter 40 (2010): 10136–10153.

Bruggeman, Véronique, Michael G. Faure, and Karine Fiore. "The Government as Reinsurer of Catastrophe Risks?" Geneva Papers on Risk and Insurance: Issues and Practice 35 (3) (2010): 369–390.

Bruggeman, Véronique, Michael G. Faure, and Tobias Heldt. "Insurance against Catastrophe: Government Stimulation of Insurance Markets for Catastrophic Events." Duke Environmental Law and Policy Forum 23 (185) (2012): 185–241.

Brydie, J. R., E. H. Perkins, D. Fisher, M. Girard, M. Valencia, M. Olson, and T. Rattray. "The Development of a Leak Remediation Technology for Potential Non-Wellbore Related Leaks from CO_2 Storage Sites." Energy Procedia 63 (2014): 4601–4611.

Buonasorte, G., G. M. Cameli, A. Fiordelisi, M. Parotto, and I. Perticone. "Results of Geothermal Exploration in Central Italy (Latium-Campania)." In Proceedings of the World Geothermal Congress, 18–31. Florence, Italy, 1995.

Burrows, Paul. "Combining Regulation and Legal Liability for the Control of External Costs." International Review of Law and Economics 19 (2) (1999): 227–244.

"The Business of Survival: What Is the Secret of Corporate Longevity?" Economist, December 16, 2004.

Cafaggi, Fabrizio. "New Foundations of Transnational Private Regulation." Revista de Derecho Privado 26 (2014): 185–217.

Calabresi, Guido. The Costs of Accidents: A Legal and Economic Analysis. New Haven: Yale University Press, 1970.

Campbell, Angela J. "Self-Regulation and the Media." Federal Communications Law Journal 51 (3) (1999): 711–772.

Carey, J. William. "Geochemistry of Wellbore Integrity in CO_2 Sequestration: Portland Cement-Steel-Brine-CO_2 Interactions." Reviews in Mineralogy and Geochemistry 77 (1) (2013): 505–539.

Chadwick, R. A., R. Arts, M. Bentham, O. Eiken, S. Holloway, G. A. Kirby, J. M. Pearce, J. P. Williamson, and P. Zweigel. "Review of Monitoring Issues and

Technologies Associated with the Long-Term Underground Storage of Carbon Dioxide." *Geological Society of London, Special Publications* 313 (1) (2009): 257–275.

Chelius, James R. "Liability for Industrial Accidents: A Comparison of Negligence and Strict Liability Systems." *Journal of Legal Studies* 5 (2) (1976): 293–309.

Chialvo, Ariel A., Lukas Vlcek, and David R. Cole. "Acid Gases in CO_2-Rich Subsurface Geologic Environments." *Reviews in Mineralogy and Geochemistry* 77 (1) (2013): 361–398.

Chiaramonte, Laura, Amie Lucier, Hannah Ross, and Mark Zoback. "Geomechanics and Seal Integrity for the Geologic Sequestration of CO_2." In *Carbon Capture and Storage: R&D Technologies for Sustainable Energy Future*, ed. Malti Goel, Baleshwar Kumar, and S. Nirmal Charan, 111–118. Oxford: Alpha Science, 2008.

Chu, Steven. "Carbon Capture and Sequestration." *Science* 325 (5948) (2009): 1599.

Coase, Ronald H. "The Federal Communications Commission." *Journal of Law and Economics* 2 (1959): 1–40.

Coase, Ronald H. *The Problem of Social Cost*. Basingstoke, UK: Palgrave Macmillan, 1960.

Coase, Ronald, William Meckling, and Jora Minasian. *Problems of Radio Frequency Allocation*. Santa Monica, CA: Rand Corporation, 1995.

Coate, Stephen. "Altruism, the Samaritan's Dilemma, and Government Transfer Policy." *American Economic Review* 85 (1) (1995): 46–57.

Coghlin, T. G. "Protection and Indemnity Clubs." *Lloyd's Maritime and Commercial Law Quarterly* 11 (1984): 403–416.

Coglianese, Cary, and David Lazer. "Management-Based Regulation: Prescribing Private Management to Achieve Public Goals." *Law and Society Review* 37 (4) (2003): 691–730.

Coman, Hannah. "Balancing the Need for Energy and Clean Water: The Case for Applying Strict Liability in Hydraulic Fracturing Suits." *Boston College Environmental Affairs Law Review* 39 (2012): 131–160.

Cooter, Robert. "Prices and Sanctions." *Columbia Law Review* 84 (6) (1984): 1523–1560.

Cooter, Robert, and Thomas Ulen. *Law and Economics*. Upper Saddle River, NJ: Pearson, 2004.

Cousy, H. "Recent Developments in Environmental Insurance." In *Recent Economic and Legal Developments in European Environmental Policy*, ed. F. Abraham, K. Deketelaere, and T. Stuyck, 235–237. Leuven: Leuven University Press, 1995.

Cypser, Darlene A., and Scott D. Davis. "Liability for Induced Earthquakes." *Journal Environmental Law and Litigation* 9 (1994): 551–589.

Dana, David A., and Hannah J. Wiseman. "A Market Approach to Regulating the Energy Revolution: Assurance Bonds, Insurance, and the Certain and Uncertain Risks of Hydraulic Fracturing." *Iowa Law Review* 99 (2014): 1523–1593.

Dari-Mattiacci, Giuseppe. "Tort Law and Economics." In *Economic Analysis of Law: A European Perspective*, ed. Aristides Hatzis. Cheltenham: Edward Elgar, 2006.

Davies, Lincoln L., Kirsten Uchitel, and John Ruple. "Understanding Barriers to Commercial-Scale Carbon Capture and Sequestration in the United States: An Empirical Assessment." *Energy Policy* 59 (2013): 745–761.

Davies, Lincoln, Kirsten Uchitel, John Ruple, and Heather Tanana. *Carbon Capture and Sequestration: A Regulatory Gap Assessment*. Salt Lake City: University of Utah, 2012.

De Figueiredo, Mark Anthony. "The Liability of Carbon Dioxide Storage." PhD diss., MIT, 2007.

De Figueiredo, Mark Anthony, D. M. Reiner, and H. J. Herzog. "Framing the Long-Term in Situ Liability Issue for Geologic Carbon Storage in the United States." *Mitigation and Adaptation Strategies for Global Change* 10 (4) (2005): 647–57.

De Smedt, K. *Environmental Liability in a Federal System: A Law and Economics Analysis*. Antwerp: Intersentia, 2007.

Dobbs, Dan B. *The Law of Torts*. Minneapolis, MN: West Group, 2000.

Dooley, James J., Chiara Trabucchi, and Lindene Patton. "Design Considerations for Financing a National Trust to Advance the Deployment of Geologic CO_2 Storage and Motivate Best Practices." *International Journal of Greenhouse Gas Control* 4 2) (2010): 381–87.

Dow, James C. *Nuclear Energy and Insurance*. Livingston, UK: Witherby and Co, 1989.

Dowding, T. *Global Developments in Captive Insurance*. London: FT Energy, Pearson Professional Ltd, 1997.

Dubin, Jeffrey A., and Geoffrey S. Rothwell. "Subsidy to Nuclear Power through Price-Anderson Liability Limit." *Contemporary Economic Policy* 8 (3) (1990): 73–79.

Duncan, Ian J., Jean-Philippe Nicot, and Jong-Won Choi. "Risk Assessment for Future CO_2 Sequestration Projects Based CO_2 Enhanced Oil Recovery in the US." *Energy Procedia* 1 (1) (2009): 2037–2042.

Eames, Fred, and Scott Anderson. "The Layered Approach to Liability for Geological Sequestration of CO_2." *Environmental Law Reporter News and Analysis* 43 (2013): 10653–10655.

Eccles, Jordan K., and Lincoln Pratson. "Global CO_2 Storage Potential of Self-Sealing Marine Sedimentary Strata." *Geophysical Research Letters* 39 (no. L19604) (2012): 1–7.

Ehrlich, Isaac, and Richard A. Posner. "An Economic Analysis of Legal Rulemaking." *Journal of Legal Studies* 3 (1) (1974): 257–286.

Endres, Alfred, Cornelia Ohl, and Bianca Rundshagen. "'Land unter!'-Ein institutionenökonomischer Zwischenruf." *ListForum* 29 (2003): 284–294.

Endres, Alfred, and Reimund Schwarze. *Allokationswirkungen einer Umwelthaftpflichtversicherung.* Berlin: Springer, 1992.

Enright, Paul L. "The Six-Minute Walk Test." *Respiratory Care* 48 (8) (2003): 783–785.

Epstein, Richard A. "Catastrophic Responses to Catastrophic Risks." *Journal of Risk and Uncertainty* 12 (2–3) (1996): 287–308.

Estep, Samuel D. "Radiation Injuries and Statistics: The Need for a New Approach to Injury Litigation." *Michigan Law Review* 59 (2) (1960): 259–304.

Faure, Michael G. "Economic Models of Compensation for Damage Caused by Nuclear Accidents: Some Lessons for the Revision of the Paris and Vienna Conventions." *European Journal of Law and Economics* 2 (1) (1995): 21–43.

Faure, Michael G. "Compensation for Non-Pecuniary Losses from an Economic Perspective." In *European Tort Law, Liber Amicorum for Helmut Koziol,* ed. U. Magnus and J. Spier, 143–159. Frankfurt am Main: Peter Lang, 2000.

Faure, Michael G. "Environmental Damage Insurance in the Netherlands." *Environmental Liability* 10 (1) (2002): 31–41.

Faure, Michael G. *Deterrence, Insurability, and Compensation in Environmental Liability: Future Developments in the European Union.* Vienna: Springer, 2003.

Faure, Michael G. "Economic Criteria for Compulsory Insurance." 31 *Geneva Papers on Risk and Insurance* 31 (2006): 149–168.

Faure, Michael G. "Financial Compensation for Victims of Catastrophes: A Law and Economics Perspective." *Law and Policy* 29 (3) (2007): 339–367.

Faure, Michael G. "Designing Incentives Regulation for the Environment." Maastricht Faculty of Law Working Paper 2008–2007 (2008).

Faure, Michael G. "Environmental Liability." In *Tort Law and Economics,* ed. Michael G. Faure. Cheltenham: Edward Elgar, 2009.

Faure, Michael G. "Economic Analysis of Tort Law." In *Tort Law and Economics,* ed. Michael G. Faure. Cheltenham: Edward Elgar, 2010.

Faure, Michael G. "Regulatory Strategies in Environmental Liability." In *The Regulatory Function of European Private Law*, ed. F. Cafaggi, H. Watt, and H. Muir, 129–187. Cheltenham: Edward Elgar, 2011.

Faure, Michael G. "Effectiveness of Environmental Law: What Does the Evidence Tell Us?" *William and Mary Environmental Law and Policy Review* 36 (2) (2012): 293–336.

Faure, Michael G. "Designing Incentives Regulation for the Environment." In *Global Environmental Commons: Analytical and Political Challenges in Building Governance Mechanisms*, ed. E. Brousseau, T. Dedeurwaerdere, P.-A. Jouvet, and M. Willinger, 275–307. Oxford: Oxford University Press, 2012.

Faure, Michael G. "Climate Change Adaptation and Compensation." In *Research Handbook on Climate Change Adaptation Law*, ed. Jonathon Verschuuren, 110–141. Cheltenham: Edward Elgar, 2013.

Faure, Michael G., and Karine Fiore. "An Economic Analysis of the Nuclear Liability Subsidy." *Pace Environmental Law Review* 26 (2009): 419–447.

Faure, Michael G., and T. Hartlief. *Insurance and Expanding Systemic Risks*. Paris: OECD, 2003.

Faure, Michael G., Ingeborg M. Koopmans, and Johannes C. Oudijk. "Imposing Criminal Liability on Government Officials under Environmental Law: A Legal and Economic Analysis." *Loyola of Los Angeles International and Comparative Law Journal* 18 (1995): 529–569.

Faure, Michael, and Jing Liu. "New Models for the Compensation of Natural Resources Damage." *Kentucky Journal of Equine, Agricultural and Natural Resources Law* 4 (2011): 261–314.

Faure, Michael G., and Jing Liu. "The Tsunami of March 2011 and the Subsequent Nuclear Incident at Fukushima: Who Compensates the Victims?" *William and Mary Environmental Law and Policy Review* 37 (2012): 129–218.

Faure, Michael G., and Marjan Peeters. *Climate Change Liability*. Cheltenham: Edward Elgar, 2011.

Faure, Michael, and Marieke Ruegg. "Environmental Standard Setting through General Principles of Environmental Law." In *Environmental Standards in the European Union in an Interdisciplinary Framework*, 39–60. Antwerp: Maklu, 1994.

Faure, Michael G., and Göran Skogh. "Compensation for Damages Caused by Nuclear Accidents: A Convention as Insurance." *Geneva Papers on Risk and Insurance: Issues and Practice* 65 (1992): 499–513.

Faure, Michael G., and S. Ubachs. "Comparative Benefits and Optimal Use of Environmental Taxes." *Critical Issues in Environmental Taxation* 1 (2003): 29.

Faure, Michael G., and Roger Van den Bergh. "Negligence, Strict Liability and Regulation of Safety under Belgian Law: An Introductory Economic Analysis." *Geneva Papers on Risk and Insurance* 12 (43) (1987): 95–114.

Faure, Michael, and Roger Van den Bergh. "Liability for Nuclear Accidents in Belgium from an Interest Group Perspective." *International Review of Law and Economics* 10 (3) (1990): 241–254.

Faure, Michael, and Roger Van den Bergh. "Restrictions of Competition on Insurance Markets and the Applicability of EC Antitrust Law." *Kyklos* 48 (1) (1995): 65–85.

Faure, Michael, and Roger Van den Bergh. "Competition on the European Market for Liability Insurance and Efficient Accident Law." *Maastricht Journal of European and Comparative Law* 9 (2002): 279–306.

Faure, Michael G., and Tom Vanden Borre. "Compensating Nuclear Damage: A Comparative Economic Analysis of the US and International Liability Schemes." *William and Mary Environmental Law and Policy Review* 33 (1) (2008): 219–286.

Faure, Michael G., and Stefan E. Weishaar. "The Role of Environmental Taxation: Economics and the Law." In *Handbook of Research on Environmental Taxation*, 399–421. Cheltenham: Edward Elgar, 2012.

Finsinger, Jörg, and Mark V. Pauly. "The Double Liability Rule." *Geneva Papers on Risk and Insurance Theory* 15 (2) (1990): 159–169.

Flatt, Victor Byers. "Paving the Legal Path for Carbon Sequestration." *Duke Environmental Law & Policy Forum* 19 (2009): 211–246.

Flynn, David P., and Susan M. Marriott. "Carbon Sequestration: A Liability Pathway to Commercial Viability." *Natural Resources and Environment* 24 (2009): 37–40.

Folger, Peter. *Carbon Capture and Sequestration: Research, Development, and Demonstration at the US Department of Energy.* Washington, DC: Congressional Research Service, 2012.

Frey, Bruno S. "Morality and Rationality in Environmental Policy." *Journal of Consumer Policy* 22 (4) (1999): 395–417.

Friehe, Tim. "Precaution v. Avoidance: A Comparison of Liability Rules." *Economics Letters* 105 (3) (2009): 214–216.

Garrett, Justine, and Brendan Beck. *Carbon Capture and Storage: Legal Regulatory Review Edition*, 2nd ed. Paris: OECD/International Energy Agency, 2011.

Garret, T., and R. Sobel. "The Political Economy of FEMA Disaster Payment." *Economic Inquiry* 41 (3) (2003): 496–509.

Gerard, David, and Elizabeth J. Wilson. "Environmental Bonds and the Challenge of Long-Term Carbon Sequestration." *Journal of Environmental Management* 90 (2) (2009): 1097–1105.

Gilead, Israel. "Tort Law and Internalization: The Gap between Private Loss and Social Cost." *International Review of Law and Economics* 17 (4) (1997): 589–608.

Gittleson, Kim. "Can a Company Live Forever?" BBC News, January 19, 2012. http://www.bbc.com/news/business-16611040.

Glachant, Matthieu. "The Use of Regulatory Mechanism Design in Environmental Policy: A Theoretical Critique." In *Sustainability and Firms: Technological Change and the Changing Regulatory Environment,* 179–188. Cheltenham: Edward Elgar, 1998.

Gollier, C. "Some Aspects of the Economics of Catastrophe Risk Insurance." *Catastrophic Risks and Insurance* 8 (2005): 13–30.

Gouveia, F. J., and S. J. Friedmann. *Timing and Prediction of CO_2 Eruptions from Crystal Geyser, UT.* Washington, DC: Department of Energy, 2006.

Grajzl, Peter, and Andrzej Baniak. "Industry Self-Regulation, Subversion of Public Institutions, and Social Control of Torts." *International Review of Law and Economics* 29 (4) (2009): 360–374.

Grajzl, Peter, and Peter Murrell. "Allocating Lawmaking Powers: Self-Regulation vs. Government Regulation." *Journal of Comparative Economics* 35 (3) (2007): 520–545.

Gron, A., and A. O. Sykes. "A Role for Government." *Regulation* 25 (2002): 44–51.

Gron, A., and A. O. Sykes. "Terrorism and Insurance Markets: A Role for the Government as Insurer?" *Indiana Law Review* 36 (2003): 447–463.

Gunningham, Neil, Peter N. Grabosky, and Darren Sinclair. *Smart Regulation: Designing Environmental Policy.* Oxford: Clarendon Press, 1998.

Gunningham, Neil, Martin Phillipson, and Peter Grabosky. "Harnessing Third Parties as Surrogate Regulators: Achieving Environmental Outcomes by Alternative Means." *Business Strategy and the Environment* 8 (4) (1999): 211–224.

Gunningham, Neil, and Darren Sinclair. "Regulatory Pluralism: Designing Policy Mixes for Environmental Protection." *Law and Policy* 21 (1) (1999): 49–76.

Haake, Ch. H., and K. B. Marsh. "The Trouble with Angels: Carbon Capture and Storage Hurdles and Solutions." World Climate Change Report. Washington, DC: Bureau of National Affairs, 2009.

Haan-Kamminga, Avelien. "Long-Term Liability for Geological Carbon Storage in the European Union." *Journal of Energy and Natural Resources Law* 29 (3) (2011): 309–31.

Haan-Kamminga, Avelien, Martha Roggenkamp, and Edwin Woerdman. "Legal Uncertainties of Carbon Capture and Storage in the EU: The Netherlands as an Example." *Carbon and Climate Law Review* 4 (3) (2010): 240–249.

Haazen, O. A., and J. Spier. "Feitelijke ontwikkelingen en verzekerbaarheid." In *De uitdijende reikwijdte van de aansprakelijkheid uit onrechtmatige daad*, ed. A. T. Boltand and J. Spier, 25–26. Zwolle: W. E. J. Tjeenk Willink, 1996.

Hannah, Leslie. "Survival and Size Mobility among the World's Largest 100 Industrial Corporations, 1912–1995." *American Economic Review* 88 (2) (1998): 62–65.

Hannah, Leslie. "Marshall's Trees and the Global Forest: Were Giant Redwoods Different?" In *Learning by Doing in Markets, Firms, and Countries*, ed. Naomi R. Lamoreaux, Daniel M. G. Raff and Peter Temin, 253–294. Chicago: University of Chicago Press, 1999.

Harrington, Scott E. "Rethinking Disaster Policy." *Regulation* 23 (1) (2000): 40–46.

Havercroft, Ian, and Richard Macrory. *Legal Liability and Carbon Capture and Storage: A Comparative Perspective.* London: Global CCS Institute, University College London, Faculty of Law, 2014.

Hawkins, David, George Peridas, and John Steelman. "Twelve Years after Sleipner: Moving CCS from Hype to Pipe." *Energy Procedia* 1 (1) (2009): 4403–4410.

Hayano, Delissa. "Guarding the Viability of Coal and Coal-Fired Power Plants: A Road Map for Wyoming's Cradle to Grave Regulation of Geologic CO_2 Sequestration." *Wyoming Law Review* 9 (2009): 139–164.

Hellevang, Helge. "Carbon Capture and Storage (CCS)." In *Petroleum Geoscience: From Sedimentary Environments to Rock Physics*, ed. K. Bjørlykke. Berlin: Springer-Verlag, 2015.

Heremans, T. *Professional Services in the EU Internal Market: Quality Regulation and Self-Regulation.* Oxford: Hart, 2012.

Herzog, Howard. "Carbon Dioxide Capture and Storage." In *The Economics of Climate Change*, ed. D. Helms and C. Hepburn, 263–283. Oxford: Oxford University Press, 2010.

Heyes, Anthony G., and Catherine Liston-Heyes. "Subsidy to Nuclear Power through Price-Anderson Liability Limit: Comment." *Contemporary Economic Policy* 16 (1) (1998): 122–124.

Heyes, Anthony, and Catherine Liston-Heyes. "Capping Environmental Liability: The Case of North American Nuclear Power." *Geneva Papers on Risk and Insurance: Issues and Practice* 25 (2) (2000): 196–202.

Hoffman, Nathan R. "The Feasibility of Applying Strict-Liability Principles to Carbon Capture and Storage." *Washburn Law Journal* 49 (2009): 527–562.

House, Kurt Zenz, Daniel P. Schrag, Charles F. Harvey, and Klaus S. Lackner. "Permanent Carbon Dioxide Storage in Deep-Sea Sediments." *Proceedings of the National Academy of Sciences of the United States of America* 103 (33) (2006): 12291–12295.

Hylton, Keith N. "When Should We Prefer Tort Law to Environmental Regulation?" *Washburn Law Journal* 41 (2001): 515.

Hylton, Keith N. "A Positive Theory of Strict Liability." *Review of Law and Economics* 4 (1) (2008): 153–181.

Ide, S. Taku, S. Julio Friedmann, and Howard J. Herzog. "CO_2 Leakage through Existing Wells: Current Technology and Regulations." In *Proceedings of the Eighth International Conference on Greenhouse Gas Control Technologies, IEA Greenhouse Gas Programme.* 2006.

Ingelson, Allan, Anne Kleffner, and Norma Nielson. "Long-Term Liability for Carbon Capture and Storage in Depleted North American Oil and Gas Reservoirs: A Comparative Analysis." *Energy Law Journal* 31 (2010): 431.

Jacobs, Wendy B. "Carbon Capture and Sequestration." In *Global Climate Change and U.S. Law*, 2nd ed., ed. Jody Freeman and Michael Gerrard, 581–620. Chicago: ABA, 2014.

Jaffee, Dwight M., and Thomas Russell. "Should Governments Support the Private Terrorism Insurance Market?" Paper presented at WRIEC Conference, Salt Lake City, August 2005.

Jenkins, Charles, Andy Chadwick, and Susan D. Hovorka. "The State of the Art in Monitoring and Verification—Ten Years On." *International Journal of Greenhouse Gas Control* 40 (2015): 312–349.

Jost, Peter. "J. "Limited Liability and the Requirement to Purchase Insurance." *International Review of Law and Economics* 16 (2) (1996): 259–276.

Kahan, Marcel. "Causation and Incentives to Take Care under the Negligence Rule." *Journal of Legal Studies* 18 (1989): 427.

Kaplow, Louis. "Incentives and Government Relief for Risk." *Journal of Risk and Uncertainty* 4 (2) (1991): 167–175.

Karapanou, Vaia, and Louis Visscher. "Towards a Better Assessment of Pain and Suffering Damages." *Journal of European Tort Law* 1 (1) (2010): 48–74.

Kharaka, Yousif K., David R. Cole, James J. Thordsen, Kathleen D. Gans, and R. Burt Thomas. "Geochemical Monitoring for Potential Environmental Impacts of Geologic Sequestration of CO_2" In *GeoChemistry of Geologic CO_2 Sequestration*, ed. Donald J. DePaolo, David R. Cole, Alexandra Navrotsky, and Ian C. Bourg, 399–430. Washington DC: Geochemical Society, 2013.

Klass, Alexandra B., and Elizabeth J. Wilson. "Climate Change and Carbon Sequestration: Assessing a Liability Regime for Long-Term Storage of Carbon Dioxide." *Emory Law Journal* 58 (2008): 103–179.

Klass, Alexandra B., and Elizabeth J. Wilson. "Carbon Capture and Sequestration: Identifying and Managing Risks." *Issues in Legal Scholarship* 8 (1) (2009): 1–30.

Koide, H., Y. Shindo, Y. Tazaki, M. Iijima, K. Ito, N. Kimura, and K. Omata. "Deep Sub-Seabed Disposal of CO_2—The Most Protective Storage." *Energy Conversion and Management* 38 (1997): S253–S258.

Kole, Allison. "Carbon Capture and Storage: How Bad Policy Is By-Passing Environmental Safeguards." *Journal Environmental and Sustainability Law* 20 (2013): 101–150.

Kolstad, Charles D., Thomas S. Ulen, and Gary V. Johnson. "Ex Post Liability for Harm vs. Ex Ante Safety Regulation: Substitutes or Complements?" *American Economic Review* 80 (1990): 888–901.

Kornhauser, Lewis A., and Richard L. Revesz. "Sharing Damages among Multiple Tortfeasors." *Yale Law Journal* 98 (5) (1989): 831–884.

Kornhauser, Lewis A., and Richard L. Revesz. "Apportioning Damages among Potentially Insolvent Actors." *Journal of Legal Studies* 19 (2) (1990): 617–651.

Kunreuther, Howard. "Mitigating Disaster Losses through Insurance." *Journal of Risk and Uncertainty* 12 (2–3) (1996): 171–187.

Kunreuther, Howard C., and Paul K. Freeman. "Insurability, Environmental Risks and the Law." In *The Law and Economics of the Environment*, ed. Anthony Heyes, 302–317. Cheltenham: Edward Elgar, 2001.

Kunreuther, Howard C., and E. Michel-Kerjan. "Challenges for Terrorism Risk Insurance in the United States." *Journal of Economic Perspectives* 18 (2004): 201–214.

Kunreuther, Howard C., and Mark Pauly. "Rules Rather Than Discretion: Lessons from Hurricane Katrina." *Journal of Risk and Uncertainty* 33 (1–2) (2006): 101–116.

Kunreuther, Howard C., R. Hogarth, and J. Meszaros. "Insurer Ambiguity and Market Failure." *Journal of Risk and Uncertainty* 7 (1) (1993): 71–87.

Lahnstein, Christian. "A Market-Based Analysis of the Financial Assurance Issues of Environmental Liability, Taking Special Account of Germany, Austria, Italy and Spain." In *Deterrence, Insurability and Compensation in Environmental Liability: Future Developments in the European Union*, ed. Michael G. Faure, 303–330. Vienna: Springer, 2003.

Landes, William M., and Richard A. Posner. "Tort Law as a Regulatory Regime for Catastrophic Personal Injuries." *Journal of Legal Studies* 13 (3) (1984): 417–434.

Lepore, Matthew J., and Derek L. Turner. "Legislating Carbon Sequestration: Pore Space Ownership and Other Policy Considerations." *Colorado Lawyer* 40 (10) (2011): 61–68.

Levmore, Saul, and Kyle D. Logue. "Insuring against Terrorism–and Crime." *Michigan Law Review* 102 (2) (2003): 268–327.

Lewicki, Jennifer L., Jens Birkholzer, and Chin-Fu Tsang. "Natural and Industrial Analogues for Leakage of CO_2 from Storage Reservoirs: Identification of Features, Events, and Processes and Lessons Learned." *Environmental Geology* 52 (3) (2007): 457–467.

Li, Qi, Zhishen Wu, and Xiaochun Li. "Prediction of CO_2 Leakage during Sequestration into Marine Sedimentary Strata." *Energy Conversion and Management* 50 (3) (2009): 503–509.

Linnerooth-Bayer, Joanne, and Reinhard Mechler. "Disaster Safety Nets for Developing Countries: Extending Public–Private Partnerships." *Environmental Hazards* 7 (1) (2007): 54–61.

Liu, Jing. *Compensating Ecological Damage: Comparative and Economic Observations.* Antwerp: Intersentia, 2013.

Lucas, Alastair R., William A. Tilleman, and Elaine Lois Hughes. *Environmental Law and Policy.* Toronto: Emond Montgomery Publication, 2003.

Lux, K.-H. "Design of Salt Caverns for the Storage of Natural Gas, Crude Oil, and Compressed Air: Geomechanical Aspects of Construction, Operation, and Abandonment." In *Underground Gas Storage: Worldwide Experiences and Future Development in the UK and Europe,* ed. D. J. Evans and R. A. Chadwick. London: British Geological Society, Keyworth, 2009.

Lynch, Richard D., Edward J. McBride, Thomas K. Perkins, and Michael E. Wiley. "Dynamic Kill of an Uncontrolled CO_2 Well." *Journal of Petroleum Technology* 37 (8) (1985): 1267–1275.

Mace, M. J., Chris Hendriks, and Rogier Coenraads. "Regulatory Challenges to the Implementation of Carbon Capture and Geological Storage within the European Union under EU and International Law." *International Journal of Greenhouse Gas Control* 1 (2) (2007): 253–260.

Mackaay, Ejan. *Economics of Information and Law.* Paris: Groupe de recherche en consommation, 1980.

Magnus, U. "Germany." In *Compensation for Victims of Catastrophes: A Comparative Legal Approach,* ed. Michael G. Faure and T. Hartlief, 119–144. Vienna: Springer, 2006.

Makuch, Z., S. Georgieva, and B. Oraee-Mirzamani. *Carbon Capture and Storage: Regulating Long-Term Liability*. London: Centre for Environmental Policy, Imperial College London, 2011.

Manceau, Jean-Charles, Dimitrios G. Hatzignatiou, Louis De Lary, Niels Bo Jensen, Kristin Flornes, Thomas Le Guénan, and Arnaud Réveillère. "Methodologies and Technologies for Mitigation of Undesired CO_2 Migration in the Subsurface." In *The Proceedings of the Seventh Trondheim CCS Conference*. 2013.

Marshall, John M. "Moral Hazard." *American Economic Review* 66 (5) (1976): 880–890.

Marston, Philip M., and Patricia A. Moore. "From EOR to CCS: The Evolving Legal and Regulatory Framework for Carbon Capture and Storage." *Energy Law Journal* 29 (2008): 421–490.

McGarr, A., Barbara Bekins, Nina Burkardt, J. Dewey, Paul Earle, W. Ellsworth, Shemin Ge, S. Hickman, A. Holland, E. Majer, J. Rubinstein, and A. Sheehan. "Coping with Earthquakes Induced by Fluid Injection." *Science* 347 (6224) (2015): 830–831.

McLaren, James, and James Fahey. "Key Legal and Regulatory Considerations for the Geosequestration of Carbon Dioxide in Australia." *Australian Resources and Energy Law Journal* 24 (2005): 45–73.

Miceli, Thomas J. "On Negligence Rules and Self-Selection." *Review of Law and Economics* 2 (3) (2006): 349–361.

Miller, I. I. I., and C. James. "The FTC and Voluntary Standards: Maximizing the Net Benefits of Self-Regulation." *Cato Journal* 4 (1984): 897–903.

Miyazaki, Brent. "Well Integrity: An Overlooked Source of Risk and Liability for Underground Natural Gas Storage: Lessons Learned from Incidents in the U.S.A." In *Underground Gas Storage: Worldwide Experiences and Future Development in the UK and Europe*, ed. D. J. Evans and R. A. Chadwick. London: British Geological Society, Keyworth, 2009.

Monast, Jonas J., Brooks R. Pearson, and Lincoln F. Pratson. "A Cooperative Federalism Framework for CCS Regulation." *Environmental and Energy Law and Policy Journal* 7 (2012): 1–46.

Monti, Alberto. "Environmental Risk: A Comparative Law and Economics Approach to Liability and Insurance." *European Review of Private Law* 9 (1) (2001): 51–79.

Monti, Alberto. "Public-Private Initiatives to Cover Extreme Events." In *Extreme Events and Insurance: 2011 Annus Horribilis*, ed. Chr. Courbage and W. R. Stahel, 27–38. Geneva: Geneva Association, 2012.

Much, Susanna. "The Emerging International Regulation of Carbon Storage in Sub-Seabed Geological Formations." In *Shipping, Law and the Marine Environment in the 21st Century: Emerging Challenges for the Law of the Sea—Legal Implications and Liabilities*, ed. Richard Caddel and Rhidian Thomas, 255–75. Witney: Lawtext Publishing, 2013.

Nell, Martin, and Andreas Richter. "The Design of Liability Rules for Highly Risky Activities—Is Strict Liability Superior When Risk Allocation Matters?" *International Review of Law and Economics* 23 (1) (2003): 31–47.

Niezen, G. J. "Aansprakelijkheid voor milieuschade in de Europese Unie." In *Ongebonden Recht Bedrijven*, 165–69. Berlin: Kluwer, 2000.

Nuclear Power: Insurance and the Pooling System, Special Edition of the Nuclear Pools' Bulletin, 1992.

Nordbotten, Jan Martin, Michael A. Celia, and Stefan Bachu. "Injection and Storage of CO_2 in Deep Saline Aquifers: Analytical Solution for CO_2 Plume Evolution during Injection." Transport in Porous Media 58 (3) (2005): 339–360.

Nussim, Jacob, and Avraham D. Tabbach. "A Revised Model of Unilateral Accidents." *International Review of Law and Economics* 29 (2) (2009): 169–177.

Ogus, Anthony I. "Quantitative Rules and Judicial Decision-Making." In *The Economic Approach to Law*, ed. P. Burrows and C. Veljanovski. 210–225. London: Butterworths, 1981.

Ogus, Anthony I. "Standard Setting for Environmental Protection: Principles and Processes." In *Environmental Standards in the European Union*, ed. Michael G. Faure, J. Vervaele, and A. Weale, 25–37. Antwerp: Maklu, 1994.

Ogus, Anthony. "Rethinking Self-Regulation." *Oxford Journal of Legal Studies* 15 (1) (1995): 97–108.

Ogus, Anthony. "Self-Regulation." In *Encyclopedia of Law and Economics*, ed. B. Bouckaert and G. De Geest, 587–602. Cheltenham: Edward Elgar, 2000.

Ogus, Anthony. "Regulation Revisited." *Public Law* 2 (2009): 332–346.

Ogus, Anthony. "The Paradoxes of Legal Paternalism and How to Resolve Them." *Legal Studies* 30 (1) (2010): 61–73.

Oldenburg, Curtis M., and André J. A. Unger. "On Leakage and Seepage from Geologic Carbon Sequestration Sites." *Vadose Zone Journal* 2 (3) (2003): 287–296.

Pacces, Alessio M., and Louis T. Visscher. "Methodology of Law and Economics." In *Law and Method: Interdisciplinary Research into Law*, ed. Bart van Klink and Sanne Taekema, 85–107. Tübingen: Möhr Siebeck, 2011.

Parchomovsky, Gideon, and Alex Stein. "Torts and Innovation." *Michigan Law Review* 107 (2008): 285–315.

Partain, Roy Andrew. "Moerman versus Pierson: The Nexus of Occupancy in Animals Ferae Naturae and Liability in Tort." *Soongsil Law Review* 28 (2012): 241–290.

Partain, Roy Andrew. "Avoiding Epimetheus: Planning Ahead for the Commercial Development of Offshore Methane Hydrates." *Sustainable Development Law and Policy* 15 (2015): 16–25.

Partain, Roy Andrew. "Is a Green Paradox Spectre Haunting International Climate Change Laws and Conventions?" *UCLA Journal of Environmental Law and Policy* 33 (1) (2015): 61–134.

Pendergrass, John. "Long-Term Stewardship of Geological Sequestration of CO_2." *Environmental Law Reporter: News and Analysis* 43 (8) (2013): 10659–10661.

Philipsen, Niels. *Regulation of and by Pharmacists in the Netherlands and Belgium: An Economic Approach*. Antwerp: Intersentia, 2003.

Pigou, Arthur Cecil. *The Economics of Welfare*. London: Macmillan, 1920.

Plaat, Hans. "Underground Gas Storage: Why and How." In *Underground Gas Storage: Worldwide Experiences and Future Development in the UK and Europe*, ed. D. J. Evans and R. A. Chadwick, 25–37. London: British Geological Society, Keyworth, 2009.

Polborn, Mattias K. "Mandatory Insurance and the Judgment-Proof Problem." *International Review of Law and Economics* 18 (2) (1998): 141–146.

Polinsky, A. M. *Strict Liability versus Negligence in a Market Setting*. Cambridge, MA: National Bureau of Economic Research, 1980.

Pollak, Melisa F., and Elizabeth J. Wilson. "Regulating Geologic Sequestration in the United States: Early Rules Take Divergent Approaches." *Environmental Science and Technology* 43 (9) (2009): 3035–3041.

Posner, Richard A. *Economic Analysis of Law*. Boston: Little, Brown, 1998.

Posner, Richard A. *Economic Analysis of Law*, 6th ed. New York: Aspen, 2003.

Posner, Richard A. *Catastrophe: Risk and Response*. Oxford: Oxford University Press, 2004.

Priest, George L. "The Current Insurance Crisis and Modern Tort Law." *Yale Law Journal* 96 (7) (1987): 1521–1590.

Priest, George L. "The Government, the Market, and the Problem of Catastrophic Loss." *Journal of Risk and Uncertainty* 12 (2–3) (1996): 219–237.

Radetzki, Marcus, and Marian Radetzki. "Private Arrangements to Cover Large-Scale Liabilities Caused by Nuclear and Other Industrial Catastrophes." *Geneva Papers on Risk and Insurance. Issues and Practice* 25 (2) (2000): 180–195.

Rankin, Adam Gardner. "Geologic Sequestration of CO_2: How EPA's Proposal Falls Short." *Natural Resources Journal* 49 (2009): 883–942.

Raschky, Paul A., and Hannelore Weck-Hannemann. "Charity Hazard—A Real Hazard to Natural Disaster Insurance?" *Environmental Hazards* 7 (4) (2007): 321–329.

Reader, Thomas W. "Is Self-Regulation the Best Option for the Advertising Industry in the European Union? An Argument for the Harmonization of Advertising Laws through the Continued Use of Directives." *University of Pennsylvania Journal of International Business Law* 16 (1995): 181–215.

Reisinger, Will, Nolan Moser, and Trent A. Dougherty. "Reconciling King Coal and Climate Change: A Regulatory Framework for Carbon Capture and Storage." *Vermont Journal of Environmental Law* 11 (2009): 1–43.

Reitsma, S. M. S. "Nuclear Insurance Pools: History and Development." In *Nuclear Accidents-Liabilities and Guarantees*, 341–347. OECD-IAEA, 1993.

Reitz, John C. "How to Do Comparative Law." *American Journal of Comparative Law* 46 (4) (1998): 617–636.

Reitze, Arnold W. Jr., and Marie Bradshaw Durrant. "State and Regional Control of Geological Carbon Sequestration." *Environmental Law Reporter News and Analysis* 41 (2011): 10348–10373.

Rhodes, Roderick Arthur William. "The New Governance: Governing without Government." *Political Studies* 44 (4) (1996): 652–667.

Richardson, Benjamin J. "Financial Institutions for Sustainability." *Environmental Liability* 8 (2) (2000): 52–64.

Richardson, Benjamin. *Environmental Regulation through Financial Organisations: Comparative Perspectives on the Industrialised Nations*. London: Kluwer Law International, 2002.

Rochon, Emily, Jo Kuper, Erika Bjureby, Paul Johnston, R. Oakley, D. Santillo, Nina Schulz, and G. von Goerne. *False Hope: Why Carbon Capture and Storage Won't Save the Climate*. Amsterdam: Greenpeace International, 2008.

Rose-Ackerman, Susan. *Rethinking the Progressive Agenda: The Reform of the American Regulatory State*. New York: Free Press, 1992.

Rose-Ackerman, Susan. "Public Law versus Private Law in Environmental Regulation: European Union Proposals in the Light of United States Experience." *Review of European Community and International Environmental Law* 4 (4) (1995): 312–320.

Rose-Ackerman, S., and M. Geisteld. "The Diversions between Social and Private Incentives to Sue: A Comment on Shavell, Menell and Kaplow." *Journal of Legal Studies* 16 (1987): 483–491.

Rubinfeld, Daniel L. "The Efficiency of Comparative Negligence." *Journal of Legal Studies* 16 (2) (1987): 375–394.

Schäfer, Hans-Bernd. "The Bundling of Similar Interests in Litigation: The Incentives for Class Actions and Legal Actions taken by Associations." *European Journal of Law and Economics* 9 (2000): 183–223.

Schäfer, Hans-Bernd, and Frank Müller-Langer. "Strict Liability versus Negligence." In *Tort Law and Economics*, ed. Michael G. Faure, 3–45. Cheltenham: Edward Elgar, 2009.

Schäfer, Hans-Bernd, and Andreas Schönenberger. "Strict Liability versus Negligence." In *Encyclopedia of Law and Economics*, 597–624. Cheltenham: Edward Elgar, 2000.

Schmitz, P. W. "On the Joint Use of Liability and Safety Regulation." *International Review of Law and Economics* 20 (2000): 371–382.

Schrag, Daniel P. "Storage of Carbon Dioxide in Offshore Sediments." *Science* 325 (5948) (2009): 1658–1659.

Schremmer, Joe. "Avoidable Fraccident: An Argument against Strict Liability for Hydraulic Fracturing." *University of Kansas Law Review* 60 (2011): 1215–1257.

Schwartz, Alan. "Statutory Interpretation, Capture, and Tort Law: The Regulatory Compliance Defense." *American Law and Economics Review* 2 (1) (2000): 1–57.

Schwartz, Alan, and Louis L. Wilde. "Intervening in Markets on the Basis of Imperfect Information: A Legal and Economic Analysis." *University of Pennsylvania Law Review* 127 (3) (1979): 630–682.

Schwartz, Gary T. "Mixed Theories of Tort Law: Affirming Both Deterrence and Corrective Justice." *Texas Law Review* 75 (1996): 1801–1835.

Schwarze, Reimund, and Gert G. Wagner. "In the Aftermath of Dresden: New Directions in German Flood Insurance." *Geneva Papers on Risk and Insurance: Issues and Practice* 29 (2) (2004): 154–168.

Scott, Colin, Fabrizio Cafaggi, and Linda Senden. "The Conceptual and Constitutional Challenge of Transnational Private Regulation." *Journal of Law and Society* 38 (1) (2011): 1–19.

Sengul, Mahmut. "Reservoir-Well Integrity Aspects of Carbon Capture and Storage." In *Carbon Capture and Storage: R&D Technologies for Sustainable Energy Future*, ed. Malti Goel, Baleshwar Kumar, and S. Nirmal Charan, 208. Oxford: Alpha Science, 2008).

Shavell, Steven. "On Moral Hazard and Insurance." *Quarterly Journal of Economics* 93 (4) (1979): 541–562.

Shavell, Steven. "Strict Liability versus Negligence." *Journal of Legal Studies* 9 (1) (1980): 1–25.

Shavell, Steven. "The Social versus the Private Incentive to Bring Suit in a Costly Legal System." *Journal of Legal Studies* 11 (2) (1982): 333–339.

Shavell, Steven. "Liability for Harm versus Regulation of Safety." *Journal of Legal Studies* 13 (2) (1984): 357–374.

Shavell, Steven. "A Model of the Optimal Use of Liability and Safety Regulation." *Rand Journal of Economics* 15 (2) (1984): 271–280.

Shavell, Steven. "Criminal Law and the Optimal Use of Nonmonetary Sanctions as a Deterrent." *Columbia Law Review* 85 (6) (1985): 1232–1262.

Shavell, Steven M. "The Judgment Proof Problem." *International Review of Law and Economics* 6 (1986): 43–58.

Shavell, Steven. "The Optimal Use of Nonmonetary Sanctions as a Deterrent." *American Economic Review* 77 (4) (1987): 584–592.

Shavell, Steven. *Economic Analysis of Accident Law*. Cambridge: Harvard University Press, 1987.

Shover, Neal. "Zelfregulering door ondernemingen: Ontwikkeling, beoordeling en bange voorgevoelens." *Tijdschrift voor Criminologie* 51 (2) (2008): 9.

Singh, Ajay K. "R&D Challenges for CO_2 Storage in Coal Seams." In *Carbon Capture and Storage: R&D Technologies for a Sustainable Energy Future*, 139–149. Hyderabad, Alpha Science International, 2008.

Skinner, Les. "CO_2 Blowouts: An Emerging Problem: Well Control and Intervention." *World Oil* 224 (1) (2003): 38–42.

Skogh, Göran. "The Transactions Cost Theory of Insurance: Contracting Impediments and Costs." *Journal of Risk and Insurance* 56 (4) (1989): 726–732.

Skogh, Göran. "Mandatory Insurance: Transaction Costs Analysis of Insurance." In *Encyclopedia of Law and Economics, II, Civil Law and Economics*, ed. B. Bouckaert and G. De Geest, 521–537. Cheltenham: Edward Elgar, 2000.

Smit, Berend, Jeffrey A. Reimer, Curtis M. Oldenburg, and Ian C. Bourg. *Introduction to Carbon Capture and Sequestration*. London: Imperial College Press, 2014.

Smith, Richard E., Emily A. Canelo, and Anthony M. Di Dio. "Reinventing Reinsurance Using the Capital Markets." *Geneva Papers on Risk and Insurance: Issues and Practice* 22 (82) (1997): 26–37.

Solomon, Semere. "Carbon Dioxide Storage: Geological Security and Environmental Issues–Case Study on the Sleipner Gas Field in Norway." *Bellona Report* (2007).

Solomon, Semere, Beate Kristiansen, Aage Stangeland, Tore A. Torp, and Olav Kårstad. "A Proposal of Regulatory Framework for Carbon Dioxide Storage in Geological Formations." International Risk Governance Council Workshop, 2007.

Spier, Jaap, and A. T. Bolt. "De uitdijende reikwijdte van het aansprakelijkheidsrecht?" In *De uitdijende reikwijdte van de aansprakelijkheid uit onrechtmatige daad, Preadvies Nederlandse Juristenvereniging*. Deventer: Kluwer, 1996.

Stigler, George J. "The Economics of Information." *Journal of Political Economics* 69 (3) (1961): 213–225.

Swayne, Nicola, and Angela Phillips. "Legal Liability for Carbon Capture and Storage in Australia: Where Should the Losses Fall?" *Environmental and Planning Law Journal* 29 (3) (2012): 189–216.

Testa, Stephen M., and James A. Jacobs. *Oil Spills and Gas Leaks: Environmental Response, Prevention, and Cost Recovery*. New York: McGraw-Hill, 2014.

Texas General Land Office, Railroad Commission of Texas, Texas Commission on Environmental Quality, Bureau of Economic Geology. *Injection and Geologic Storage Regulation of Anthropogenic Carbon Dixoide: A Preliminary Joint Report to the Legislature Required Under Sections 9 and 10 of SB 1387, 81st Leg, 2009*. 2010. http://www.rrc.tx.us/forms/reports/notices/SB1387-FinalReport.pdf.

Tietenberg, Tom H. "Indivisible Toxic Torts: The Economics of Joint and Several Liability." *Land Economics* 65 (4) (1989): 305–319.

Tietenberg, Thomas H., and Lynne Lewis. *Environmental and Natural Resource Economics*. Reading, MA: Addison-Wesley, 2000.

Trabucchi, Chiara, Michael Donlan, and Sarah Wade. "A Multi-Disciplinary Framework to Monetize Financial Consequences Arising from CCS Projects and Motivate Effective Financial Responsibility." *International Journal of Greenhouse Gas Control* 4 (2) (2010): 388–395.

Trabucchi, Chiara, and L. Patton. "Storing Carbon: Options for Liability Risk Management, Financial Responsibility." World Climate Change Report 170. Washington, DC: Bureau of National Affairs, 2008.

Trauberman, Jeffrey. "Statutory Reform of Toxic Torts: Relieving Legal, Scientific, and Economic Burdens on the Chemical Victim." *Harvard Environmental Law Review* 7 (1983): 177–296.

Trebilcock, Michael J. "The Social Insurance-Deterrence Dilemma of Modern North American Tort Law: A Canadian Perspective on the Liability Insurance Crisis." *San Diego Law Review* 24 (1987): 929–1046.

Trebilcock, Michael, and Ralph A. Winter. "The Economics of Nuclear Accident Law." *International Review of Law and Economics* 17 (2) (1997): 215–243.

Tyran, Jean-Robert, and Peter Zweifel. "Environmental Risk Internalization through Capital Markets (ERICAM): The Case of Nuclear Power." *International Review of Law and Economics* 13 (4) (1993): 431–444.

Van den Bergh, Roger. "De maatschappelijke wenselijkheid van gedragscodes vanuit rechtseconomisch perspectief." *Weekblad voor Privaatrecht, Notariaat en Registratie* 6772 (2008): 792–798.

Van den Bergh, R., and L. T. Visscher. "Optimal Enforcement of Safety Law." In *Mitigating Risk in the Context of Safety and Security: How Relevant Is a Rational Approach?* ed. R. V. de Mulder. Rotterdam: Erasmus University Rotterdam, 2008.

Vanden Borre, Tom. "Shifts in Governance in Compensation for Nuclear Damage, 20 Years after Chernobyl." In *Shifts in Compensation for Environmental Damage*, ed. Michael G. Faure and A. Verheij, 261–311. Vienna: Springer, 2007.

Van Eijk-Graveland, J. C. *Verzekerbaarheid van Opzet in het Schadeverzekeringsrecht.* Zwolle: Tjeenk Willink, 1999.

Van Huizen, F. "Enkele Begrenzingen van de (Beroeps-) Aansprakelijkheidsverzekering." In *Drie Treden*, 325–339. De Ruiter-Bundel, 1995.

Van Velthoven, B. "Empirics of Tort." In *Tort Law and Economics*, ed. Michael G. Faure. 453–498. Cheltenham: Edward Elgar, 2010.

Viscusi, W. Kip. "The Dimensions of the Product Liability Crisis." *Journal of Legal Studies* 20 (1) (1991): 147–177.

Visscher, Louis T. "Tort Damages. " In *Tort Law and Economics, Encyclopedia of Law and Economics*, ed. Michael G. Faure, 153–200. Cheltenham: Edward Elgar, 2009.

Wagner, Fred. "Risk Securitization: An Alternative of Risk Transfer of Insurance Companies." *Geneva Papers on Risk and Insurance. Issues and Practice* 23 (1998): 574–607.

Wagner, Fred. "Tort Law and Liability Insurance." In *Tort Law and Economics*, ed. Michael G. Faure, 377–405. Cheltenham: Edward Elgar, 2010.

Wang, Hui. *Civil Liability for Marine Oil Pollution Damage: A Comparative and Economic Study of the International, US and Chinese Compensation Regimes.* Alphen aan den Rijn: Kluwer Law International, 2011.

Wansink, J. H. *De Algemene Aansprakelijkheidsverzekering.* Zwolle: W. E. J. Tjeenk Willink, 1994.

Wansink, J. H. *De aansprakelijkheidsverzekering en de dekking van long-tail-risico's.* Aansprakelijkheid en Verzekering, 1995.

Warner, Robin. *Protecting the Oceans beyond National Jurisdiction: Strengthening the International Law Framework.* Leiden: Brill, 2009.

Wibisana, A. "The Myths of Environmental Compensation in Indonesia: Lessons from the Sidoarjo Mudflow." In *Regulating Disasters, Climate Change and Environmental Harm: Lessons from the Indonesian Experience*, ed. Michael G. Faure and A. Wibisana, 277–354. Cheltenham: Edward Elgar, 2013.

Wils, Wouter P. J. "Insurance Risk Classifications in the EC: Regulatory Outlook." *Oxford Journal of Legal Studies* 14 (3) (1994): 449–467.

Wilson, Elizabeth J., and Mark A. de Figueiredo. "Geologic Carbon Dioxide Sequestration: An Analysis of Subsurface Property Law." *Environmental Law Reporter News and Analysis* 36 (2) (2006): 10114.

Wilson, Elizabeth J., S. Julio Friedmann, and Melisa F. Pollak. "Research for Deployment: Incorporating Risk, Regulation, and Liability for Carbon Capture and Sequestration." *Environmental Science and Technology* 41 (17) (2007): 5945–5952.

Wilson, Elizabeth J., Timothy L. Johnson, and David W. Keith. "Regulating the Ultimate Sink: Managing the Risks of Geologic CO_2 Storage." *Environmental Science and Technology* 37 (16) (2003): 3476–3483.

Wilson, Elizabeth J., Alexandra B. Klass, and Sara Bergan. "Assessing a Liability Regime for Carbon Capture and Storage." *Energy Procedia* 1 (1) (2009): 4575–4582.

Wilde, M. *Civil Liability for Environmental Damage: Comparative Analysis of Law and Policy in Europe and the US*, 2nd ed. Alphen aan den Rhijn: Kluwer Law International, 2013.

Zanchetta, Barbara Pozzo. "The Liability Problem in Modern Environmental Statutes." *European Review of Private Law* 4 (2) (1996): 111–144.

Zech, Jürgen. "Will the International Financial Markets Replace Traditional Insurance Products?" *Geneva Papers on Risk and Insurance. Issues and Practice* 23 (89) (1998): 490–495.

Zeckhauser, Richard. "The Economics of Catastrophes." *Journal of Risk and Uncertainty* 12 (2–3) (1996): 113–140.

Zhou, Di, Zhongxian Zhao, Jie Liao, and Zhen Sun "A Preliminary Assessment on CO_2 Storage Capacity in the Pearl River Mouth Basin Offshore Guangdong China." *International Journal of Greenhouse Gas Control* 5 (2) (2011): 308–317.

Index